Inland Waters - Dynamics and Ecology

Edited by Adam Devlin, Jiayi Pan and Mohammad Manjur Shah

Published in London, United Kingdom

IntechOpen

Supporting open minds since 2005

Inland Waters - Dynamics and Ecology
http://dx.doi.org/10.5772/intechopen.87463
Edited by Adam Devlin, Jiayi Pan and Mohammad Manjur Shah

Assistant to the Editor
Aliyu Ahmad Nuhu

Contributors
Kenneth W. Cummins, Claudio Valdovinos Zarges, Pablo Fierro, Viviana Olmos, Abdul-Rahman Dirisu, John Ovie Olomukoro, Natividade Vieira, Ana Sofia Soares, Carlos Pinheiro, Matthew Bingham, Jason Kinnell, Wu Bo, Jinmu Zhang, Yindi Zhao, Humberto Alves Barbosa, Franklin Javier Paredes, T. V. Lakshmi Kumar, Manoj Kumar Thakur, Catarina Buriti, Scott A. Wells, Oscar Donde, Caren Atalitsa, Sani Dauda Ahmed, Sampson Kwaku Agodzo, Kwaku Amaning Adjei, Gevorg Simonyan, Satarupa Ghosh, Snigdha Chatterjee, Prasanna Pal, Ghora Shiva Prasad, Belda Q. Mosepele, Kethlatlogile Mosepele, Uirá Oliveira

First published in London, United Kingdom, 2021 by IntechOpen
IntechOpen is the global imprint of INTECHOPEN LIMITED, registered in England and Wales, registration number: 11086078, 5 Princes Gate Court, London, SW7 2QJ, United Kingdom
Printed in Croatia

British Library Cataloguing-in-Publication Data
A catalogue record for this book is available from the British Library

Additional hard and PDF copies can be obtained from orders@intechopen.com

Inland Waters - Dynamics and Ecology
Edited by Adam Devlin, Jiayi Pan and Mohammad Manjur Shah
p. cm.
Print ISBN 978-1-83968-294-0
Online ISBN 978-1-83968-295-7
eBook (PDF) ISBN 978-1-83968-296-4

For EU product safety concerns:
IN TECH d.o.o., Prolaz Marije Krucifikse Kozulić 3, 51000 Rijeka, Croatia, info@intechopen.com or visit our website at intechopen.com.

Meet the editors

Dr. Adam T. Devlin is a physical oceanographer who analyzes the fluctuations in tides and sea level on a global and regional level, with a special interest in the dynamics of the Western Pacific Ocean. He also studies the patterns of water level and meteorological changes using historical records and modern satellite data to better predict future flood scenarios and impacts on coastal populations and ecosystems. Dr. Devlin holds a Ph.D. in Applied Physics from Portland State University, an MS in Applied Physics from Appalachian State University, and a BS in Physics from the University of Oregon. He worked at the Chinese University of Hong Kong from 2016 to 2019. Dr. Devlin joined the School of Geography and Environment, Jiangxi Normal University as a distinguished research professor in May 2019.

Professor Jiayi Pan received his Ph.D. in Physical Oceanography from the Institute of Oceanology, Chinese Academy of Sciences in 1996. He was a postdoctoral researcher at the University of Delaware from 1999 to 2003, then worked at the University of Southern Mississippi, Oregon Health and Science University, and Portland State University in 2003-2004, 2004-2005, and 2005-2009, respectively. He joined the Institute of Space and Earth Information Science, the Chinese University of Hong Kong as an associate professor in February 2009. Professor Jiayi Pan has been engaged in research on ocean remote sensing, air-sea interaction, estuarine dynamics, and internal wave dynamics for many years, and was the principal investigator (PI) or Co-PI for a number of research projects. In July 2019, he joined the School of Geography and Environment, Jiangxi Normal University as a distinguished professor.

Dr. Mohammad Manjur Shah obtained his Ph.D. from Aligarh Muslim University in 2003. He is a pioneer in the field of insect parasitic nematodes in the North East Part of India. He has presented his findings in several conferences and published his articles in various reputed international journals. He completed post-doctoral fellowship twice under the Ministry of Science and Technology, Government of India before joining the Yusuf Maitama Sule University, Kano, Nigeria in 2015. Apart from the present book, he has already edited five books with IntechOpen. He is also the reviewer of several journals of international repute. He has been listed in various biographies published from the USA and UK. At present he is working as an Associate Professor in Biology at the Yusuf Maitama Sule University, Kano, Nigeria.

Contents

Preface

Inland waters, lakes, rivers, and their connected wetlands are the most important and the most vulnerable sources of freshwater on the planet. The ecology dependent on these systems includes a wide range of flora and fauna, as well as most human populations and civilization. The study of inland waters includes analyses of the biology, chemistry, physical dynamics, morphology, geology, and geography of inland aquatic systems. Inland water investigations may include examinations of lakes, rivers, ponds, streams, reservoirs, wetlands, and groundwater, as well as the ecological and anthropogenic factors that define and influence such systems. In particular, inland waters and wetlands are highly susceptible to chemical and biological pollutants from natural or human sources, changes in watershed dynamics due to the establishment of dams and reservoirs, and land use changes from agriculture and industry. This book provides a comprehensive review of issues important to the understanding of inland waters and discusses many worldwide inland water systems. The main topics of this text are water quality investigation, analyses of the ecology of inland water systems, remote sensing observation and numerical modelling methods, and biodiversity investigations.

This book is organized into four main sections. The first section is entitled "Water Quality of Inland Waters", which is concerned with factors that affect the status of freshwater system quality. Chapter 1 of the first section investigates the contamination by microplastics found in salt works used for table salt in the Portuguese coastal waters, which reveals that microplastic contamination can be a serious concern. Chapter 2 of the first section adopts a novel approach to consider the entropy of inland water systems in the analysis of pollution and analyzes and compares many different freshwater quality indices in the waters of Armenia.

The second section of the book is entitled "Ecological Factors Affecting Inland Waters", which presents different issues related to the ecological health of wetlands and lakes. Chapter 1 of the second section analyzes how the impacts of global climate change can affect the production of inland freshwater fisheries, which in turn can impact the economy of those who depend on these industries. This chapter also provides suggestions for efficient management of fisheries under increased climate change scenarios. Chapter 2 of the second section is about techniques and methodologies of designing constructed wetlands for river diversion as possible improvements to water quality. Chapter 3 of the second section discusses the dynamics of hazardous algae blooms in Lake Erie, and the economic impacts on tourism in the region. Chapter 4 in the second section discusses various forms of constructed wetland structures and how this may help local aquatic plants respond and adapt to contamination by pollutants such as antibiotic resistant genes.

The third section of this book is focused on observational and modelling methodology, entitled "Remote Sensing and Modelling of Inland Waters". Chapter 1 of the third section examines how thermal stratification in lakes and reservoirs can be categorized and provides an in-depth analysis of the governing equations for modelling the hydrodynamics and water quality in stratified inland water systems, including particular case studies. Chapter 2 of the third section analyzes precipitation in inland water systems as observed from satellites using the CHIRPS method. Chapter 3 of the third section presents a detailed methodology for using remote sensing to extract narrow water features from satellite imagery using a morphological linear enhancement technique.

The fourth and final section of the book deals with biological issues in inland water systems and is entitled " Biodiversity of Inland Waters". Chapter 1 of the fourth section presents a detailed zoological study of invertebrates found in the stream and river systems of North America, and discusses relevant factors affecting these populations. Chapter 2 of the fourth section reviews the biodiversity dynamics in the Okavango Delta in Botswana, discussing biotic and abiotic factors, flood dynamics, and other factors important to this wetland system. Chapter 3 of the fourth section discusses the biodiversity and environmental integrity of rivers in Nigeria, involving a detailed and thorough statistical analysis of various ecological factors. Finally, Chapter 4 of the fourth section reviews the present status of knowledge of invertebrates found in rivers, lakes and wetlands in western South America, and provides a characterization of the diverse groups of insects, mollusks, crustaceans and other smaller groups spread over two major invertebrate phyla.

The chapters of this book provide detailed and varied information about many of the important factors affecting inland water systems, with a good variety of systems worldwide discussed. It is hoped that the methods and analyses presented here will inspire new in-depth investigations and other future studies. We are grateful to the authors, and to the publishing staff of IntechOpen for their contributions to this textbook and their hard work.

Adam T. Devlin and Jiayi Pan
Jiangxi Normal University,
Nanchang, Jiangxi, China

Mohammad Manjur Shah
Yusuf Maitama Sule University,
Nigeria

Section 1

Water Quality of Inland Waters

Chapter 1

Microplastic Pollution in Portuguese Saltworks

Ana Sofia Soares, Carlos Pinheiro, Uirá Oliveira and Maria Natividade Vieira

Abstract

Currently, microplastics are dispersed everywhere; from our oceans to our rivers, sediments, organisms, air, and even food resources. Therefore, this study aimed to assess the degree of contamination present in the Portuguese traditional table salts depending on their origin and type of salt. Fourteen samples were selected: seven from *fleur de sel* and seven from coarse salts, corresponding to seven distinct regions of the Portuguese territory. The concentration of microplastics, depending on salts' origin, ranged between 595 and 5090 MPs/kg, in sea salt, and in Rio Maior's well salt it varied from 3325 to 6430 MPs/kg. By salt type, the concentration of microplastics in the *fleur de sel* was 2320–6430 MPs/kg and in the coarse salt was 595–3985 MPs/kg. In the analyzed table salt, the most abundant anthropogenic particles were fibers (64%) and fragments (35%). The most predominant colors were transparent, blue, and black. The concentration of microplastics did not vary significantly ($p > 0.05$) between *fleur de sel* samples within different regions. However, statistically significant differences were found between coarse salt samples from the various regions. The results, gathered from this study, demonstrate the high contamination within artisanal Portuguese table salts, thus, becoming crucial to develop more future research, leading to a better understanding of the health risks associated with salt consumption.

Keywords: contamination, food security, microplastic, Portugal, table salt

1. Introduction

Plastics have played a fundamental role in man's daily life since the invention of Bakelite, the first synthetic polymer created by the chemist Leo Baekeland in 1907. Newly synthetic polymers were developed in the following years. These were capable of resisting higher temperatures without degrading, thus allowing the creation of a revolutionary era, within all commercial sectors (mainly industrial, health, and domestic), that shaped the habits of the future generations [1, 2]. The mass production of plastic began in the 1950s and has since evolved exponentially. For instance, in 2017, worldwide plastic production surpassed 348 million tons [3, 4]. Nevertheless, plastics have become persistent pollutants in the most diverse environments (atmospheric air, sources of freshwater, brackish, and saltwater, and soils) mainly due to their industrial characteristics (durability, hydrophobic composition, and plasticity) but also due to their improper handling over the years [1, 2, 5].

There are several ways in which plastic debris can reach the oceans, such as: (1) terrestrial sources (e.g., rivers and estuaries), (2) several industrial sectors,

(3) treated and untreated urban effluents, (4) and human activities (e.g., fishing) [6]. For example, our oceans receive, annually, a total of 4.8 to 12.7 million tons of plastic, from which 1.15 to 2.41 million tons come directly from freshwater streams [5, 7, 8]. Several studies have been carried out over the last two decades regarding the problem of plastics and their degradation into smaller particles. Since the 1970s, many scientists have suspected the harmful effects of these anthropogenic particles on the environment [9–12]. However, it was not until the 1990s that microplastics were considered and recognized as emerging pollutants. Therefore, scientists have been developing identification methods to better understand this problem's magnitude [13].

Microplastics (MPs) are characterized as plastic materials or fragments of small proportions, with measurements inferior to 5 mm [14]. MPs existing in aquatic environments may originate from primary sources (where they are designed and produced for a specific purpose) or from secondary sources (resulting from the degradation of larger plastic debris) [15]. In the oceans, the most prevalent synthetic polymers are polypropylene (PP), polyethylene (PE), polystyrene (PS), polyvinyl chloride (PVC) and polyethylene terephthalate (PET) [16]. Today we are often surrounded by the presence of microplastics as they are everywhere; from our oceans, rivers, sediments, living organisms, atmospheric air and even food resources [14, 17]. These synthetic micropolymers have the ability to accumulate in our food chain, thus reaching various types of organisms. Several studies have been conducted in recent years to understand the potential impacts of these anthropogenic compounds on human health [6, 18].

In Portugal, the population is culturally linked to the sea and its resources, thus becoming more exposed to pollution present in the Atlantic Ocean. In recent years, several studies have been published about the accumulation of microplastics on beaches [19–21], estuaries [22, 23] and aquatic organisms [22, 24, 25] in Portugal. Moreover, many of these MPs had high concentrations of polychlorinated biphenyls (PCBs), pesticides (e.g., DDTs), polycyclic aromatic hydrocarbons (PAHs), and other persistent contaminants adsorbed to their surface [20, 26–28]. Additionally, all these chemical compounds are capable of causing problems in various aquatic ecosystems (such as marine, estuarine, lotic and lentic ecological communities). Hence, Yang et al. [14] hypothesized that saltwater-based commercial table salts were possibly contaminated, since oceans are polluted by plastic debris. In their work, the presence of MPs was analyzed in numerous types of Chinese sea salt brands, where it was found 7–681 MPs/kg of sea salt. After this first study was published, six others were able to characterize and identify (with relevant results) the presence of MPs in commercial table salts, with their concentrations varying mostly between sites [3, 5, 8, 29–31].

Considering that sodium (Na^+) is an essential element for our well-being [3] and the above information, it can be assumed that MP intake may vary depending on the country's gastronomic culture and environmental pollution. For example, WHO suggests a maximum daily intake of 5 g of salt, however, the Portuguese population consumes, on average, 3 g more than the recommended [32]. We can then assume that the Portuguese people are subject to a greater exposure of microplastics via table salt consumption. Therefore, this study aims to analyze the available Portuguese table salts and determine the degree of microplastic contamination, depending on their origin and type of salt.

2. Methods and materials

2.1. Study area and sample collection

A total of seven geographically distinct saltworks were previously identified and selected for this study: Aveiro (Av), Figueira da Foz (FF), Rio Maior (RM), Tejo (Tj), Olhão (Ol), Tavira (Tv) and Castro Marim (CM) (**Figure 1**). In each region, samples of *fleur de sel* and artisanal coarse salt were collected (n_t = 14), in triplicate, between February–April 2019 under the same brand.

Depending on their origin, salts can have different designations. In all the selected saltworks, except for RM, both *fleur de sel* and coarse salt are produced through the solar evaporation of brackish and/or marine waters. Thus, salt originated from these solar saltworks is commonly named sea salt. However, in RM's saltworks the salt is produced from the storm- and groundwater that leaches the halite deposits, present in the region, to form a brine. This one is then collected in open solar saltworks and undergoes the same evaporation processes as the remaining. Therefore, this type of salt is designated as "well salt" [33].

The packages obtained from *fleur de sel* and artisanal coarse salt were sold in plastic bags of 200–250 and 1000–1500 g, respectively. Previous studies have already shown that plastic packaging does not influence the concentration of MPs in the final results [5, 8, 31].

2.2. Preventive methods of contamination

A protocol, based on published studies [5, 14, 29], was developed to prevent microplastic contamination throughout the study. Protective gloves and white coats were implemented when handling any material or reagent, since numerous plastic fibers can be found in our hands and clothing, which can have in its composition synthetic fibers (e.g., polyester, nylon...) [34]. Moreover, all equipment was cleaned with previously filtered deionized water and 70% (v/v) ethanol. Finally, materials were protected and/or sealed [29].

Figure 1.
Location of Portugal's main saltworks and respective sampling sites.

The deionized water used during the study was previously filtered with the aid of a vacuum pump using a 1 μm porosity cellulose nitrate membrane (Whatman®; CAT: 7190-002), thus, removing possible microplastic contamination. This filtered water was used to clean the equipment and dissolve the tested table salts, thus, preventing any contamination from external sources via microplastics [14].

2.3. Microplastic extraction and identification

The microplastic extraction from both salt types was performed according to Iñiguez et al. [5], with some modifications. Briefly, 200 g of salt (*fleur de sel* or coarse salt) was homogeneously dissolved in 1 L of previously filtered deionized water. Next, the solution was placed in a centrifuge for 1 h at 1900 rpm in order to isolate the denser (in)organic material from the supernatant [5]. The supernatant was subsequently collected and filtered with the aid of a vacuum pump using a 5 μm porosity cellulose nitrate filter (Whatman®; CAT: 7195-004). According to Branca [35], the plastic particles are expected to be suspended in the supernatant due to density differences. Nevertheless, preliminary tests were performed to assess whether certain microplastics were "trapped", or not, in the precipitate. No microplastics were found in all of the precipitates. After filtration, the membranes were carefully collected, preserved and sealed in Petri dishes and allowed to dry at room temperature. Each Petri dish was previously brushed with petroleum jelly to fix the membrane and avoid, consequently, the plastic microparticles displacement during microscopic or stereoscopic manipulations. All filtrations and hand-manipulations were performed inside a fume chamber, hence, avoiding any kind of atmospheric contamination [29].

The visual identification of microplastics was performed using a stereoscope (Carl Zeiss Stemi DV4 CLS120X) and a camera. A total of 59 membrane filters were recorded since some replicas required more than one membrane due to clogging. The open software ImageJ (v1.80) was used to analyze each collected picture. The classification of microplastics was made according to their shape and color [3] (**Table 1**).

Shape	Color
• Fibers, thin plastics and often cylindrical in shape.	• Beige
• Films, thin and flat plastics.	• Black
• Fragments, microplastics with an irregular shape and/or surface.	• Blue
• Microspheres, perfectly round plastic particles.	• Brown
• Styrofoam, a lightweight polymer with a sponge-like texture.	• Gray
	• Green
	• Multicolor
	• Orange
	• Pink
	• Red
	• Transparent
	• Violet
	• White
	• Yellow

Table 1.
Microplastics classification according to their shape and color as used in this study (adapted from [3]).

2.4. Data analysis

A one-way ANOVA, followed by Tukey's HSD multiple comparison test, was performed in order to discover significant statistical differences between the abundance of microplastics of each region, for the same salt type (*fleur de sel* and coarse salt). Homogeneity and normality tests were applied to the data to validate the tests. Additionally, several independent sample *t*-student tests were employed to understand the differences between salt types, in each saltworks region. All analyses were performed with a significance level of 0.05. All statistical tests were performed using the SPSS v25 software.

3. Results

3.1. Presence of microplastics in Portuguese salt

In this study, all samples (n_t = 14) of artisanal *fleur de sel* and coarse salt were contaminated by microplastics. Overall, 23,175 anthropogenic plastic microparticles were analyzed, of which approximately 64% corresponded to fibers, 35% to irregular fragments, and only <1% to films (**Figure 2A**). Indeed, fibers were the

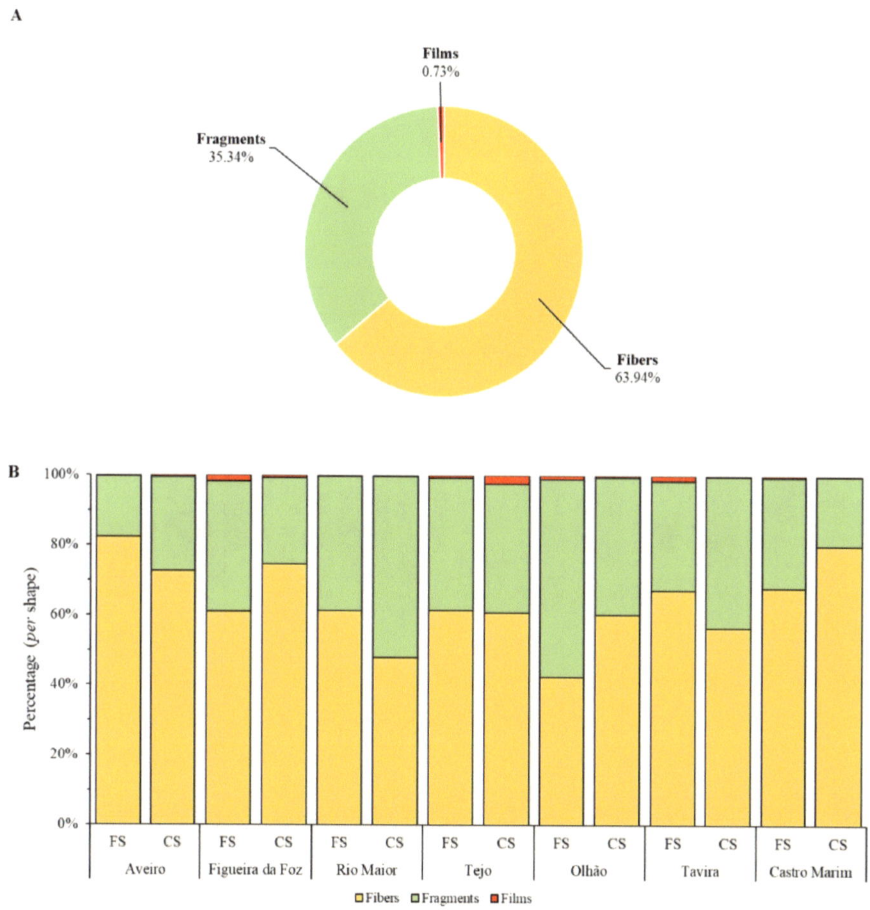

Figure 2.
*Percentage of the different shapes of microplastics identified (A) throughout the study and (B) in each saltworks region. Microspheres and Styrofoam microparticles were not found. FS—*Fleur de sel; *CS—Coarse salt.*

most predominant type of microplastic (>50%) present in all the studied regions, except in Rio Maior and Olhão (**Figure 2B**). Here, a higher percentage of fragments (51.8 and 56.5%, respectively) was found in coarse salt and *fleur de sel*, respectively. Films were also present in all samples, although in lower percentages (<2.5%) when compared to the other microplastic shapes (**Figure 2B**). No spherical and Styrofoam microparticles were identified in all samples.

The most common colors found were transparent (54.9%), blue (15.3%), and black (14.6%), corresponding to roughly 85% of the analyzed microplastic polymers (**Figure 3A**). Gray and green accounted for about 4.6 and 3.2%, respectively. The remaining 7.5% includes microplastics with the colors designated as yellow, white, beige, brown, orange, pink, red, and violet (**Figure 3A**).

According to **Figure 3B**, transparent was the most predominant color (>40%) throughout all sampled regions. From the remaining colors, blue and black proved to be the most prevalent in all regions when compared to the remaining colors (**Figure 3B**). No multicolored microplastics were found in this study.

Moreover, it was possible to observe a high amount of lesser dense organic matter throughout the membrane filter analysis (e.g., small invertebrates, *Artemia* sp.

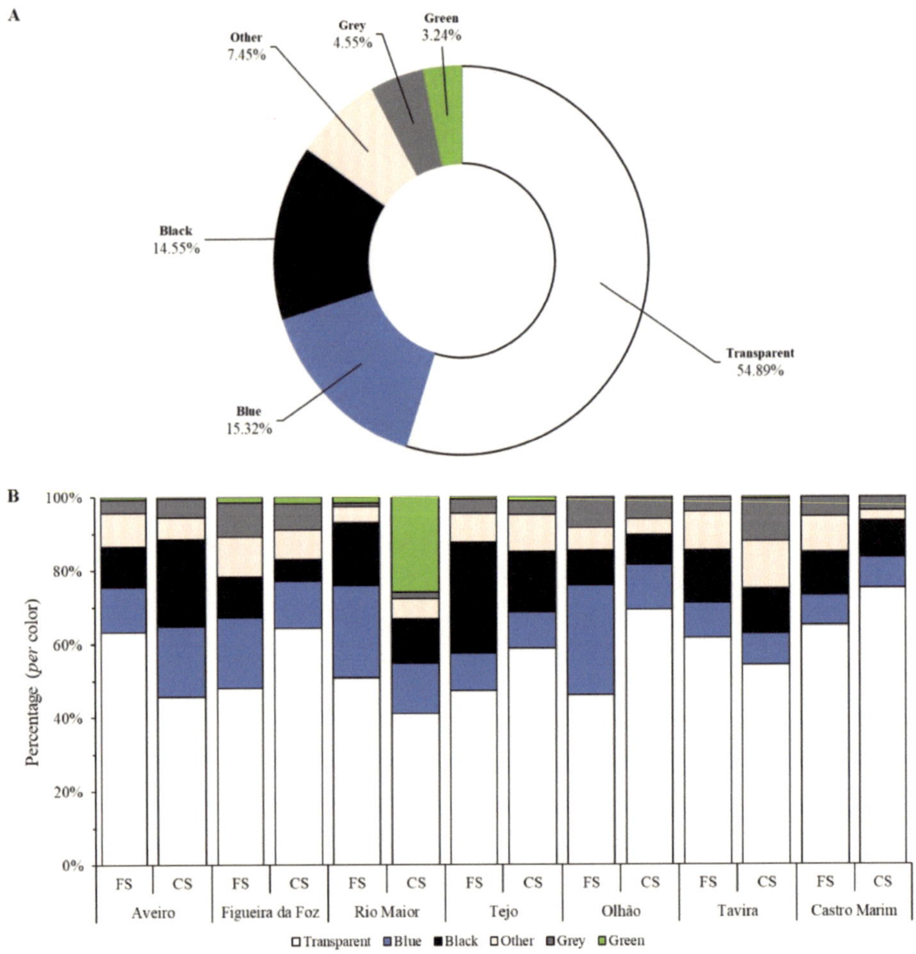

Figure 3.
*Percentage of the different colors of microplastics identified (A) throughout the study and (B) in each saltworks region. "Other" represents the sum between the following colors: beige, brown, orange, pink, red, violet, yellow, and white. No multicolor microplastics were identified. FS—*Fleur de sel*; CS—Coarse salt.*

cysts, feathers, vegetal organic matter...; **Figure 4**), which demonstrates the degree of impurities present in these types of salt.

3.2. *Fleur de sel* vs. coarse salt

Figure 5 shows the significant differences between the different types of salt (artisanal *fleur de sel* and coarse salt) in each saltworks' region. According to **Figure 5**, significantly higher values ($p < 0.05$) of microplastics in the *fleur de sel* were found in all studied regions, when compared to coarse salt (**Av** – $t_{(4)} = 2.816, p = 0.048$; **FF** – $t_{(4)} = 4.000, p = 0.016$; **Tj** – $t_{(4)} = 7.376, p = 0.002$; **Ol** – $t_{(4)} = 2.990, p = 0.040$; **Tv** – $t_{(4)} = 5.249, p = 0.006$), except for Rio Maior's ($t_{(4)} = 1.457, p = 0.219$) and Castro Marim's ($t_{(4)} = 2.152, p = 0.098$) saltworks (see **Table 2** for mean and standard deviation values). Moreover, regarding artisanal coarse salt, statistically significant differences were detected between regions according to the one-way ANOVA ($F_{(6.14)} = 18.752, p = .000$). Tukey's *post hoc* test revealed that Rio Maior artisanal coarse salt contains a significantly higher amount of microplastics than the coarse salt from the other regions studied (Av, $p = 0.012$; FF, $p = 0.000$; Tj, $p = 0.000$; Ol, $p = 0.000$; Tv, $p = 0.000$; CM, $p = 0.001$). However, there was no statistically significant differences ($p > 0.05$) between the different *fleur de sel* samples from the various studied saltworks ($p = 0.170$).

Additionally, the region who presented the highest quantities of microplastics in artisanal *fleur de sel* and coarse salt was Rio Maior (4830.0 ± 1408.0 and 3611.7 ± 338.4, respectively). The lowest amount of *fleur de sel* and coarse salt was found in the two most southeastern saltworks of Portugal: Castro Marim (2798.3 ± 595.3) and Tavira (666.7 ± 72.5), respectively (**Table 2**).

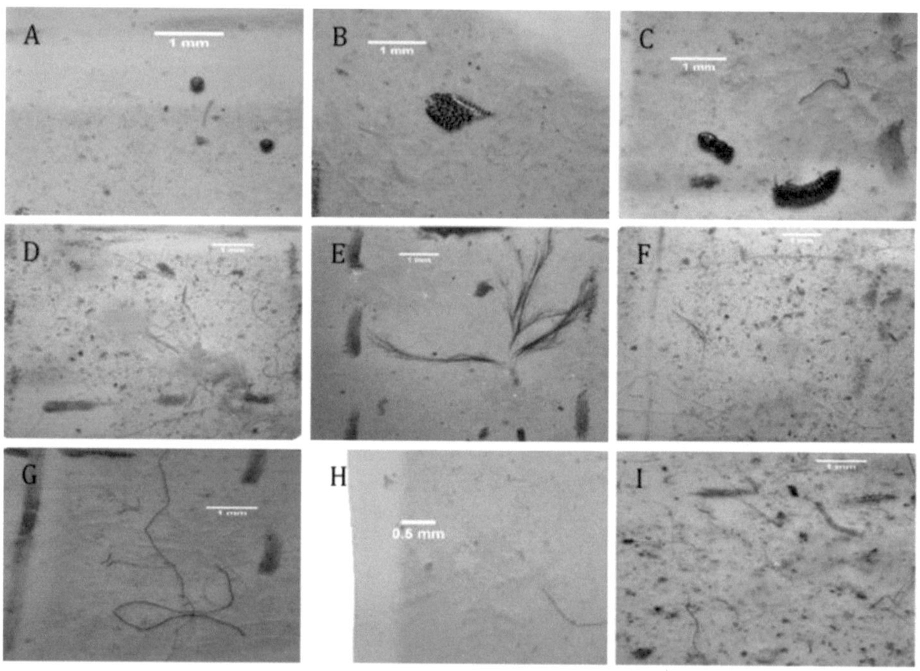

Figure 4.
Different types of organic and inorganic particles found in artisanal table salts: (A) Artemia *sp. cysts, present in most of* fleur de sel *samples; (B) an invertebrates' chitin; (C) insects body; (D, F, I) overview of a membrane filter being analyzed; (E) birds' feather; (G) a mesoplastic, with more than 5 mm; (H) blue microplastic fragment, identified in most samples across regions.*

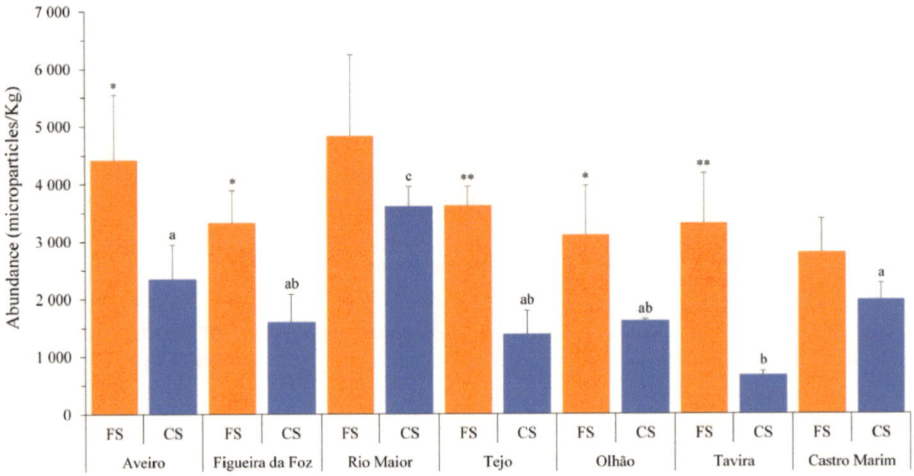

Figure 5.
Comparison between the abundance of plastic microparticles present in the artisanal fleur de sel *(FS) and coarse salt (CS) from the various saltworks selected for this study. Each value represents the mean ± standard deviation. Asterisks represent statistically significant differences between* fleur de sel *and coarse salt according to Student's* t-test. *—$p < 0.05$; **—$p < 0.01$. *Different letters represent statistically significant differences ($p < 0.05$) between samples of artisanal coarse salt across the various regions studied. There are no statistically significant differences ($p > 0.05$) between the samples of* fleur de sel.

	Microplastic particles *per* kilogram (Kg)							
	Fleur de sel				Coarse salt			
	Min	**Max**	**Mean**	**SD**	**Min**	**Max**	**Mean**	**SD**
Av	3120	5050	4420.0	1126.0	1670	2735	2351.7	591.9
FF	2675	3730	3320.0	565.4	1085	2040	1603.3	482.7
RM	3780	6430	4830.0	1408.0	3985	3900	3611.7	338.4
Tj	3255	3900	3621.7	331.5	1030	1830	1390.0	406.0
Ol	2395	4070	3108.3	864.7	1585	1630	1615.0	21.2
Tv	2510	4235	3310.0	869.3	595	740	666.7	72.5
CM	2320	3465	2798.3	595.3	1700	2270	1978.3	285.2

Table 2.
Minimum (min) and maximum (max) values of the number of microplastic particles per *kilogram (kg), and respective mean and standard deviation (SD) for the different types of artisanal table salt* (fleur de sel *and coarse salt) collected in all regions.*

4. Discussion

The main purpose of this study was to assess the level of anthropogenic contamination via microplastic particles of two types of table salts present in the Portuguese territory. Seven different locations were selected, where both artisanal *fleur de sel* and coarse salt were obtained for analysis, from the same brand.

In Portugal, the vast majority of the salt production starts by capturing seawater, due to tidal changes, into several successive ponds with different widths and heights. Next, it undergoes through various evaporation processes, due to the wind and solar actions, improving the salt crystals' precipitation. The coarse salt is then collected, roughly washed, and packaged. On the other hand, *fleur de sel* is only the first surface layer of formed salt produced in saltworks. This type of table salt is collected and immediately packaged without being previously cleaned [36]. Nevertheless, some types of salts may undergo sanitization, as well as a refining process, before packaging [8]. Therefore, to understand the level of contamination existing in the Portuguese saltworks, all the salts acquired for this study were of artisanal origin, i.e., no refinement or industrial treatment was applied.

Once samples were analyzed, it was evident the presence of a high MP concentration compared to other studies already carried out in several countries (**Table 3**). These values may be a direct proof of the lack of refinement or treatment processes that undergo in Portuguese saltworks, thus, reflecting the traditional methods still used nowadays. Most studies related to the presence of microplastics in table salt samples either used refined, treated salts and/or salts from which there is no information on their treatment [5, 8, 14, 30, 31]. Hence, it becomes important to study and understand how refining processes influence the abundance of microplastics in table salts.

Overall, *fleur de sel* presented always higher contamination values of MPs than those found in coarse salts. These salt "scales" are formed at the crystallizers' surface and, as such, greater air contamination of plastic particles is expected [2]. Also, since it does not undergo any cleaning process up to its packaging, it is expected a higher concentration of MPs, when compared to artisanal coarse salt.

Moreover, laboratory errors should be considered while manipulating samples. For instance, MPs were only analyzed using a stereoscope, with most of them measuring less than 500 μm. However, several studies indicate that from this size

Abundance of MPs *per* Kg($n_{samples}$)			Predominant		References
Sea salt	Well salt	*Fleur de sel*	Shapes	Colors	
550–681 (5)	7–204 (5)	—	Fr/Fb	BLa/G/BLu/W	[14]
50–280 (16)	115–185 (5)	—	Fb	BLa/G/ BLu/W	[5]
0–10 (17)	0 (5)	—	Fr/Fb/Fl	ND	[3]
16–84 (5)	—	—	Fb/Fr/Fl	ND	[29]
46.7–806 (11)	113–367 (1)	—	Fb/Fr	BLu/G/R/T	[30]
22–19,800 (12)	—	—	Fr/Fb	BLa/T/G/BLu	[31]
0–13,629 (28)	0–148 (9)	—	Fr/Fb	W/T/BLa/BLu	[8]
595–5090 (6)	3325–6430 (1)	2320–6430 (7)	Fb/Fr	T/BLu/BLa	This study

The microplastic particles are ordered from most to least predominant.
Fr—Fragments; Fb—Fibers; Fl—Films; BLa—Black; BLu—Blue; G—Green; R—Red; T—Transparent; W—White; ND—No data.

Table 3.
Comparison between the concentrations of MPs found in several published articles and the current study.

an underestimation of the real value may occur. Hidalgo-Ruiz et al. [37], stated that the visual identification of MPs is a valid method for dimensions superior to 500 μm, while MPs smaller than this threshold need to be analyzed using stricter methodologies. A visual identification-only approach can lead to underestimations ranging from 20% [38] to 70% [37], with that percentage increasing inversely with MPs' size [31].

Among the studies conducted so far, two presented similar results to those found in the Portuguese table salts: Renzi and Blašković [31] analyzed sea salt from Italy and Croatia and found between 22 and 19,800 MPs/Kg; and Kim et al. [8], that discovered 0 to 13,629 MPs/Kg and 0 to 148 MPs/Kg in sea salt and well salt, respectively, from worldwide salt samples. In this study, values ranged from 595 to 5090 MPs/Kg in sea salt and from 3325 to 6430 MPs/kg in Rio Maior's well salt.

Regarding *fleur de sel*, values ranged from 2320 to 6430 MPs/Kg and, for coarse salt, between 595 and 3985 MPs/Kg. Until now, only Portuguese salt has had a higher concentration of MPs in well salt [33], when compared to sea salt. Unfortunately, there is a lack of studies in other regions with similar characteristics to Rio Maior's saltworks, which do not allow for further data analysis. Concerning the MPs classification (shape and color), this study presented similar results to Gündoğdu [29], Iñiguez et al. [5], and Kosuth et al. [30], with fibers and fragments being the most predominant overall, as well as, black, blue, and transparent microplastics.

Therefore, we can argue that regardless of its source, microplastic contamination in table salts is an emerging concern, mainly due to their public health implications and environmental pollution. Indeed, microplastics are consumed not only through table salts but also via tap water, beer, honey [30], atmospheric air [2] and a wide variety of seafood. In fact, bivalves are the most studied group of animals since most species are filterers and become easily contaminated [39–41]. Microplastics can be hazardous already by themselves in the environment, however, they also function as transporters/emitters of persistent organic pollutants (e.g., bisphenol A, organochlorides) due to their adsorption ability. Hence, it is important to investigate and evaluate possible transmission risks in which microplastics may affect public health [5].

The World Health Organization recommends a maximum daily salt (Na^+) intake of 5.0 g/day. Nevertheless, the Portuguese population generally consumes 8.0 g/day, due to a strong and rooted gastronomic culture. Therefore, with the values obtained in this study, Portuguese people would consume, on average, approximately 7443 or 12,325 MPs/year depending on the table salts origin (sea salt and well salt, respectively). If we calculate by the salt type, an average Portuguese person can consume 5551 or 10,769 MPs/year if they consume only coarse salt or *fleur de sel*, respectively. However, these values are only theoretical and that in reality the amount MPs ingested is relatively smaller, since WHO also takes into account the salt found in food additives and processed food for the maximum daily salt intake threshold [3].

5. Conclusions

The present study was the first to assess the presence/absence of microplastic polymers in *fleur de sel*, a type of salt formed at the surface of saltworks' crystallizers. Overall, microplastics were found in all samples of Portuguese table salt, regardless of their origin and type, with higher contamination values being found in Rio Maior's "well salt".

Moreover, more studies are needed, since salt is a very common and well-rooted ingredient in Portuguese gastronomic culture. In fact, there are other available table salts to the population that were not studied yet. Additionally, new management tools need to be applied to decrease the concentration of impurities (e.g., insects' feathers, exoskeletons, etc.,...) and contaminants, such as microplastics.

Nowadays, research needs to be increasingly strengthened regarding microplastics contamination so that better regulations can exist together with a broader understanding. Also, these regulations need to be supported by government agencies in order to implement actions that reduce the emission of plastics into the environment and, thus, preventing environmental and human impairments.

Acknowledgements

This study was funded by the "Fundação para a Ciência e a Tecnologia, I.P. (FCT), Portugal, with national funds (FCT/MCTES, "orçamento de Estado", project reference PTDC/MAR-PRO/1851/2014), and the European Regional Development Fund (ERDF) through the COMPETE 2020 program (POCI-01-0145-FEDER-016885) through the project "PLASTICGLOBAL—Assessment of plastic-mediated chemicals transfer in food webs of deep, coastal and estuarine ecosystems under global change scenarios" that is also funded by the Lisboa 2020 program (LISBOA-01-0145-FEDER-016885). The study was also supported by the Strategic Funding UID/Multi/04423/2013 through national funds provided by FCT and ERDF in the framework of the program Portugal 2020 to CIIMAR.

Conflict of interest

The authors declare no conflict of interest.

Author details

Ana Sofia Soares[1], Carlos Pinheiro[2,3], Uirá Oliveira[2,3] and Maria Natividade Vieira[2,3]*

1 Abel Salazar Institute of Biomedical Sciences, Porto, Portugal

2 Faculty of Sciences, University of Porto, Porto, Portugal

3 Interdisciplinary Centre of Marine and Environmental Research, Matosinhos, Portugal

*Address all correspondence to: mnvieira@fc.up.pt

References

[1] Barnes DK, Galgani F, Thompson RC, Barlaz M. Accumulation and fragmentation of plastic debris in global environments. Philosophical Transactions of the Royal Society London B Biological Sciences. 2009;**364**:1985-1998. DOI: 10.1098/rstb.2008.0205

[2] Lusher A. Microplastics in the marine environment: Distribution, interactions and effects. In: Bergmann M, Gutow L, Klages M, editors. Marine Anthropogenic Litter. 1st ed. Cham: Springer; 2015. pp. 245-307. DOI: 10.1007/978-3-319-16510-3_10

[3] Karami A, Golieskardi A, Keong Choo C, Larat V, Galloway TS, Salamatinia B. The presence of microplastics in commercial salts from different countries. Science Reports. 2017;7:1-11. DOI: 10.1038/srep46173

[4] Plastic Europe. An Analysis of European Plastics Production, Demand and Waste Data [Internet]. 2018. Available from: https://www.plasticseurope.org/application/files/6315/4510/9658/Plastics_the_facts_2018_AF_web.pdf

[5] Iñiguez ME, Conesa JA, Fullana A. Microplastics in Spanish table salt. Science Reports. 2017;7:1-7. DOI: 10.1038/s41598-017-09128-x

[6] Barboza LGA, Dick Vethaak A, Lavorante B, Lundebye AK, Guilhermino L. Marine microplastic debris: An emerging issue for food security, food safety and human health. Marine Pollution Bulletin. 2018;**133**:336-348. DOI: 10.1016/j.marpolbul.2018.05.047

[7] Jambeck JR, Geyer R, Wilcox C, Siegler TR, Perryman M, Andrady A, et al. Plastic waste inputs from land into the ocean. Science. 2015;**347**:768-771. DOI: 10.1126/science.1260352

[8] Kim JS, Lee HJ, Kim SK, Kim HJ. Global pattern of microplastics (MPs) in commercial food-grade salts: Sea salt as an indicator of seawater MP pollution. Environmental Science & Technology. 2018;**52**:12819-12828. DOI: 10.1021/acs.est.8b04180

[9] Austin HM, Stoops-Glas PM. The distribution of polystyrene spheres and nibs in Block Island sound during 1972-1973. Chesapeake Science. 1977;**18**:89-92. DOI: 10.2307/1350372

[10] Carpenter EJ, Smith KL. Plastics on the Sargasso Sea surface. Science. 1972;**175**:1240-1241. DOI: 10.1126/science.175.4027.1240

[11] Colton JB, Burns BR, Knapp FD. Plastic particles in surface waters of the Northwestern Atlantic. Science. 1974;**185**:491-497. DOI: 10.1126/science.185.4150.491

[12] Kartar S, Milne RA, Sainsbury M. Polystyrene waste in the Severn estuary. Marine Pollution Bulletin. 1973;**4**:144. DOI: 10.1016/0025-326X(73)90010-6

[13] Critchell K, Bauer-Civiello A, Benham C, Berry K, Eagle L, Hamann M, et al. Plastic pollution in the coastal environment: Current challenges and future solutions. In: Wolanski E, Day JW, Elliot M, Ramachandran R, editors. Coasts and Estuaries. 1st ed. Amsterdam, Netherlands: Elsevier; 2019. pp. 595-609. DOI: 10.1016/B978-0-12-814003-1.00034-4

[14] Yang D, Shi H, Li L, Li J, Jabeen K, Kolandhasamy P. Microplastic pollution in table salts from China. Environmental Science & Technology. 2015;**49**:13622-13627. DOI: 10.1021/acs.est.5b03163

[15] OSPAR. Assessment Document of Land-Based Inputs of Microplastics in the Marine Environment [Internet]. 2017. Available from: https://www.ospar.org/documents?v=38018

[16] Guzzetti E, Sureda A, Tejada S, Faggio C. Microplastic in marine organism: Environmental and toxicological effects. Environmental Toxicology and Pharmacology. 2018;**64**:164-171. DOI: 10.1016/j.etap.2018.10.009

[17] Li J, Qu X, Su L, Zhang W, Yang D, Kolandhasamy P, et al. Microplastics in mussels along the coastal waters of China. Environmental Pollution. 2016;**214**:177-184. DOI: 10.1016/j.envpol.2016.04.012

[18] Pinheiro C, Oliveira U, Vieira MN. Occurrence and impacts of microplastics in freshwater fish. Journal of Aquaculture and Marine Biology. 2017;**5**:00138. DOI: 10.15406/jamb.2017.05.00138

[19] Antunes J, Frias J, Sobral P. Microplastics on the Portuguese coast. Marine Pollution Bulletin. 2018;**131**:294-302. DOI: 10.1016/j.marpolbul.2018.04.025

[20] Antunes JC, Frias JGL, Micaelo AC, Sobral P. Resin pellets from beaches of the Portuguese coast and adsorbed persistent organic pollutants. Estuarine, Coastal and Shelf Science. 2013;**130**:62-69. DOI: 10.1016/j.ecss.2013.06.016

[21] Martins J, Sobral P. Plastic marine debris on the Portuguese coastline: A matter of size? Marine Pollution Bulletin. 2011;**62**:2649-2653. DOI: 10.1016/j.marpolbul.2011.09.028

[22] Bessa F, Barria P, Neto JM, Frias J, Otero V, Sobral P, et al. Occurrence of microplastics in commercial fish from a natural estuarine environment. Marine Pollution Bulletin. 2018;**128**:575-584. DOI: 10.1016/j.marpolbul.2018.01.044

[23] Rodrigues SM, Almeida CMR, Silva D, Cunha J, Antunes C, Freitas V, et al. Microplastic contamination in an urban estuary: Abundance and distribution of microplastics and fish larvae in the Douro estuary. Science of the Total Environment. 2019;**659**:1071-1081. DOI: 10.1016/j.scitotenv.2018.12.273

[24] Frias JP, Otero V, Sobral P. Evidence of microplastics in samples of zooplankton from Portuguese coastal waters. Marine Environmental Research. 2014;**95**:89-95. DOI: 10.1016/j.marenvres.2014.01.001

[25] Neves D, Sobral P, Ferreira JL, Pereira T. Ingestion of microplastics by commercial fish off the Portuguese coast. Marine Pollution Bulletin. 2015;**101**:119-126. DOI: 10.1016/j.marpolbul.2015.11.008

[26] Frias JP, Sobral P, Ferreira AM. Organic pollutants in microplastics from two beaches of the Portuguese coast. Marine Pollution Bulletin. 2010;**60**:1988-1992. DOI: 10.1016/j.marpolbul.2010.07.030

[27] Mizukawa K, Takada H, Ito M, Geok YB, Hosoda J, Yamashita R, et al. Monitoring of a wide range of organic micropollutants on the Portuguese coast using plastic resin pellets. Marine Pollution Bulletin. 2013;**70**:296-302. DOI: 10.1016/j.marpolbul.2013.02.008

[28] Ogata Y, Takada H, Mizukawa K, Hirai H, Iwasa S, Endo S, et al. International Pellet Watch: Global monitoring of persistent organic pollutants (POPs) in coastal waters. 1. Initial phase data on PCBs, DDTs, and HCHs. Marine Pollution Bulletin. 2009;**58**:1437-1446. DOI: 10.1016/j.marpolbul.2009.06.014

[29] Gündoğdu S. Contamination of table salts from Turkey with microplastics. Food additives & contaminants. Part A, Chemistry, analysis, control, exposure & risk assessment. 2018;**35**:1006-1014. DOI: 10.1080/19440049.2018.1447694

[30] Kosuth M, Mason SA, Wattenberg EV. Anthropogenic contamination of tap water, beer, and sea salt. PLoS One. 2018;**13**:1-18. DOI: 10.1371/journal.pone.0194970

[31] Renzi M, Blaskovic A. Litter & microplastics features in table salts from marine origin: Italian versus Croatian brands. Marine Pollution Bulletin. 2018;**135**:62-68. DOI: 10.1016/j.marpolbul.2018.06.065

[32] Santos M, Nascimento AC, Santiago S, Gama AC, Calhau MA. O sal na alimentação dos portugueses. Instituto Nacional de Saúde. 2016;**8**:17-20. DOI: 10400.18/4125

[33] Calado C, Brandão JM. Salinas interiores em Portugal: O caso das marinhas de Rio Maior. Geonovas. 2009;**22**:45-54. DOI: 10400.9/926

[34] Napper IE, Thompson RC. Release of synthetic microplastic plastic fibres from domestic washing machines: Effects of fabric type and washing conditions. Marine Pollution Bulletin. 2016;**112**:39-45. DOI: 10.1016/j.marpolbul.2016.09.025

[35] Branca DL. Uniformity of low density fibrous structures and the effects of manufacturing processes on apparent density [thesis]. Syracuse: State University of New York College of Environmental Science and Forestry; 2007

[36] Rodrigues CM, Bio A, Amat F, Vieira MN. Artisanal salt production in Aveiro/Portugal-an ecofriendly process. Saline Systems. 2011;**7**:3. DOI: 10.1186/1746-1448-7-3

[37] Hidalgo-Ruz V, Gutow L, Thompson RC, Thiel M. Microplastics in the marine environment: A review of the methods used for identification and quantification. Environmental Science & Technology. 2012;**46**:3060-3075. DOI: 10.1021/es2031505

[38] Eriksen M, Mason S, Wilson S, Box C, Zellers A, Edwards W, et al. Microplastic pollution in the surface waters of the Laurentian Great Lakes. Marine Pollution Bulletin. 2013;**77**:177-182. DOI: 10.1016/j.marpolbul.2013.10.007

[39] Davidson K, Dudas SE. Microplastic ingestion by wild and cultured Manila clams (*Venerupis philippinarum*) from Baynes Sound, British Columbia. Archives of Environmental Contamination and Toxicology. 2016;**71**:147-156. DOI: 10.1007/s00244-016-0286-4

[40] Foekema EM, De Gruijter C, Mergia MT, van Franeker JA, Murk AJ, Koelmans AA. Plastic in North Sea fish. Environmental Science & Technology. 2013;**47**:8818-8824. DOI: 10.1021/es400931b

[41] van Cauwenberghe L, Janssen CR. Microplastics in bivalves cultured for human consumption. Environmental Pollution. 2014;**193**:65-70. DOI: 10.1016/j.envpol.2014.06.010

Chapter 2

Systemic-Entropic Approach for Assessing Water Quality of Rivers, Reservoirs, and Lakes

Gevorg Simonyan

Abstract

Water is a nonrenewable resource, and its unsustainable use almost everywhere has led to a decrease in water quality. The development of water quality indices and the introduction of indexing methods used in assessing the quality of surface waters (pollution) are particularly relevant in recent years. As a result of anthropogenic pollution of the aquatic environment, the entropy of the system changes, which is not always taken into account in hydrochemical studies. This chapter analyzes dozens of freshwater quality indicators existing in science literature and presents the advantages of the water quality indicators developed by the author and colleagues: the geoecological evolving organized index (GEVORG), and the Armenian Water Quality Index. Water quality analyses have been tested for most of the rivers, reservoirs, and lakes of Armenia. It was found that the Armenian Water Quality Index has a linear relationship with most water quality indexes, and an inverse relationship with the Canadian Water Quality Index. The quality of river and reservoir water has been assessed according to the new standards for background concentrations.

Keywords: water quality index, GEVORG index, Armenian Water Quality Index, rivers, reservoirs and lakes

1. Introduction

Water resources play a vital role in various sectors of economy, such as industrial activities, agriculture, forestry, fisheries, hydropower, and other creative activities [1].

The study of the ecological status of rivers, reservoirs, and lakes in Armenia is important for assessing the quality of their water, as well as for their further rational use.

The suitability of water sources for human consumption is studied using a water quality index (WQI), which is one of the most effective ways to describe water quality. WQI uses water quality data and helps in changing policies that are formulated by various environmental monitoring agencies [2].

WQI was first developed by Horton (HWQI) [3] in the United States, and is based on the 10 most frequently used water quality variables, such as dissolved oxygen (DO), pH, coliform, conductivity, alkalinity, chloride, etc., which are widely used and accepted in European, African, and Asian countries. Horton placed grading scales and weights for determining factors to give the relative importance of each parameter for assessing water quality. Furthermore, a new WQI similar to Horton's index has also been developed by the group of Brown in 1970 [4], which is based on weights to individual parameter. Recently, many modifications have been considered for WQI (BWQI) concept by various scientists and experts. Canadian Council of Ministers of the Environment has developed a WQI, Canadian Water Quality Index (CWQI), which can be applied by many water agencies in various countries with slight modification [5]. In 1995, the Canadian Ministry of the Environment developed the British Columbia Water Quality Index [6]. The Oregon Water Quality Index (OWQI) takes into account eight water quality variables (temperature, DO, pH, BOD, total phosphorus, total solids, fecal *E. coli*, ammonia, and nitrate nitrogen). The Delphi method has been used to select variables [7]. Malaysia Water Quality Index (MWQI) developed by the Department of the Environment of Malaysia was successfully applied for measuring water quality of 462 rivers in Malaysia. The calculation includes six water parameters: DO, BOD, COD, ammonia nitrogen, suspended solids, and pH [7]. The Bascaron Water Quality Index was developed by Bascarón specifically for Spain [8]. In 1976, the Scottish Engineering Department improved and developed the Scottish Water Quality Index [9]. An effective gradation index for diagnosing a generalized river quality has been developed and illustrated with the case study of the Keya River in Taiwan (TWGI) [10]. The Universal Water Quality Index for Turkey was developed by Boyacioglu [11] based on water quality standards set by the Council of European Communities. Sargaonkar and Deshpande [12] developed Overall Index of Pollution (OIP) for Indian rivers based on measurements and subsequent classification of pH, turbidity, dissolved oxygen, BOD, hardness, total dissolved solids, total coliforms, arsenic, and fluoride.

Some indexes and their variables are given in **Table 1**.

For the evaluation of the degree of water contamination, the comprehensive indicators are used, which make it possible to evaluate the contamination of water at the same time on a wide range of quality indicators. Water Contamination Index (WCI), CWQI, and specific combinatorial water quality index (SCWQI) are used for the evaluation of surface water quality in Republic of Armenia [5, 13, 14]. It must be noted that most developed complex characteristics of water bodies are in one way or another connected with the existing maximum allowable concentration (MAC) [15, 16].

According to the Water Framework Directive (WFA) (2000/60/EC) developed by the European Union (EU), all European surface waters should be in good ecological condition after 2015, and water bodies with poor quality water should be improved through targeted measuring. Each EU Member State has developed schemes for water quality classification according to WFD [17]. For example, in France, the SEQ-system is used for the classification of river water quality, consisting of three sections. To classify water quality, 15 descriptors are separated into 156 indicators, taking into account similar factors and effects. The evaluation is carried out using the boundary value table, which defines the boundaries of classes.

The index values are calculated based on parameters, which are classified into five classes based on the water usability. Germany's chemical quality classification scheme consists of four main classes and three subclasses, with a similar biological classification. The grades obtained are mapped through color codes.

Water quality assessment in the Danube River Basin according to the EU WFD (2000/60/EC) program is carried out according to separate indicators [17, 18]. In this classification scheme, indicators are classified into five classes. Class I is referred to as "reference" or background concentration; class II is a target value that should be followed; classes III–V are part of the "non-executable" classification scheme and their values are usually 2–5 times higher than the target value.

Parameters	HWQI	BWQI	MWQI	OWQI	OIR	TWQI
DO	+	+	+	+	+	+
BOD_5		+	+	+	+	+
COD			+			
pH	+	+	+	+	+	+
t°	+	+		+		+
Conductivity	+					
Carbon chloroform extract	+					
Turbidity		+			+	+
Hardness					+	
Suspended solids			+			+
Alkalinity	+					
Obvious pollution	+					
Sanitation facility	+					
Fecal *E. coli*				+		
Total coliforms					+	
Fecal coliforms count	+	+				+
Nitrate		+				
Nitrate nitrogen				+		
Ammonia nitrogen			+			+
Ammonia				+		
Total phosphorus		+		+		
Total dissolved solids				+	+	
Total solid content		+				
Chloride	+					
Fluoride					+	
As					+	
Cd						+
Cu						+
Cr						+
Pb						+
Zn						+

Table 1.
Some indexes and their variables.

According to the EU WFW Rural Water Quality Assessment, due to the lack of biological monitoring, assessment was made only with the use of chemical indicators of water quality. Natural background concentrations of hydrochemical indices were taken into account. The determination of background concentrations according to the EU WFD was performed using a statistical method using the logarithmic probability distribution function. The expected background status of the reference state is the absence or insignificance of anthropogenic pressure. It is closely connected with background concentration (BC). Background concentration is the value of the water quality indicator concentration before exposure to any source of pollution.

The Government of the Republic of Armenia ("Decree No. 75-N of March 27, 2011") established a new system for assessing surface water quality in Armenia for each water quality indicator for each watercourse [19]. The advantages of the new water quality standards in Armenia are that, firstly, the classification of environmental norms is based on natural BC, and secondly, the choice of indicators was made taking into account the load on the surface waters of the Republic of Armenia (based on 43 water indicators). The calculations of the BC took place in the RA rivers in 2005–2010 hydrochemical monitoring.

In recent years, for a comprehensive assessment of surface water quality, we have proposed the geoecological evolving organized index (GEVORG) or entropy water quality index (EWQI) and the Armenian Water Quality Index (AWQI) [20, 21].

Using EWQI and AWQI, a comprehensive assessment of surface water quality was carried out [22–26], and a structural analysis of the state of biological systems at the level of proteins, ribonucleic acid, and cell [27, 28]; of the state of trees [29]; and of the state of naftide systems [30] was made.

The aim of this work is to assess the water quality of the rivers, reservoirs, and Lake Sevan using the Armenian Water Quality Index and for the WFD using BC.

2. Materials and methods

2.1 Study area

2.1.1 Rivers

Dzknaget River is a river in the Gegharkunik and Tavush regions of Armenia. It is located in the eastern slopes of the Pambak Mountains and 1 km south of Tsovagyugh in the north-western corner of Lake Sevan. The river's length is 22 km. In this river, the fish caviar of Lake Sevan are debugged. Partly because of this reason, the river was named after a *river of fish*. There are two monitoring posts: No. 60–0.5 km above Semyonovka and No. 61—at the mouth of the river [31–33].

Masrik is a river in the Gegharkunik region of Armenia. It starts from the slopes of the eastern Sevan Mountains and flows into Lake Sevan in the north of the village of Tsovak. Its length is 45 km. The catchment area is 682 km^2, and the annual runoff is 131 million m^3. There is a monitoring post, No. 63—at the river's mouth [31–33].

Sotk (Zod), a river in the Gegharkunik region, is the right tributary of Masrik. It starts from the western slopes of the eastern Sevan ridge at a height of 2670 m. The length of the river is 21 km, the catchment basin area is 59.5 km^2. In the upper and middle streams, it flows through the V-shaped valley. Average annual expenditure is 0.28 m^3/s. Its water is used for irrigation. There are two monitoring posts: No. 64 —0.5 km from the mine top and No. 65—at the mouth of the river [31–33].

Vardenis River, is a river in the Gegharkunik region, in the Lake Sevan basin. It starts from the northern slopes of the central part of the Vardenis Range, at an altitude of 3215 m. The river's length is 28 km, the catchment basin area is 116 km^2. River valley is V-shaped in the upper and middle currents, extending below it, leaving the semi-desert plain and north of Lake Vardenik into Lake Sevan. Its water is used for irrigation. There is a monitoring post, No. 70—at the mouth of the river [31–33].

Martuni River, is a river in the Gegharkunik region, in the Lake Sevan basin. It starts from the northern slopes of the Vardenis Ridge, at an altitude of 3300 meters. Its length is 27.6 km, and the catchment basin area is 101 km^2. River valley is a

V-shaped at an upper flow, on average, a cane. There are two monitoring posts: No. 71—0.5 km from Geghahovit top and No. 72—at the mouth of the river [31–33].

Argichi is a river in the Gegharkunik region, in the basin of Lake Sevan; it starts from the northern slope of the Gndasar mountains of the Geghama mountain range, at a height of 2600 m. The river's length is 51 km, and the drainage basin area is 384 km^2. Its water is used for irrigation purposes and energy production. There is a monitoring post, No. 74—at the river's mouth [31–33].

Gavaraget is a river in the Gegharkunik region, in Lake Sevan basin. It starts from the northern slope of the Geghama mountain range, at a height of 3050 m and flows into Lake Sevan. The river's length is 50 km, the drainage basin area is 480 km^2. The river freezes in winter. Its water is used for irrigation purposes and energy production. There is a monitoring post, No. 74—at the river's mouth [31–33].

The locations and monitoring posts of all the mentioned rivers are given in **Figure 1**.

2.1.2 Reservoirs

Akhurian Reservoir is located in the Akhurian River basin in Armenia and Turkey. The reservoir has a surface area of 54 km^2, and maximum length of 20 km. It is one of the largest reservoirs in the Caucasus, with coordinates 40° 33′ 47.67″ N 43° 39′ 16.26″ E.

Lake Arpi is situated in the north-west of the Republic of Armenia. The lake is fed by meltwater and four streams, and it is the source of the Akhurian River. Being an alpine-specific ecosystem with its rare flora and fauna, it ensures ecological balance of adjacent extensive area. The reservoir-lake is 7.3 km long and 4.3 km wide, with an area of 20 km^2 and coordinates 41° 03′ 0″ N 43° 37′ 00″ E.

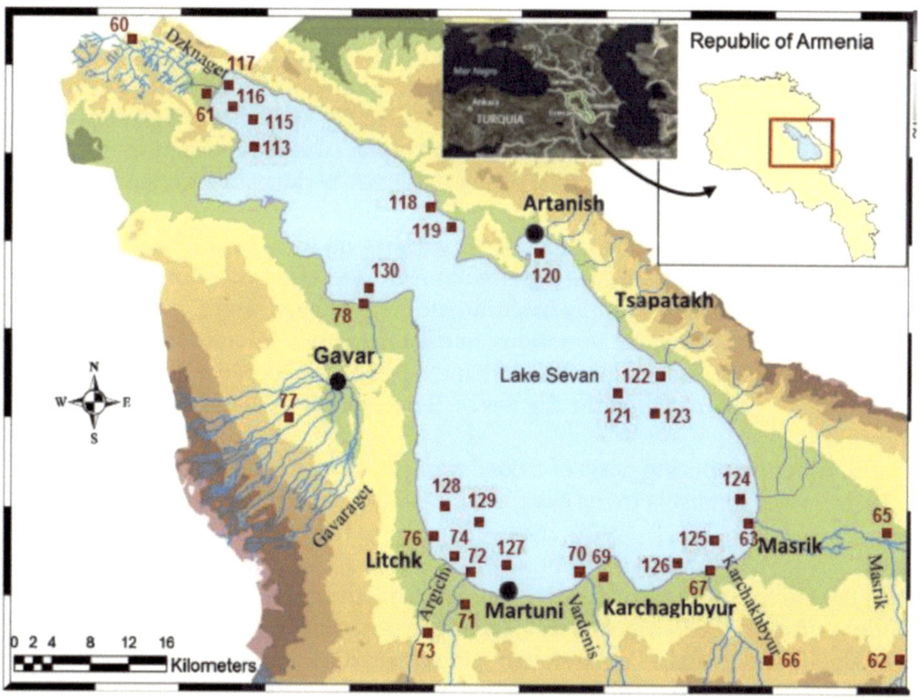

Figure 1.
Location of monitoring posts in Lake Sevan and rivers Dzknaget, Gavaraget, Argichi, Martuni, Vardenis, Masrik, and Sotq.

Yerevan Lake is an artificial reservoir located in the capital of Armenia in Yerevan. The reservoir-lake Yerevan is 7.3 km long and 5.0 km wide, with an area of 0.65 km^2 and coordinates 40° 9′ 35.04″ N 44° 28′ 36.54″ E.

Aparan Reservoir is located in the Aragatsotn region of Armenia. It has been built on the river Kassah. It has an area of 7.9 km^2 and coordinates 40° 29′ 49″ N 44° 26′ 07″ E. Its water is used for irrigation.

Kechut Reservoir is located in the Vayots Dzor region of Armenia, on the river Arpa, 3.5 km south of the resort town of Jermuk. The reservoir was built in 1981. Water from it through the conduit enters Lake Sevan to regulate the level. It has coordinates 39° 47′ 54″ N 45° 39′ 22″ E.

Azat Reservoir is located in Armenia, in the Ararat region, above the village of Lanjazat, at an altitude of 1050 m above sea level. The reservoir was built on the Azat River. The volume of the lake is 70 million m^3. Its water is used to irrigate the Ararat Plain. It has coordinates 40° 04′ 00″ N 44° 36′ 00″ E [21, 31–33].

2.1.3 Lake Sevan

Lake Sevan is located in the north-eastern part of the Armenian Highland, in the Gegharkunik Region. Sevan is considered to be one of the three ancient and biggest lakes (Vana and Urmia) of the Armenian Kingdom. It was called the "blue eyes" of Armenia and is surrounded by Geghama, Vardenis, Pambak, Sevan, and Areguni mountain chains. The blue beauty of Armenia is situated at an altitude of 1900 m above sea level and the total surface area is about 5000 km^2. It was famously known as "Geghama Tsov (in English sea), Gegharkunyats Tsov." It is the world's second-highest lake with freshwater after the Titicaca in South America and is the largest in the South Caucasus. The lake's length is 70 km, and maximum width is 55 km. It has an area of 1240 km^2 (1360 km^2 before the level is lowered). Twenty-eight rivers flow into the lake, the largest of which reaches a length of 50 km. Only one river flows from Sevan-Hrazdan, which flows into the Araks. The mineralization of water is about 700 mg/l. Lake is of tectonic barren nature. The basin of the same name is of tectonic origin, and the dam was formed due to the outflow of the Holocene lavas. The lake consists of two unequal parts called Big and Small. The Sevan's Peninsula is located in the north-western part of the lake and it is famous for its medieval monasteries and khachkars (cross-stones). Sevanavank is a monastery complex situated on the peninsula.

Small and Big Sevan: Small Sevan is very deep—up to 83 m and has rugged banks. It is in this part that the greater volume of lake water is concentrated. In the Big Sevan, the bottom is flat, the banks are not very rugged, and the depth does not exceed 30 meters. There are 26 stations on the Lake Sevan (monitoring posts), from No. 115 to No. 119; also stations No. 130 and 131 are located in the Small Sevan and those from No. 120 to 129 in the Big Sevan. These monitoring posts are shown in **Figure 1**.

The water monitoring posts of Lake Sevan are located at: No. 115—3.5 km distance from the peninsula to the east; No. 116—70° azimuth from the peninsula, from the surface; No. 117—a distance of 1 km from the Dzknaget river, from the surface; No. 117—a distance of 1 km from the Dzknaget river, at depth of 20 m; No. 118—0.5 km south-west from the village Shorzha, from the surface; No. 119—6 km south-west from the village Shorzha, from the surface; No. 119′—6 km south-west from the village Shorzha, at a depth of 20 m; No. 120—2 km from the village Artanish with 135° azimuth, from the surface; No. 120′—2 km from the village Artanish with 135° azimuth, at depth of 20 m; No.121—10 km from the village Pambak with 270° azimuth, from the surface; No. 121′—10 km from the village Pambak with 270° azimuth, at a depth of 20 m; No. 122—2.2 km from the village Pambak with 255° azimuth from the surface; No. 122′—2.2 km from the village Pambak with 255° azimuth, at a depth of 20 m; No. 123—13 km from the village Pambak with 235° azimuth, from the surface; No. 123′—13 km from the village Pambak with 235° azimuth, at a depth of 20 m; No. 124—1 km from the village Tsovak to the north-west from the surface; No. 125—1 km from the mouth of the river Karchaghbyur to the west, from the surface; No. 126—at Arpa-Sevan tunnel exit; No. 127—1.5 km from the city of Martuni, to the north, from the surface; No. 128—15 km from the village Eranos with 90° azimuth, from the surface; No. 128′—15 km from the village Eranos with 90° azimuth, at a depth of 20 m; No. 129—24 km from the village Eranos with 90° azimuth, from the surface; No. 129′—24 km from the village Eranos with 90° azimuth, from the surface, at a depth of 20 m; No. 130—7 km north-west of the village of Noratus, from the surface; No. 131–7.5 km north of the village of Chkalovka, from the surface; No. 131′—7.5 km north of the village of Chkalovka, at a depth of 20 m [21, 31–33].

2.2 Index determination

2.2.1 Canadian Water Quality Index (CWQI)

CWQI provides a consistent method, which has been formulated by Canadian jurisdictions, for conveying water quality information to both the management and public [5]. Moreover, a committee has been established under the Canadian Council of Ministers of the Environment WQI, which can be applied by numerous water agencies in various countries with slight modification. This method has been developed to evaluate surface water for protection of aquatic life in accordance to specific guidelines. The parameters related with various measurements may vary from one station to the other and sampling protocol requires at least four parameters, sampled at least four times. The calculation of index scores in CWQI method can be obtained by using the following relation:

$$CWQI = 100 - \frac{\sqrt{F_1^2 + F_2^2 + F_3^2}}{1.732}, \tag{1}$$

where scope (F_1) represent the percentage of variable that do not meet their objectives at least once during the time period under consideration (“failed variables”), relative to the total number of variables measured frequency (F_2) is the number of times by which the objectives do not meet; and amplitude (F_3) is the amount by which the objectives do not meet.

Therefore, five categories have been suggested for classification of water quality, which are summarized in **Table 2**.

2.2.2 Water contamination index (WCI)

WCI was established by the USSR Goskomgidromet (State Committee of Hydrometeorology) [13] and belongs to the category of indicators most often used to assess the quality of water bodies. This index is a typical additive coefficient and represents the average percentage of exceeding the MAC for a strictly limited number of individual ingredients:

$$WCI = \frac{1}{n}\sum_{i=1}^{n}\frac{C_i}{MAC_i}, \tag{2}$$

CWQI value	Rating of water quality	Water quality classes
95–100	Excellent water quality	1
80–94	Good water quality	2
60–79	Fair water quality	3
45–59	Marginal water quality	4
0–44	Poor water quality	5

Table 2.
Classes of water quality depending on the value of CWQI.

WCI value	Rating of water quality	Water quality classes
up to 0.2	Very clean	I
0.2–1.0	Clean	II
1.0–2.0	Moderately polluted	III
2.0–4.0	Contaminated	IV
4.0–6.0	Dirty	V
6.0–10.0	Very dirty	VI
>10.0	Extremely dirty	VII

Table 3.
Classes of water quality depending on the value of WCI.

where C_i is the concentration of the component (in some cases the value of the physicochemical parameter) and n is the number of indicators used for calculating the index, $n = 6$ (pH, biological oxygen demand of BOD_5 dissolved oxygen in water, petroleum products, nitrite ions (NO_2^-), and ammonium ion (NH_4^+)). Seven categories have been proposed for the classification of water quality, which are listed in **Table 3**.

2.2.3 Specific combinatory water quality index (SCWQI)

In accordance with RD 52.24.643-2002, "The method for the integrated assessment of the degree of contamination of surface waters by hydrochemical indicators" the calculation of the specific combinatorial water quality index has been introduced [14]. To assess the quality of water of rivers and water bodies, it is divided into several contamination classes. The classes are based on the intervals of the specific combinatory water pollution index, depending on the number of critical pollution indicators. At least 15 indicators are analyzed. The required list includes: dissolved oxygen in water, BOD_5, chemical oxygen consumption—COD, phenols, petroleum products, nitrite ions (NO_2^-), nitrate ions (NO_3^-), ammonium ion (NH_4^+), iron total (Fe^{2+} and Fe^{3+}), copper (Cu^{2+}), zinc (Zn^{2+}), nickel (Ni^{2+}), manganese (Mn^{2+}), chlorides, and sulfates. The value of SCWQI is determined by the frequency and the multiplicity of the MPC exceeding by several indicators and can vary in waters of different degrees of contamination from 1 to 16 (for pure water is 0). The highest index value corresponds to the worst water quality. Taking into account the number of bullpen, it allows dividing the surface waters into five classes, depending on the degree of their contamination. The third and fourth classes for more detailed water quality assessment are respectively divided into two and four categories.

2.2.4 Geoecological evolving organized index and Armenian index of water quality

An open system can exchange energy, material, and, which is not less important, information from environment. The system consumes information from the environment and provides information to environment for acting and interacting with environment. Shannon [34, 35] was the first who related concepts of entropy and information. He has suggested that entropy is the amount of information attributable to one basic message source, generating statistically independent reports. The information entropy for independent random event *x* with *N* possible states is calculated by the following equation:

$$H = -\sum_{i=1}^{N} p_i \log_2 p_i, \tag{3}$$

where P_i is the probability of frequency of occurrence of an event.

Different processes in hydroecological systems can occur both with increase and decrease in of entropy. Pollution of water systems can be represented as a system of the hydrochemical parameters (elements), the concentration of which exceeds the MAC. Then, in the equation, Shannon P_i is the probability of the number of cases of MAC excess of i-substance or indicator of water of total cases of *MAC-N*, $P_i = n_i/N$.

For determination of the values of the EWQI and AWQI of environmental quality, the following computational algorithm is used [17–19]:

1. To determine the number of cases of MAC excess of i-substance or indicator of water

2. Estimate the total amount of cases at the maximum allowable concentration (N)—$N = \sum n$.

3. Compute $\log_2 N$, $n\log_2 n$ and $\sum n\log_2 n$.

4. Determine geoecological syntropy (*I*) and entropy (*H*):

$$I = \sum n \log_2 n/N \tag{4}$$

and

$$H = \log_2 N – I. \tag{5}$$

5. Then GEVORG index (G) is determined:

$$G = H/I \tag{6}$$

6. Further, the total amount multiplicity of MAC exceedances is estimated: (M)– $M = \sum m$.

7. Then, $\log_2 M$ is computed.

8. Finally, Armenian Water Quality Index is obtained:

$$AWQI = G + 0.1 \log_2 M. \tag{7}$$

Therefore, five categories have been suggested for classification of the water qualities, which are summarized in **Table 4**.

GEVORG value	AWQI value	Rating of water quality	Water quality classes
< 0.7	< 1.1	Excellent water quality	1
0.7–1.0	1.1–1.4	Good water quality	2
1.0–1.4	1.4–1.8	Fair water quality	3
1.4–1.7	1.8–2.1	Marginal water quality	4
> 1.7	> 2.1	Poor water quality	5

Table 4.
Classes of water quality depending on the value of EWQI and AWQI.

Water quality class	Assessment	Water quality
1		Excellent
2		Good
3		Moderate
4		Unsatisfactory
5		Bad

Table 5.
Water quality classification by EU WFD.

2.2.5 Water quality classification by EU WFD

The calculations of the BC took place in the RA rivers in 2005–2010 hydrochemical monitoring.

According to the decision of the Government of the Republic of Armenia, "On establishing standards for ensuring water quality for each area of water basin management," there are five classes: "Excellent" (1st grade), "Good" (2nd grade), "Moderate" (3rd grade), "Unsatisfactory" (4th grade), and "Bad" (5th grade). Each class is indicated by color (**Table 5**). A general assessment of the chemical quality of water is performed by the class of the lowest quality indicator. So if different quality indicators of a surface water body fall into different quality classes, the final classification is considered the worst. The following principle applies: "If someone is in bad shape, then everyone is in poor condition" or the principle "someone is out, everyone is out."

3. Results and discussion

3.1 Results for rivers

In this work, we present data on the study of water quality of rivers in 2009–2019. Since 2013, in Armenia, the quality of river water has been assessed by the new standards for background concentrations.

The quality of the waters of the Dzknaget, Sotk, Masrik, Vardenis, Martuni, Argichi, and Gavaraget rivers is comprehensively evaluated by the indices: AWQI, EWQI, WCI, CWQI, and SCWQI.

The values of the WQIs are shown in **Table 6**.

With the help of the computer program "Origin-6," an analysis of the linear relationship between AWQI and other WQIs is done: AWQI = $a + b$ (WQI).

Sampling points	EWQI	AWQI	WCI	CWQI	SCWQI
60	0.415	0.650	0.77	90.38	0.8
61	0.856	1.208	0.92	83.98	1.48
63	0.604	0.993	2.21	78.71	1.74
64	0.321	0.559	0.64	88.74	1.31
65	0.642	0.989	1.2	75.25	1.86
70	0.370	0.625	0.82	90.63	1.20
71	0.625	0.899	0.66	90.52	0.68
72	0.333	0.584	0.95	86.62	1.40
74	0.303	0603	1.51	81.7	1.04
77	0.955	1.325	1.62	83.8	1.38
78	0.625	1.077	3.86	70.14	2.15

Table 6.
Water quality indices of rivers (2009).

$$AWQI = (0.196 \pm 0.060) + (1.217 \pm 0.095)\blacksquare EWQI, R = 0.97914, N = 9$$
$$AWQI = (0.717 \pm 0.142) + (0.127 \pm 0.085)\blacksquare WCI, R = 0.46584, N = 9$$
$$AWQI = (0.539 \pm 0.287) + (0.251 \pm 0.196)\blacksquare SCWQI, R = 0.41219, N = 9$$
$$AWQI = (2.685 \pm 0.957) - (0.021 \pm 0.011)\blacksquare CWQI, R = 0.55362, N = 9$$

Analysis of obtained data indicates that AWQI has liner dependence on WCI, SCWQI, and EWQI and an inverse dependence on CWQI. This result is based on the fact that the scale of the Canadian index of quality of water begins from 100, and scales of indexes of impurity of water, and EWQI, WQI, and SCWQI, start from scratch.

The quality of the water in the rivers was also evaluated according to the new standards of background concentrations (see **Table 7**).

In 2013–2019, the waters of the Dzknaget, Martuni, Sotk, Gavarvget rivers (monitoring post 77) and Martuni (monitoring post 71) were found to be of "moderate" or "good" quality. The water at the mouth of the Vardenis and Gavarvget

River	Sampling points		2013	2014	2015	2016	2017	2018	2019
Dzknaget	60	0.5 km above Semyonovka							
	61	River mouth							
Masrik	63	River mouth							
Sotq	64	0.5 km from the mine top							
	65	River mouth							
Vardenis	70	River mouth							
Martuni	71	0.5 km from Geghhovit							
	72	River mouth							
Argichi	74	River mouth							
Gavaraget	77	0.5 km from Tsakhkvan							
	78	River mouth							

Table 7.
Water quality classes of analyzed rivers.

rivers had an average and "unsatisfactory" quality for ammonium ions and phosphate. The water at the mouth of the Martuni River in 2014 was of "poor" quality for ammonium and phosphate ions, and the water at the mouth of the Masrik River in 2017–2019 was also of "poor" quality for vanadium.

3.2 Results for reservoirs

In this chapter [26], we studied the quality of water in the years 2009–2012 of the reservoirs of the lakes of Arpi, Yerevan, Akhuryan, Azat, Aparaan, and ketchut using AWQI, EWQI WCI, and SCWQI, and CWQI. An analysis of the data shows that AWQI has a linear relationship with WCI, SCWQI, and EWQI and an inverse relationship with CWQI.

In this work, we presented data on the study of water quality in reservoirs in 2013–2019. Since 2014, in Armenia, the quality of reservoir water has been assessed by the new standards for background concentrations.

In 2013, it was found out that the reservoirs of lakes Arpi, Yerevan and Akhuryan regularly increased the MACs of nitrite ions, ammonium, copper, vanadium, aluminum, chromium, manganese and iron. For example, in reservoir Akhuryan for NO^{2-}, Al, V, Cu, Mn, and Cr the number of cases of an increase in the MAC is 5, 8, 8, 8, 6 and 7 times, respectively. The amount of excess cases of MPC – $N = 42$, $\sum n \log_2 n = 118.76$, $I = 118.76/42 = 2.8276$, $H = \log_2 42 – 2.8276 = 2.5616$, AWGI = $G = 2.5616/ 2.8276 = 0.9059$. The total amount of the multiplicity of MAC

Reservoir	Lake Arpi		Akhuryan		Aparan		Lake Yerevan	
Positions	109		110		111		112	
Indicator	n	$n\log_2 n$	n	$n\log_2 n$	n	$n\log_2 n$	n	$n\log_2 n$
BOD_5	0	0	0	0	5	11.61	0	0
NH_4^+	0	0	0	0	0	0	10	33.2
NO_2^-	0	0	5	11.61	0	0	12	43
Al	6	15.51	8	24	8	24	2	2
V	6	15.51	8	24	8	24	12	43
Cu	6	15.51	8	24	6	15.51	11	38
Mn	5	11.61	6	15.51	8	24	11	38
Se	0	0	0	0	0	0	8	24
Cr	5	11.61	7	19.64	0	0	12	43
N	28		42		35		78	
$\sum$nlog2n	69.75		118.76		99.12		264.2	
I	2.491		2.8276		2.8320		3.3871	
H	2.313		2.5616		2.2943		2.8946	
EQWI	0.9288		0.9059		0.8101		0.8546	
$M = \Sigma m$	33.6		24.5		15.2		49.3	
$\log_2 M$	5.067		4.612		3.924		5.620	
AQWI	1.4355		1.3671		1.2025		1.4166	

Table 8.
Entropic and Armenian water quality indexes for reservoirs of Lake Arpi, Akhuryan, Aparan, and Lake Yerevan (2013).

Reservoir	Azat		Ketchut	
Positions	**113**		**114**	
Indicator	**n**	**$n\log_2 n$**	**n**	**$n\log_2 n$**
BOD_5	4	8	0	0
Al	0	0	6	15.51
V	6	15.51	11	38
Cu	0	0	4	8
Mn	3	4.752	7	19.64
N	13		28	
$\sum$nlog2n	28.262		81.15	
I	2.173		2.898	
H	1.5253		1.9066	
EQWI	0.7019		0.6579	
M = Σm	17.9		9	
$\log_2 M$	4.159		3.168	
AQWI	1.1178		0.9746	

Table 9.
EQWI and AQWI for reservoirs of Azat and Ketchut (2013).

Reservoirs	2014	2015	2016	2017	2018	2019
Lake Arpi						
Lake Yerevan						
Akhuryan						
Aparan						
Azat						
Ketchut						

Table 10.
Water quality classes of analyzed reservoirs.

exceedances-M $= \sum$ m =24.5, $\log_2 M$ = 4.6123, AWQI = 0.9059 + 0.4612 = 1.3671. The values of the indices EWQI and AWQI are given in **Tables 8** and **9**.

An analysis of the data shows that AWQI has a linear relationship with EWQI. AWQI = $-(0.054 \pm 0.205) + (1.613 \pm 0.251)$•EWQI; R^2 = 0.95.457; N = 6.

The quality of the water in the reservoirs was also evaluated according to the new standards of background concentrations (see **Table 10**).

In 2014, the water of the Arpi Lake reservoir was of "moderate" quality in terms of phosphate ion and COD, and the water of the Akhuryan reservoir was "moderate" in terms of ammonium, nitrite, and phosphate ions. The water of the Azat reservoir was also of "moderate" quality in terms of phosphate ion, the water of the Aparan reservoir was "good" in terms of phosphate ion, and the water of Yerevan lake was of "poor" quality. The waters of the Kechut and Aparan reservoirs were of "good" quality. In 2015, the water of the reservoir of Lake Arpi had a "moderate" quality in terms of COD and the water of the Akhuryan reservoir had a "moderate" quality in terms of phosphate ion and COD. The water in Yerevan Lake had a "moderate" quality in terms of ammonium, nitrite, and phosphate ions and COD.

The water of the Azat reservoir was also of "moderate" quality according to COD, and the water of the Kechut and Aparan reservoir was of "good" quality. In 2016, the water of the reservoirs of Lake Arpi and Azat was of "good" quality, and the waters of the Akhuryan, Aparan, and Kechut reservoirs were "moderate" in COD. The water of Yerevan Lake had a "poor" quality in terms of ammonium, nitrate, nitrite, and phosphate ions and COD.

In 2017, the water of the reservoirs of Lake Arpi, Akhuryan, Aparan, Kechut, and Azat was of "good" quality, and the water of Yerevan Lake was of "unsatisfactory" quality for ammonium and nitrite ions. In 2018, the water in the reservoir of Lake Arpi had "moderate" quality in terms of phosphate ion and suspended solids, and the water in the Akhuryan reservoir had "moderate" quality in ammonium and phosphate ions, as well as in COD and BOD_5. The water of the Aparan reservoir also had "moderate" COD quality. The water of Yerevan Lake had "unsatisfactory" quality for ammonium and nitrite ions. The waters of the Kechutsky and Azat reservoirs were of "good" quality. In 2019, the water in the reservoir of Lake Arpi had "moderate" quality in terms of phosphate ion and suspended solids, and the water in the Akhuryan reservoir had "moderate" quality in terms of COD and suspended solids. The water of Yerevan Lake had "unsatisfactory" quality in terms of nitrite ion. The waters of the Kechut, Aparan, and Azat reservoirs were of "good" quality.

According to WQI values, the water quality in the Aparan, Azat, and Kechut reservoirs has "good" and "excellent" grade. The water quality of the reservoirs of Akhuryan, Lake Arpi and in Yerevan Lake, on the contrary, is "poor" from 3rd to 2nd class, and restricts the use of water for irrigation purposes. The poor water quality of the Lake Arpi reservoir is associated with an increase in the amount of metals. The reduced water quality of the Akhuryan reservoir and Yerevan lake is associated with pollution from the main settlements in the river basin, respectively, in Gyumri and Yerevan, with municipal wastewater.

3.3 Results for Lake Sevan

The purpose of this section is to assess the water quality of Lake Sevan using the Armenian Water Quality Index and other indicators of water quality, as well as to identify the causes of the appearance of blue-green algae that contribute to growth.

In July 2019, an increase in the blue-green algae of Anaben was recorded in Lake Sevan. These algae were first found in Lake Sevan in the middle of the last century,

Year	Mg	V	Cr	Cu	Se	BOD_5
2009	1.1–1.4	5.0–7.0	2.0–3.0	2.0–3.0	3.0–4.0	—
2010	1.1–1.3	5.0	2.0	2.0–3.0	2.0	2.0–3.0
2011	1.1	5.0	2.0	—	2.0	1.1–1.9
2012	1.1–1.2	6.2–6.4	—	2.1	2.1–2.6	—
2013	1.1–1.2	5.0–5.7	1.8	—	2.5	—
2014	1.1	3.8–5.6	1.2–3.4	—	1.2–19	1.2–1.5
2015	1.2–1.7	2.9–5.9	1.2–3.9	—	1.2–5.0	1.2–1.4
2016	1.2	3.8–8.0	1.2–1.7	1.4–1.5	1.2–6.5	1.2
2017	1.2	5.0–5.9	2.0–3.8	1.3–7.3	5.7–7.0	1.2–1.4
2018	1.2–1.3	3.7–6.6	1.7–3.3	1.2–2.9	1.4–2.7	1.2
2019	—	4.5–8.9	1.2–6.6	1.2–3.5	3.1–3.3	—

Table 11.
Excess concentration from MAC (times).

Positions	115		116		117		118		119		130	
Indicator	**n**	**nlog2n**	**n**	**nlog2n**	**n**	**nlog2n**	**n**	**nlog2n**	**n**	**nlog2n**	**n**	**nlog2n**
Mg	0	0	5	11.6	5	11.6	6	15.5	4	8	5	11.6
Cu	0	0	0	0	0	0	1	0	0	0	0	0
V	8	24	8	24	8	24	8	24	8	24	8	24
Cr	5	11.6	6	15.5	6	15.5	6	15.5	5	11.6	6	15.5
Br	8	24	8	24	8	24	8	24	8	24	8	24
Se	8	24	8	24	7	19.64	7	19.64	6	15.5	7	19.64
N	29		35		34		36		31		34	
∑nlog2n	83.6		99.1		94.74		98.64		83.1		94.74	
I	2.882		2.831		2.786		2.74		2.68		2.786	
H	1.974		2.295		2.298		2.427		2.271		2.298	
G	0.686		0.811		0.826		0.886		0.847		0.825	
M = ∑m	11.8		13		14		17.1		14.1		14.3	
log2M	3.56		3.7		3.81		4.093		3.815		3.836	
AWQI	1.041		1.181		1.207		1.294		1.228		1.209	

Table 12.
Entropic and Armenian water quality indexes for Small Lake Sevan (2009).

and the first manifestation of flowers was recorded in 1964 and repeated in different volumes at different times. Large-scale flourishing was observed in 2018.

It has been established that the maximum permissible concentration of vanadium, copper, chromium, magnesium, BOD_5, and selenium is regularly exceeded in the waters of Lake Sevan (see **Table 11**).

For example, in position No. 118 of Lake Sevan, number of MAC increasing cases for V, Br, Se, Cr, Mg, and Cu has been changed 8, 8, 7, 6, 6 and 1 times, respectively [36]. The amount of excess cases of MAC – $N = 36$, $\sum n \log_2 n = 98.64$, $I = 98.64/36 = 2.740$, $H = \log_2 36 – 2.740 = 2.427$, $G = 2.740/2.427 = 0.886$. The total amount of the multiplicity of MAC exceedances -$M = \sum m = 17.1$, $\log_2 M = 4.093$, AWQI = 0.886 + 0.409 = 1.294. The calculation algorithm and the values of the EWQI and AWQI indices of other position of Smally Sevan are given in **Table 12**.

Quality of Lake Sevan water was also comprehensively evaluated by other indexes: WCI, CWQI, and SCWQI. Values of the WQIs are given in **Table 13**.

It is shown that water quality by the EWQI and AWQI of the 2nd pollution class, by the WCI and CWQI of the 3rd pollution class, and by SPWQI is mainly the 2nd and in some cases up to the 3rd class of pollution.

With the help of the computer program "Origin-6", the analysis of the linear relationship between AWQI and other WQIs was provided: AWQI = $a + b$ (WQI).

Analysis of obtained data indicates that AWQI has liner dependence with WCI, SCWQI, and EWQI, and an inverse dependence with CWQI.

A satisfactory correlation is obtained when all the positions of the Lake Sevan are considered together.

- AWQI = (0.739 ± 0.074) + (0.313 ± 0.047) ▪ WCI,R = 0.80233, N = 26
- AWQI = (1.047 ± 0.127) + (0.096 ± 0.069) ▪ SCWQI, R = 0.27301, N = 26
- AWQI = (0.203 ± 0.038) + (1.225 ± 0.046) ▪ EWQI, R = 0.98339, N = 26

Sampling points	EWQI	AWQI	WCI	CWQI	SCWQI
115	0.686	1.041	1.35	68.65	1.75
116	0.811	1.181	1.41	70.37	1.65
117	0.826	1.207	1.42	70.95	1.56
117'	0.831	1.201	1.43	70.21	1.81
118	0.867	1.276	1.87	65.98	1.88
119	0.847	1.228	1.48	69.12	1.81
119'	0.923	1.334	1.90	66.35	2.11
120	0.939	1.340	1.67	67.24	1.65
120'	0.921	1.332	1.92	65.50	1.96
121	0.803	1.187	1.60	66.79	1.94
121'	0.799	1.192	1.56	66.89	1.83
122	0.820	1.204	1.50	66.65	2.19
122'	0.820	1.214	1.49	68.68	1.59
123	0.811	1.205	1.50	67.05	1.69
123'	0.819	1.212	1.55	66.58	1.75
124	0.819	1.212	1.53	68.16	1.85
125	0.819	1.222	1.52	66.49	1.59
126	0.825	1.229	1.54	67.27	1.82
127	0.833	1.237	1.55	67.14	1.78
128	0.883	1.312	1.75	66.02	2.11
128'	0.825	1.209	1.44	68.25	2.03
129	0.815	1.198	1.44	67.60	1.93
129'	0.815	1.198	1.52	67.96	1.79
130	0.825	1.209	1.43	68.64	1.63
131	0.834	1.217	1.43	68.48	1.88
131'	0.825	1.208	1.46	68.45	1.95

Table 13.
Water quality indices of Lake Sevan (2009).

- AWQI = (2.637 ± 0.513) – (0.021 ± 0.008) ▪ CWQI, R = 0.49061, N = 26:

For the Small Lake Sevan:

- AWQI = (0.787 ± 0.213) + (0.275 ± 0.143) ▪ WCI, R = 0.65192, N = 7
- AWQI = (0.965 ± 0.438) + (0.131 ± 0.252) ▪ SCWQI, R = 0.22730, N = 7
- AWQI = (0.189 ± 0.053) + (1.234 ± 0.064) ▪ EWQI, R = 0.99321, N = 7
- AWQI = (2.097 ± 1.361) – (0.013 ± 0.019) ▪ CWQI, R = 0.28427, N = 7:

For the Big Lake Sevan:

- AWQI = (0.529 ± 0.181) + (0.452 ± 0.116) ▪ WCI, R = 0.81003, N = 10
- AWQI = (1.292 ± 0.172) + (0.031 ± 0.092) ▪ SCWQI, R = 0.11946, N = 10
- AWQI = (0.252 ± 0.082) + (1.174 ± 0.099) ▪ EWQI, R = 0.97297, N = 10
- AWQI = (2.776 ± 0.1.935) − (0.023 ± 0.028) ▪ CWQI, R = 0.27118, N = 10:

A good correlation is also obtained when the underlying layers are considered together

- AWQI = (0.778 ± 0.054) + (0.287 ± 0.034) ▪ WCI, R = 0.9545, N = 9
- AWQI = (0.849 ± 0.208) + (0.205 ± 0.111) ▪ SCWQI, R = 0.57343, N = 9
- AWQI = (0.209 ± 0.061) + (1.217 ± 0.072) ▪ EWQI, R = 0.98775, N = 9
- AWQI = (2.998 ± 0.753) − (0.026 ± 0.011) ▪ CWQI, R = 0.66353, N = 9:

Thus, a correlation between AWQI and other WQIs was established. Analysis of obtained data indicates that AWQI has liner dependence on WCI, SCWQI, EWQI and an inverse dependence on CWQI. This result is based on the fact that the scale of the Canadian index of quality of water begins from 100, and scales of indexes of impurity of water, EWQI, WQI, and SCWQI, start from scratch. It has been established that the maximum permissible concentrations of copper, vanadium, chromium, magnesium, and selenium regularly increase in the waters of Lake Sevan. It has been found that the Armenian Water Quality Index demonstrates a linear dependence on the water contamination index, a specific combinatorial water quality index, and an index of geoecological evolving organization and an inverse relationship to the Canadian Water Quality Index. It is shown that water quality by the geoecological evolving organized index and Armenian Water Quality Index of the 2nd pollution class, by the water contamination index and Canadian Water Quality Index of the 3rd pollution class, and by specific combinatorial water quality index is mainly the 2nd and in some cases up to the 3rd class of pollution.

Over the past 10 years, the water level in Sevan has risen by 3 meters, leaving under water trees, stubble and buildings that have not yet been cleaned. The ecosystem of Lake Sevan is also polluted due to debris entering the lake. In addition to sewage systems from dozens of settlements in Lake Sevan, sewage and agricultural and wastewater from service and recreation facilities operating on the shores of Lake Sevan are also discharged into Sevan.

It should be noted that in 2019 there was little rain. For example, in May, 36 million m^3 of precipitation was recorded in the lake, which is close to the historically minimal (33 million m^3) precipitation. Due to the strong wind force, evaporation in the spring was twice as high as normal.

According to the results of research conducted by the Ministry of Nature Protection in 2018, the concentrations of phosphate and ammonium ions in Lake Sevan were high, and a sharp rise in temperature created favorable conditions for intensive flowering of the lake. The average concentration of phosphate ion in the surface and middle layers was 0.08 mg/l, and in the underlying layer—0.15 mg/l, which did not exceed the norm of the RA environment (0.3 mg/l). The average concentration of ammonium ion in the surface layer is 0.25 mg/l, and in the underlying layer 0.17 mg/l, which does not exceed the norm of the environment RA (0.5 mg/l). The average concentration of nitrate ions in the surface layer was 0.19 mg/l, and in the underlying layer—0.12 mg/l. The observed concentrations do not exceed the ecological norm of RA (11 mg/l).

4. Conclusions

The quality of the waters of the Dzknaget, Sotk, Masrik, Vardenis, Martuni, Argichi, and Gavaraget rivers and the lakes of Arpi, Yerevan, Akhuryan, Azat, Aparan and Kechut reservoirs comprehensively evaluated by the indices: AWQI, EWQI, WCI, CWQI, and SCWQI.

The quality of rivers and reservoirs water has been assessed by the new standards for background concentrations.

The water at the mouth of the Martuni River in 2014 was of “poor” quality for ammonium and phosphate ions, and the water at the mouth of the Masrik River in 2017–2019 was also of “poor” quality for vanadium. In 2013–2019, the waters of the Dzknaget, Martuni, Sotk, and Gavaraget rivers (monitoring post No. 77) and Martuni (monitoring post No. 71) were mostly of “good” quality. The water at the mouth of the Vardenis and Gavarvget rivers had an average and “unsatisfactory” quality for ammonium ions and phosphate.

The poor water quality of the Lake Arpi reservoir is associated with an increase in the amount of metals. Reduced water quality of the Akhuryan reservoir and Lake Yerevan is associated with pollution from the main settlements in the river basin, respectively, in Gyumri and Yerevan, with municipal wastewater.

For the first time, the water quality in the reservoirs of Lake Sevan was evaluated using the Armenian Water Quality Index. It was found out that the water of Lake Sevan is regularly increased MAC of vanadium, copper, chromium, magnesium, bromium, and selenium. The water quality in the Lake Sevan is poor.

It has been found that the Armenian Water Quality Index is linearly dependent on the water contamination index, the specific combinatorial water quality index, the geoecological evolving organized index, and has inverse relationship to the Canadian Water Quality Index.

Author details

Gevorg Simonyan
Yerevan State University (YSU), Yerevan, Republic of Armenia

*Address all correspondence to: sim-gev@mail.ru; gevorg.simonyan@ysu.am

References

[1] Nikanorov AM. Scientific Basis for Water Quality Monitoring. Sankt-Peterburg, Rossia: Gidrometeoizdat; 2005. p. 577 (in Russian)

[2] Stambuk-Giljanovic N. Comparison of Dalmatian water evaluation indices. Water Environment Research. 2003; **75**(5):388–405

[3] Horton RK. An index number system for rating water quality. Journal of the Water Pollution Control Federation. 1965;**37**(3):300

[4] Brown RM, Mclennald NI, Deininger RA, Tozer RG. A water quality index: Do we dare? Water and Seawage Works. 1970;**117**(10):339

[5] Canadian Council of Ministers of the Environment. Canadian water quality guidelines for the protection of aquatic life: CCME water quality index 1.0, user's manual. In: Canadian Environmental Quality Guidelines, 1999. Winnipeg: Canadian Council of Ministers of the Environment; 2001

[6] Zandbergen PA, Hall KJ. Analysis of the British Columbia water quality index for watershed managers: A case study of two small watersheds. Water Quality Research Journal of Canada. 1998;**33**(4):519–549

[7] Department of Environment Malaysia (DOE) 2013. Malaysia Environmental Quality Report on 2012 (Report No. WEQR-2012). Putrajaya. OMR press Sdn.Bhd; 2013

[8] Bascaron M. Establishment of a methodology for the determination of water quality. Boletin Informativo del Medio Ambiente. 1979;**9**:30–51

[9] Scottish Research Development Department (SRDD). Development of a Water Quality Index. Applied Research & Development, Report Number ARD3. Edinburg, UK: Engineering Division; 1976. p. 61

[10] Liou S, Lo S, Wang S. A generalized water quality index for Taiwan. Environmental Monitoring and Assessment. 2004;**96**(1):35–52

[11] Boyacioglu H. Development of a water quality index based on a European classification scheme. Water SA. 2013; **33**(1):101–106

[12] Sargaonkar A, Deshpand V. Development of an overall index of pollution for surface water based on a general classification scheme in Indian context. Environmental Monitoring and Assessment. 2003;**89**(1):43–67

[13] Temporary guidelines of complex evaluation of surface and sea water quality by hydrochemical indicators have been introduced by the instruction of the State Committee of Hydrometeorology, No. 250-1163. 1986. p. 5 (in Russian). Available from: https://files.stroyinf.ru/Index2/1/4293742/4293742635.htm

[14] RD 52.24.643-2002. The Leading Document. Methodical Instructions. A Method of a Complex Assessment of Degree of Impurity of Surface Water on Hydrochemical Indicators. Gidrometeoizdat, Sankt-Peterburg; 2002. p. 55 (in Russian)

[15] Tirkey P, Bhattacharya T, Chakraborty S. Water quality indices—Important tools for water quality assessment. International Journal of Advances in Chemistry. 2015;**1**(1):15–29

[16] Maximum Permissible Concentrations of Harmful Substances in the Water. Sanitary Water Bodies and the Requirements for the Composition and Properties of Water in Reservoirs at Points of Drinking and Cultural and Domestic Water Use. HMC, Moscow:

Ministry of Health of the USSR; 1973. p. 14 (in Russian)

[17] Directive 2000/60/EC of the European Parliament and of the Council of 23 October 2000 establishing a framework for community action in the field of water policy. OJ L327; 22 December 2000. Available from: https://eur-lex.europa.eu/legal-content/en/ALL/?uri=CELEX:32000L0060

[18] Water Quality in the Danube River Basin-2004, Yearbook. International Communications for the Danube River; 2005. Available from: https://www.icpdr.org/main/sites/default/files/TNMN_Yearbook_2005_long_version.pdf

[19] RA Government Decision N 75-N of 27 January 2011 "On Defining the Norms for Securing Water Quality of Each Basin Management Area Depending on the Features of the Specific Area" (in Armenian). Available from: http://www.irtek.am/views/act.aspx?aid=58822

[20] Pirumyan G, Pirumyan E, Simonyan G, Simonyan A. Method of determining the level of water pollution. RA Patent, №3063A; 2016

[21] Pirumyan GP, Simonyan GS, Margaryan LA. Geoecological Evaluational Integrating Index of Natural Waters and Other Systems. Yerevan: Copy Print LTD.; 2019. p. 244

[22] Simonyan AG, Pirumyan GP, Simonyan GS. Analysis of environmental status of the Kechut Artificial Reservoir and river Arpa with Armenian index of water quality. Austrian Journal of Technical and Natural Sciences. 2016;**7-8**:37–41

[23] Simonayn AG. Analysis of the environmental status of the river Voghji with Armenina index of water quality. Proceedings of YSU, Series Chemistry and Biology. 2016;**2**:20–24

[24] Simonyan GS, Simonyan AG, Pirumyan GP. Systemic-entropy approach for estimating the water quality of a river. Oxidation Communications. 2018;**41**(2):307–317

[25] Simonyan AG, Simonyan GS, Pirumyan GP. Analysis of environmental status of the Rivers Aghstev and Getik with Armenian index of water quality. European Journal of Natural History. 2016;**4**:22–27

[26] Simonyan G, Pirumyan G. Entropy-system approach to assess the ecological status of reservoirs in Armenia. Preprints. 2019:2019010260. DOI: 10.20944/preprints201901.0260.v.1

[27] Simonyan GS, Simonyan AG. Entropy approach to the assessment of chaos and the order of biological systems. Successes in Modern Natural Sciences. 2015;**9**:100–104 (in Russian)

[28] Simonyan GS. Chaos and the order of biological systems in the light of the synergistic theory of information. In: International Conference "Modern Problems of Chemical Physics", Yerevan. 2012. pp. 227–228

[29] Simonyan GS, Simonyan AG, Sayadyan ML, Sarsekova DN, Pirumyan GP. Analysis of environmental status of wood and shrub vegetation by the Armenian Index of environmental quality. Oxidation Communications. 2018;**41**(4):533–541

[30] Simonyan GS. Evaluation of the influence of nitrogen on the stability of naptic systems with the help of geo-ecological evolving organized index. Oxidation Communications. 2019; **42**(3):329–336

[31] Chilingaryan LA, Mnatsakanyan BP, Aghababyan KA, Tokmajyan HV. Watercourse of Armenian Rivers and Lakes. Yerevan, Armenia: Agropress; 2002. p. 50 (in Armenian)

[32] Sargsyan VO. Vodi Armenii [Armenian Water]. Yerevan: YSUAB; 2008. p. 208

[33] Dictionary of Physical-Geographic Objects in the Republic of Armenia. State Committee of the Real Estate Cadastre. Yerevan, Armenia, 2007. p. 136 (in Armenian)

[34] Shannon C. Works on Information Theory and Cybernetics. IL, Moscow; 1963. p. 830 (in Russian)

[35] Shannon CE. A mathematical theory of communication. The Bell System Technical Journal. 1948;**27**:379–423 and 623–656

[36] The Environmental Impact Monitoring Center of the Ministry of Nature Protection of the Republic of Armenia on the Environmental Monitoring Results of the Environmental Impact Assessment. 2009. p. 55 (in Armenian)

Section 2

Ecological Factors Affecting Inland Waters

Chapter 3

Effect of Climate Change on Aquatic Ecosystem and Production of Fisheries

Satarupa Ghosh, Snigdha Chatterjee, Ghora Shiva Prasad and Prasanna Pal

Abstract

The exploitation of nature for decades due to several anthropogenic activities has changed the climatic conditions worldwide. The environment has been polluted with an increase of greenhouse gases. The major consequences are global warming, cyclone, an increase in sea level, etc. It has a clear negative impact on the natural environment including aquatic ones. As a result, production of fish in the aquaculture system and marine system is greatly affected. Marine ecosystems like coral reefs are also destroyed. Decreased fish production has also affected the livelihood and economic condition of the fish farmers. So, corrective measures should be taken to reduce the climate changes for minimizing its effects on fish production. Using more eco-friendly substances, planting more trees, and preserving our nature are some steps to be taken. Awareness should also be generated among the common people.

Keywords: aquatic environment, economy, climate change, fish production, global warming

1. Introduction

For the last few decades, climate change, food security and their complex interaction have become a global issue [1]. With the rapid increase in human population, we have destroyed our nature and polluted the environment. The level of greenhouse gases in the atmosphere is increasing day by day. Consequently, we are facing the threats of global warming and other climatic changes like cyclone, drought, flood, etc. Change in the climatic conditions may be limited to a specific region or may occur across the whole earth. But, it is affecting all the ecosystems including the aquatic ones. Aquatic organisms are very vulnerable to climate change because the average temperature of both air and water are changing simultaneously. Climate change in the aquatic system mainly occurs through sea level and temperature rise, change in monsoon patterns, extreme weather events and water stress having both direct and indirect impacts on aquatic animals including fish stocks. It directly acts upon the physiological behavior and growth pattern of organisms, subsequently decrease reproductive capacity and finally cause mortality. Indirectly it may alter the productivity, structure, function and composition of aquatic ecosystems. All these effects finally result in decreased fish production. It disturbs the economic condition of fish farmers and hamper their normal livelihood by huge economic losses. In this chapter, we will discuss how climate change affects the production of fish and the lives of fish farmers and how it could be mitigated through proper actions.

2. Causes of climate change

The factors that can cause a change in the atmospheric system or climatic regime are called "climate forcing" or "forcing mechanisms." So, forcing mechanisms can be of two types, i.e., internal forcing mechanism and external forcing mechanisms. Internal forcing mechanisms are natural processes in the climatic system like thermohaline circulation, etc. External forcing mechanisms can also be of two types- anthropogenic mechanisms including greenhouse gas emission and the emission of several other pollutants and natural mechanisms like changes in solar output, volcanic eruptions, etc. All these mechanisms are responsible for the change of climate. But overwhelming evidence exists that anthropogenic activities are the major reason behind this dreadful condition. These are described below.

- Fossil fuel burning: Fossil fuel burning is one of the most important sources of climate change. As fossil fuels contain carbon for many years, they can release back CO_2 into the air. This is one of the direct causes of carbon emission in the air, which can cause all sorts of environmental problems including global warming.

- Livestock farming: Through livestock farming, methane (CH_4) gas is emitted into the atmosphere. As we know, CH_4 is a greenhouse gas, so capable of trapping a huge amount of heat from the sun. In that way, they can contribute to global warming in broad sense.

- Aerosols: Aerosols also represent a big problem for the climate today. Aerosols are a very small naturally occurring particle in the atmosphere. Previously the number of aerosols in the atmosphere was very less, but now the level is increasing.

- Use of fertilizers: Use of fertilizers in both agricultural and aquacultural farmland can increase the availability of food source greatly to us. To meet up the growing demand for food, the use of fertilizers have increased rapidly. Fertilizer contains a huge amount of nitrous oxide, which is responsible for a steady increase in the earth's surface temperature.

3. Changes on aquatic ecosystem due to climate change

3.1 Temperature

All the aquatic organisms including fish and aquatic invertebrates are poikilothermic in nature and the body temperature of those organisms changes with environmental temperature. So, they are very much sensitive to the change in the temperature in their external environment where they live. When the external environmental temperature goes beyond the tolerance limit of these organisms, they will go for migration to the place where their internal system allows them to regain their internal homeostasis. This procedure is termed as behavioral thermoregulation [2]. This will result in rapid migration to the cooler zones of the water body [3]. This migration allows the shifting of the aquatic animals from shallow coastal waters and semi-enclosed areas into deeper cooler waters [4]. In spite of the negative impacts of these phenomenons like coral reef destruction and increased ocean acidification, it would have some conservative approach. This phenomenon of migration can alone reduce the maximum catch potential of the tropics by 40% [4].

As the major consequences of climate change, especially increased temperature strongly affects the recruitment process [5]. Some stocks may become intolerance to the sustainable fishing effort because they experience them as overfishing due to the side effects of temperature enhancement [6]. Temperature enhancement of water, where fish live, will slow down their growth and maximum size as the temperature would increase their metabolic rate [2].

Local extinction of fish species would be noticed, among freshwater and diadromous species especially [7]. Because of the higher potential for migration, terrestrial species show a higher rate (15–37%) of overall migration than marine species [8].

The increased temperature would bring a deadly impact on reef fisheries by inducing bleaching of the coral reef [7].

3.2 Primary productivity

The levels of light and temperature determine the availability of nutrients in the water body, which in turn affects the primary productivity. Due to climate variability, reduced precipitation would lead to reduced run-off from land, which caused the starvation of wetland and mangrove and damage local fisheries. In some other places, due to increased precipitation from extreme weather events like flooding, nutrient level in the water body tremendously increased causing eutrophication and washout fertilizer causing harmful algal blooms into the water bodies, known as red tides [2, 9]. Most of the small scale fisheries locate at the lower latitude, where climate change hit the most and decline the primary productivity [10] of the fisheries sector.

4. Impact of climate change on fish production and ecosystem

4.1 Aquaculture system

Fisheries and aquaculture are largely dependent on the interactions among the various factors like the earth's climate and ocean environment. So, changing the pattern of air and sea-surface temperatures, rainfall, sea level, ocean acidity and wind-pattern will adversely affect the fisheries and aquaculture [3].

4.1.1 Marine system

Marine fish production is largely disrupted by climate change. With the change in the climatic conditions, several changes are observed in the ocean including a rise in temperature, melting of polar ice, rising sea level, change in ocean current system and acidification of seawater. Over the coming decades, the temperature of the Indian seas is going to increase by 1–3°C [11]. The species that is going to be affected first due to these conditions is plankton. It forms the basis of the food chain in the marine ecosystem. Other species including corals, fishes, sea birds will be affected simultaneously. Due to increased ocean acidification, marine organisms like oysters, shrimps and corals would unable to form their outer covering or shell through the process of calcification. Thus, the entire marine food web get affected because of the formation of cracks in the marine food chain.

4.1.2 Freshwater systems

The vulnerability of the freshwater ecosystems against climate change is very high. The size, depth and trophic status of the lake determine the vulnerability of this system against climate change. According to Field and coworkers [12], the negative impact was observed on the cold-water species and positive impact on the warm-water species. Due to acute effects of climate change, alteration of shapes and distribution is seen in the freshwater lake system and in some cases, they might be disappeared. These are the attributes of the dynamics change in precipitation, evaporation and run-off [13]. Climate change promotes long-term increases in fish-production by inducing the enhancement of the production rate of invertebrate prey logarithmically with increasing temperature. The increasing rates are 2–4 times for each 10°C increase in temperature [14]. But on the other hand, climate change will result in a change in prey-species composition. This change may cause antagonistic effects on the long-term enhancement of fish production [14]. In short-time, climate change will cause a decrease in fish-production because of timing mismatch [14]. The ability of the movement of the freshwater species is vital in determining the resistance of those species to withstand climate change [13].

4.2 Coral reef

The coral reef is an important source of income for many developing countries [15]. Coral provides habitat for more than half of all marine species. But now coral reefs of the ecosystem are in great danger. The main reasons are increasing temperature, acidity, etc.

Climate change-related impact on the coral reef can be based on three different time-scales.

- Years: Coral bleaching which increased in recent years and results in degradation of reefs.
- A few decades: Acidification increased and carbonate structures degenerate.
- Multi-decades: Weakened the structural integrity of the reefs which causes large scale composition shifts.

The coral reef is one of the most resistant ecosystems and too resilient to recover from weak chronic as well as acute stresses [16]. But according to Hughes and coworkers [17], the reef ecosystem is not able to sustain against chronic plus acute stress.

Increasing acidity causes decreasing the pH of the ocean, which results in decreased aragonite saturation that can disrupt the calcification of coral [18]. Enhanced acidity of the world's ocean is very much important and represents a long-term threat to coral reefs but the impact growth of the corals on the increasing acidity is unknown [15]. The saturation level of aragonite in deep cold water corals are 90–150 m [19]. The impact of the acidification is badly seen in these deep- cold-water corals.

If corals are decreased due to adverse impacts from climatic change, it causes a negative impact on the reef fish- biodiversity [20]. According to Grandcourt and

Cesar [21], coastal fisheries are badly affected by the warming of the climate and bleaching events. It can be concluded that coral reef destruction causes a long-term impact on the animals which depends on these reefs for their food and habitat.

4.3 Global marine biodiversity

Climate change acts as an important determinant of the distribution of biodiversity in past and future aspect [22–26]. Environmental factors reflect strong influences upon species richness of aquatic organisms [27]. Ocean warming can cause change to the marine species especially in their latitudinal range [28–30] and depth range [31]. At a larger scale, such changes can lead to local extinction and invasions and shifting to their bio-geographic pattern [28]. As a result, a huge shift in species richness can occur which is regarded as the main cause of disruption of marine biodiversity and ecosystem [2, 32, 33]. The climate in the aquatic environment can affect biodiversity, community structure and ecosystem function [34–37].

5. Economical crisis of the fish farmers due to climate change

Change in the aquatic environment has a direct impact on the lives of the fish farmers. Due to disturbed fish production, farmers face economic losses. Besides global warming, cyclone is another problem that affects the lives of the farmers. Cyclone combined with a flood and heavy rainfall creates a major problem every year for the farmers especially in the coastal states of India. It is a matter of great concern that the frequency of intense tropical cyclones has increased in the Indian ocean [38]. The factors such as warm sea temperature, high humidity and instability of atmosphere are responsible for intensifying the cyclone [39]. As a consequence of global warming, the temperature of the Indian Ocean has also increased promoting devastating cyclones. In May 2019, a cyclone named Fani hit Andhra Pradesh, Odisha and West Bengal. It caused damage to the coastal land, boats, jetties and the shelters of the fishermen and five lakh houses were destroyed in 14 districts [40]. In Odisha only, the losses were estimated to be 12,000 crores. Regarding the seafood sector, the production of shrimp was declined by 60–70% [41]. Most recently, in May 2020, cyclone Amphan hit eastern India specifically West Bengal and alos Bangladesh. This was the first super cyclone in Bay of Bengal since 1999 super cyclone that hit Odisha took the life of more than 9000 people [42]. Amphan affected the coastal areas of West Bengal including East Midnapur, North 24 Parganas, South 24 Parganas, Kolkata, Hoogly and Howrah. According to Chief Minister of West Bengal, the death toll was more than 86 and the state suffered a damage of 1 lakh crore rupees (15.38 Billion USD) [43]. Specially, the Sundarban areas were highly devastated, millions of homes were damaged breached embankments led to flood in villages. It takes years for the local residents as well as fishermen to recover from these situations. They do not have shelter to stay, do not have a boat for fishing and no money to pay back the loans that ultimately affects their psychological health sometimes leading to suicidal tendencies.

6. Adaptation and mitigation measures to reduce the effects of climate change

Consideration of future climate changes in advance and making them a part of short-term decision making is known as adaptation. This includes using more eco-friendly substances, planting more trees and preserving our nature as much as possible. On another hand, preventing the chances of climate change, before it has occurred, reducing the effects of climate change in case of occurrence is known as mitigation. Reducing the carbon footprint and related activities should be a major step. The level of environmental pollution should be decreased as soon as possible before it becomes too late to act. Some strategies that we should follow immediately are discussed below.

Adaptation of forest conservation measures: Forest plays an important role in maintaining equilibrium in our ecosystem. We should conserve and prevent the destruction of forest land through afforestation as well as reforestation and prohibit the use of forestland for nonforest purposes to meet the livelihood of local people.

Inclusion of climate-study in the school-level educational system: If we want to generate awareness in the young generation by the introduction of climate-related study along with traditional educational system with the help of governmental initiatives. This will help to grow the consciousness among the young generation from very beginning which will significantly broaden this culture at the local, state and national levels.

Slowing down of population growth: Population growth is becoming a burden especially in the case of a developing country like India. It has become a major obstruction in achieving social and economic development. So, in order to fight against climate change, population pressure over the area need to be reduced by reversing down the population growth curve in developing countries.

Integration of climate issue with economic planning: Climate protection-related policies and programs should be incorporated into the local, state and national levels in order to encourage the integration of climate issues with economic planning and management.

7. Recommendation for better management of fisheries against climate changes

- The ecosystem approach should be comprehensive, sound, integrated, compact and revised to make complete management of sand oceans of coasts, fisheries and aquaculture.
- Environmental friendly aquaculture and fishing practices to be undertaken.
- Fuel-efficient aquaculture and fishing practices to be undertaken.
- Integration of climate-proof aquaculture with other sectors.
- Over-fishing and excess fishing capacity should be eliminated through the implementation of reduced subsidy systems.
- Risk assessments should be proper and accurate at the local level.
- Exploration of the carbon sequestration process by aquatic ecosystems.

8. Steps for sustaining the fish production and economy against climate change

There is a crucial knowledge gap between fisheries, aquaculture management and climate change that need to be filled practically. In order to assess the risk of climate change to coastal communities, human and institutional capacity building should be strengthened and proper adaptation and mitigation measures should be implemented. Therefore, well managed fisheries and aquaculture could give birth to a healthy and productive ecosystem. Careful use of coastal areas and catchment areas should be cross-sectoral responsibility to encourage the building process of a healthy and productive ecosystem. Moreover, youth engagement in each and every policy and decision-making process related to aquaculture and fisheries both at continental and national levels should be institutionalized efficiently as youth are the backbone of our society.

9. Positive effects of climate change on the aquatic environment

- **Slowed down the winter death rate of aquatic organisms:** Water temperature is one of the most crucial factors in determining the survival of aquatic animals. Many years ago, especially before the drastic climate change, winters were too cold to maintain the minimum metabolic rate of the aquatic organisms and the consequent death rate has increased rapidly. As a result of climate change, the average temperature of the water body increased so rapidly that winter has now become bearable considerably. So, the number of death due to winter temperature –falling has decreased.

- **Reduction of the fuel cost of the aquatic environment:** As a result of climate change, heat energy becomes available and affordable at a cheaper rate. So, the demand for fuel in the aquatic environment has decreased and the consequent cost of fuel has also become cut down.

- **Growth in aquaculture production:** Some thermophilic organisms living in the aquatic environment demand high temperature for maintaining their metabolic rate at an optimum level. The excess heat which is introduced as a result of climate change meets the demand of those aquatic organisms. So, in that way, climate change benefited the overall aquacultural yield.

10. Conclusion

Climate change is a major threat to both aquatic and terrestrial ecosystems. In present days, a random population explosion increases fossil fuel burning, industrialization, deforestation, and profit-oriented capitalism, which can, in turn, create synergistic effects on climate change. Aquaculture sector is much impacted by temperature increase in water and air, sea level rise, and associated water intrusion as affected by global warming and climate change. This change in the aquatic environment or a decrease in fish production is directly affecting the economic sustainability of fish farmers. Thus, this situation can be corrected if necessary actions will be taken in reducing environmental pollution as soon as possible. Researchers, economists, policymakers, and farmers should act together to fight economic instability and maintain harmony with nature. One thing we should remember that we should protect nature if we want to protect ourselves from the coming threats.

Author details

Satarupa Ghosh[1*], Snigdha Chatterjee[2], Ghora Shiva Prasad[1] and Prasanna Pal[3]

1 Department of Aquatic Environment Management, West Bengal University of Animal and Fishery Sciences, Kolkata, West Bengal, India

2 Department of Agricultural Chemistry and Soil Science, Bidhan Chandra Krishi Viswavidyalaya, Mohanpur, Nadia, West Bengal, India

3 Animal Physiology Division, ICAR-National Dairy Research Institute, Karnal, Haryana, India

*Address all correspondence to: satarupasonaibfsc07@gmail.com

References

[1] Plagányi É. Climate change impacts on fisheries. Science. 2019;**363**(6430):930-931. DOI: 10.1126/science.aaw5824

[2] Roessig JM, Woodley CM, Cech JJ, Hansen LJ. Effects of global climate change on marine and estuarine fishes and fisheries. Reviews in Fish Biology and Fisheries. 2004;**14**(2):251-275

[3] Cochrane K, Young C. De, Bahri, T. Climate change implications for fisheries and aquaculture. FAO Fisheries and Aquaculture Technical Paper. 2009;**530**:212

[4] Cheung WW, Lam VW, Sarmiento JL, Kearney K, Watson REG, Zeller D, et al. Large-scale redistribution of maximum fisheries catch potential in the global ocean under climate change. Global Change Biology. 2010;**16**(1):24-35

[5] Walther GR, Post E, Convey P, Menzel A, Parmesan C, Beebee TJ, et al. Ecological responses to recent climate change. Nature. 2002;**416**(6879):389

[6] Easterling WE, Aggarwal PK, Batima P, Brander KM, Erda L, Howden SM, et al. Food, fibre and forest products. Climate Change. 2007:273-313

[7] IPCC. Climate change 2007: Synthesis report. In: Pauchauri RK, Reisinger A, editors. Contribution of Working Groups I, II, and III to the Fourth Intergovernmental Panel on Climate Change. Core Writing Team. Geneva, Switzerland: IPCC; 2007. p. 8

[8] Cheung WW, Lam VW, Sarmiento JL, Kearney K, Watson R, Pauly D. Projecting global marine biodiversity impacts under climate change scenarios. Fish and Fisheries. 2009;**10**(3):235-251

[9] Epstein PR. Is global warming harmful to health? Scientific American. 2000;**283**(2):50-57

[10] FAO. Climate change implications for fisheries and aquaculture. In: The State of Fisheries and Aquaculture 2008. Rome, Italy: FAO; 2008. pp. 87-91

[11] Ralbam V. Impact of climate change on Indian marine fisheries and option for adaptation. In: Lobster Research in India. Kochi, India: CMFRI; 2010. pp. 65-71. DOI: 10.1097/NRL.0000000000000102

[12] Field CB, Mortsch LD, Brklasich M, Forbes DL, Kovacs P, Patz JA, et al. North America. In: Parry ML, Canziani OF, Palutikof JP, Hanson PJ, Linden v d CE, editors. Climate Change 2007: Impacts, Adaptation and Vulnerability. Cambridge: Cambridge University Press; 2007. pp. 617-652

[13] Poff NL, Brinson MM, Day JWJ. Aquatic ecosystems and Global climate change: Potential Impacts on Inland Freshwater and Coastal Wetland Ecosystems in the United States. 2002. Available from: www.pewclimate.org [Retrieved: 26 June 2019]

[14] Watson RT, Zinyowera MC, Moss RH, Dokken DJ. IPCC Special Report: The Regional Impacts of Climate Change: An Assessment of Vulnerability. United Nations: IPCC; 2001

[15] Parry M, Parry ML, Canziani O, Palutikof J, Van der Linden P, Hanson C. Climate Change 2007—Impacts, Adaptation and Vulnerability: Working Group II Contribution to the Fourth Assessment Report of the IPCC. Cambridge: Cambridge University Press; 2007

[16] Buddemeier RW, Kleypas JA, Aronson RB. Potential Contributions of Climate Change to Stresses on Coral Reef Ecosystems. Coral Reefs and Global Climate Change. Virginia, USA:

Pew Center on Global Climate Change; 2004

[17] Hughes TP, Baird AH, Bellwood DR, Card M, Connolly SR, Folke C, et al. Climate change, human impacts, and the resilience of coral reefs. Science. 2003;**301**(5635):929-933. DOI: 10.1126/science.1085046

[18] Orr JC, Fabry VJ, Aumont O, Bopp L, Doney SC, Feely RA, et al. Anthropogenic Ocean acidification over the twenty-first century and its impact on calcifying organisms. Nature. 2005;**437**(7059):681-686. DOI: 10.1038/nature04095

[19] Feely RA, Fabry VJ, Guinotte JM. Ocean Acidification of the North Pacific Ocean. Vol. 2. PICES press; 2007. Retrieved from: http://www.pices.int/publications/pices_press/volume16/v16_n1/pp_22-26_Acidification_f.pdf

[20] Jones GP, McCormick MI, Srinivasan M, Eagle JV. Coral decline threatens fish biodiversity in marine reserves. Proceedings of the National Academy of Sciences. 2004;**101**:8251-8253

[21] Grandcourt EM, Cesar HS. The bio-economic impact of mass coral mortality on the coastal reef fisheries of the Seychelles. Fisheries Research. 2003;**60**(2):539-550

[22] Parmesan C, Yohe G. A globally coherent fingerprint of climate change impacts across natural systems. Nature. 2003;**421**:37-42. Crossref | CAS | PubMed |

[23] Peterson AT, Ortega-Huerta MA, Bartley J, Sánchez-Cordero V, Soberón J, Buddemeier RH, et al. Future projections for Mexican faunas under global climate change scenarios. Nature. 2002;**416**:626-629

[24] Root TL, Price JT, Hall KR, Schneider SH, Roxenzweiq C, Pounds JA. Fingerprints of global warming on wild animals and plants. Nature. 2003;**421**:57-60

[25] Rosenzweig C, Karoly D, Vicarelli M, et al. Attributing physical and biological impacts to anthropogenic climate change. Nature. 2008;**453**:353-357

[26] Thomas CD, Williams SE, Cameron A, et al. Biodiversity conservation: Uncertainty in predictions of extinction risk/effects of changes in climate and land use/climate change and extinction risk. Nature. 2004;**430**:1-2

[27] Macpherson E. Large-scale species-richness gradients in the Atlantic Ocean. Proceedings of the Royal Society of London Series B. 2002;**269**:1715-1720

[28] Hiddink JG, Hofstede RT. Climate induced increases in species richness of marine shes. Global Change Biology. 2008;**14**:453-460

[29] Mueter FJ, Litzow MA. Sea ice retreat alters the biogeography of the Bering Sea continental shelf. Ecological Applications. 2008;**18**(2):309-320

[30] Perry AL, Low PJ, Ellis JR, Reynolds JD. Climate change and distribution shifts in marine shes. Science. 2005;**308**:1912-1915

[31] Dulvy NK, Rogers SI, Jennings S, Vanessa S, Dye SR, Skjoldal HR. Climate change and deepening of the North Sea sh assemblage: A biotic indicator of warming seas. Journal of Applied Ecology. 2008;**45**:1029-1039

[32] Cheung WWL, Lam VWY, Pauly D. Modelling Present and Climate-Shifted Distribution of Marine Fishes and Invertebrates. Fisheries Centre Research Report 16(3). Vancouver: University of British Columbia; 2008

[33] Worm B, Barbier EB, Beaumont N, et al. Impacts of biodiversity loss on ocean ecosystem services. Science. 2006;**314**:787-790

[34] Genner MJ, Sims DW, Wearmouth VJ, Southall EJ, Southward AJ, Henderson PA, et al. Regional climatic warming drives long-term community changes of British marine sh. Proceedings of the Royal Society of London, Series B: Biological Sciences. 2004;**271**:655-661

[35] Hooper DU, Chapin FS III, Ewel JJ, et al. Effects of biodiversity on ecosystem functioning: A consensus of current knowledge. Ecological Monographs. 2005;**75**:3-35

[36] Sala OE, Chapin FS III, Armesto JJ, et al. Global biodiversity scenarios for the year 2100. Science. 2000;**287**:1779-1774

[37] Stachowicz JJ, Fried H, Osman RW, Whitlatch RB. Biodiversity, invasion resistance, and marine ecosystem function: Reconciling pattern and process. Ecology. 2002;**83**:2575-2590

[38] Bhatia KT, Vecchi GA, Knutson TR, Murakami H, Kossin J, Dixon KW, et al. Recent increases in tropical cyclone intensification rates. Nature Communications. 2019;**10**(1):635

[39] Fitchett J. Why the Indian Ocean is Spawning Strong and Deadly Tropical Cyclones. India: Down To Earth. 2019. Retrieved from: https://www.downtoearth.org.in/news/climate-change/why-the-indian-ocean-is-spawning-strong-and-deadly-tropical-cyclones-64489

[40] Chhotoray S. Cyclone Fani Took Us 20 Yrs behind, Snatched Our Livelihood': Distressed Chilika Fishermen Await Aid. India: News 18. 2019. Retrieved from: https://www.news18.com/news/india/cyclone-fani-has-snatched-our-livelihood-taken-us-20-years-behind-say-chilika-fishermen-2150891.html

[41] Krishnakumar PK. Shrimp Production in Odisha Hit by Cyclone Fani. The Economic Times. 2019. Retrieved from: https://economictimes.indiatimes.com/news/economy/agriculture/shrimp-production-in-odisha-hit-by-cyclone-fani/articleshow/69370625.cms?from=mdr

[42] Nandi J, Thakur J. Amphan Transforming into Super Cyclone, first after Deadly 1999 Super Cyclone in Bay of Bengal. Hindustantimes. 2020. Retrieved from https://www.hindustantimes.com/india-news/amphan-transforming-into-super-cyclone-first-after-deadly-1999-super-cyclone-in-bay-of-bengal/story-12vMmByNipIQwibsJwHJIO.html

[43] Annonymus. Mamata Pegs Cyclone Amphan Damage at Rs. 1 Lakh Crore, Toll Rises to 86. India: The Hindu. 2020. Retrieved from https://www.thehindu.com/news/national/other-states/mamata-pegs-cyclone-amphan-damage-at-1-lakh-crore-toll-rises-to-86/article31660847.ece

Chapter 4

Designing River Diversion Constructed Wetland for Water Quality Improvement

Sani Dauda Ahmed, Sampson Kwaku Agodzo and Kwaku Amaning Adjei

Abstract

Constructed wetlands are recognized as viable potential technology for reducing pollution load and improving quality of water and wastewater. The use of river diversion wetlands is gaining place for improving quality of river and stream water. However, the design criterion for this category of wetlands has not been fully established, and there is a need to optimize existing approach to enhance operational performance. This chapter presents a step-by-step approach for the design of a typical river diversion constructed wetland intended to remove some pollutants and improve river water quality. The approach focused mainly on water quality objective and outlined simple criteria, guidelines, and model equations for the design procedure of a new river diversion constructed wetland. The design of constructed wetlands is generally an iterative process based on empirical equations. Thus, this approach combines simple equations and procedure for estimating the amount of river water to be diverted for treatment so as to assist the designer in sizing the wetland system. The novel approach presented may be useful to wetland experts as some of the procedures presented are not popular in wetland studies. However, this may improve existing river diversion wetlands' design and development.

Keywords: design, river diversion, constructed wetland, water quality, rating curve, empirical equations

1. Introduction

There is no doubt that streams and rivers are important freshwater sources for man due to their influence on social and economic development of human societies. However, the quality of water in most streams and rivers is being threatened worldwide due to pollution connected with human activities [1]. The situation is worsened with increasing industrial pollution and use of fertilizers and other agrochemicals in agriculture, rapid urbanization, and continuing use of improper sanitation systems especially in developing countries [2]. Consequently, aquatic ecosystems that depend on water flows and seasonal changes within these water bodies are often threatened by poor water quality [3]. Water quality problems represent a major global challenge. For example, pollution of water bodies, especially nutrient loading, has worsened water quality in almost all rivers in Africa, Asia, and Latin

America. Therefore, future global water demands cannot be met unless concerted efforts are made to address water quality and wastewater management challenges.

Therefore, sustainable management of freshwater resources needs to aim at protecting or reducing pollution load of freshwater sources especially streams and rivers to avoid negative impacts on water quality and ecosystems. In this regard, constructed wetlands are recognized as potential technology for meeting water quality and other requirements of these important freshwater sources. The use of constructed wetlands for water quality improvement is increasing with new applications and technological possibilities [4, 5]. In recent times, the use of river diversion wetlands is gaining more relevance for improving quality of water in riverine systems [6–8]. The incorporation of constructed wetlands into management strategies for rivers and streams may help to reduce pollution load and enhance their absorbing capacity against impacts [9].

Despite the recognition of constructed wetlands as an effective and economical way of improving water quality, many of those in operation are underperforming. The shortcomings are partly attributed to limitation and inconsistencies of equations used in designing them [10–12]. Besides, most of the available design methods are either related to municipal wastewater treatment or stormwater quality improvement with the primary aim of peak flow retention to attenuate flood water which may lead to overestimation. For river diversion wetlands, specific design criteria have not been fully established, and further research is needed to optimize existing approach in order to enhance performance capabilities of these types of wetlands [7]. However, the design of constructed wetlands is generally based on empirical equations using zero- or first-order plug flow kinetics as basis for predicting pollutants' removal and improving water quality [13].

This chapter aimed to provide guidance on the design of a typical river diversion constructed wetland intended to improve quality of river water. The chapter provides an overview of factors to be considered for the wetland design, water quality characterization, wetland inflow estimation, computation of the wetland hydrodynamic parameters, wetland sizing, and configuration and guide on designing of conveying and inlet and outlet structures. The approach presented may be useful to wetland experts as some of the procedures adopted are not popular in wetland studies.

2. Types of constructed wetland systems

Basically, two main types of constructed wetlands exist. These are free water surface (FWS) flow and subsurface flow (SSF) systems. FWS flow wetlands operate with water surface open to the atmosphere, while for SSF, water flow is below the ground through a sand or gravel bed without direct contact with the atmosphere [14, 15]. Both are characterized by shallow basins usually less than 1 m deep. FWS wetlands require more land than SSF wetlands for the same pollution reduction but are easier and cheaper to design and build [16].

FWS flow wetlands are further sub-classified based on the dominant type of vegetation planted in them such as emergent, submerged, or floating aquatic plants. SSF wetlands which are often planted with emergent aquatic plants are best subclassified according to their flow direction as horizontal subsurface flow (HSSF), vertical subsurface flow (VSSF), and hybrid system [17]. Another sub-division of constructed wetland types which have emerged recently is river diversion wetlands. These are mostly FWS wetlands located near or within a stream or river system. They are distinguished according to their location as off-stream and in-stream wetlands. Off-stream wetlands are constructed nearby a river or stream where only a portion of the river flow enters the wetland. On the other hand, in-stream wetlands are constructed within the river bed, and all flows of the river enter into the wetland [18]. **Figure 1** shows a typical arrangement of both types.

Potential benefits of river diversion wetlands include merits relating to river water quality improvement, flood attenuation, increasing connectivity between rivers and floodplains, and creation of mixed habitat of flora and fauna communities [8, 19]. The systems are also cost-effective due to their simple designs and construction when compared to conventional treatment systems. Major drawbacks of these types of wetland systems relate to emissions of greenhouse gases and losses of biodiversity which may result from continued pollution loading [20]. Unlike the in-stream wetlands, a major advantage of the off-stream river diversion wetlands is that they can be used to mitigate non-point source pollution from agricultural lands before reaching the river channel. However, off-stream wetlands may require storage and flow control structures to regulate flow and a large space for layout of the wetlands which may result in high initial costs for land easements. Additionally, only part of the river flow volume can be treated at a time. On the other hand, space availability may not be a big issue for in-stream wetlands as they are constructed within the river bed, and as such the whole river flow volume can be subjected to treatment. However, it may be difficult to regulate flow especially during river peak flows and consequently retention time which is an important aspect of wetland for effective pollutant removal.

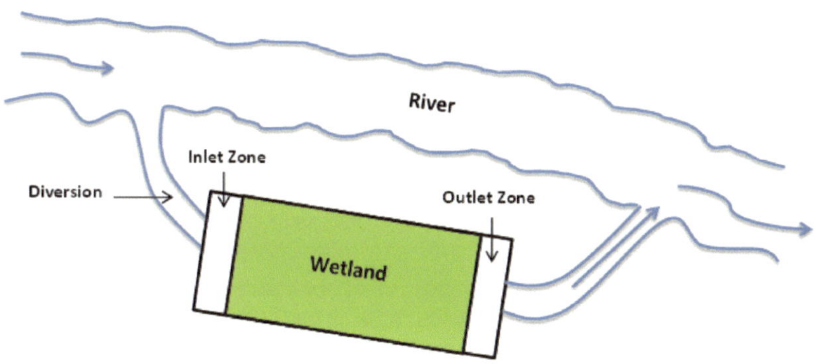

(a)

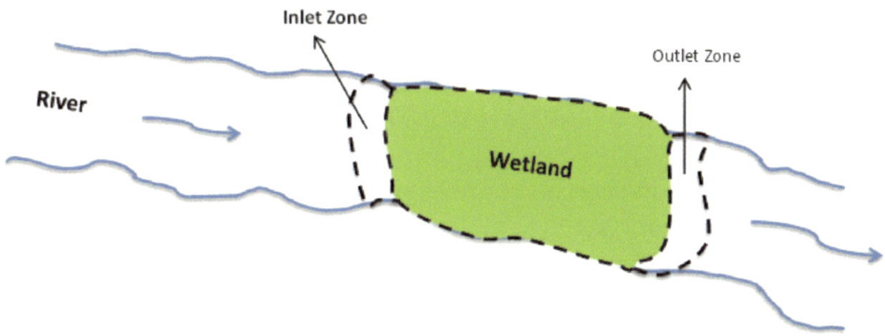

(b)

Figure 1.
Arrangement of off-stream and in-stream river diversion wetlands. (a) Off-stream river diversion wetland and (b) in-stream river diversion wetland.

3. Design consideration for river diversion constructed wetland

The design of a constructed river diversion wetland is an iterative process involving site-specific data. Prior to design and construction, site conditions must be evaluated to assess the appropriateness of the site for the proposed constructed wetland system [4]. Thus, the following are recommended as part of the design process:

- Investigation of site characteristics
- Water quality characterization
- Wetland design inflow estimation

3.1 Investigation of site characteristics

Site condition is a very important factor in the design of a constructed river diversion wetland. This is particularly necessary when a suitable site or land is not readily available as the situation often limits possible options the designer may utilize. Thus, site investigation enables the designer to have an idea of the site characteristics including size of area or land available for the design. However, where there is sufficient suitable site or land, it gives the designer the latitude and flexibility of several design options. Therefore, identifying the required area available for optimal layout of the wetland is vital for effective reduction of pollutants.

Site characteristics to be evaluated when designing and possibly constructing a river diversion wetland include:

- Proximity of the site to the river system (the site should be situated close to the source of water to be treated for easy diversion or within the river channel depending on the type (in-stream or off-stream))
- Climate (climate can affect type and size of the space required for the wetland; climatic factors that are important include rainfall, evaporation, evapotranspiration, insolation, and wind velocity)
- Topography of the land (topographic conditions such as natural depressions and slopes are important consideration; the gradient of the land should preferably have a gentle slope so that water can easily flow by gravity)
- Groundwater condition (assess groundwater levels within the site in different seasons to guide against possible contamination)
- Soil and environmental condition of the site (the site should contain soils that can be sufficiently compacted to minimize seepage to groundwater, or necessary measures should be put in place to minimize groundwater contamination)
- Distance of the site from residential buildings to avoid creating an environment that is not conducive for inhabitants

After due consideration of the above conditions, a suitable location can be selected for siting the wetland system, and the designer can then take cognizance of the space available for the system design.

3.2 Water quality characterization

Characterization of pollutant concentration of the river water to be treated is essential for sizing of a constructed river diversion wetland and in creating a clear understanding of whether the wetland can effectively treat the water or not. Thus, the constituents of the river water and their respective concentrations need to be known before beginning the design process of the constructed river diversion wetland. However, water quality is highly variable especially in rivers due to fluctuations and variability of discharge and contaminant concentration from pollution sources [21]. Thus, a clear definition of water quality is essential, and it may be necessary to take into account previous distribution of the contaminants' concentrations in the water over time [4]. According to [22], characterization of the river water quality can be done based on available data which provides information on temporal and spatial distribution of parameters of interest and their level of concentrations in the water to be treated. Water quality parameters that are characterized in most situations include biochemical oxygen demand (BOD), nitrogen, phosphorus, suspended solids, and coliform bacteria [23]. These are pollutants that originate mostly from organic sources and are considered of most interest in treatment wetland design [24]. Others include metals, phenols, pesticides, and surfactants which may also be treated. However, these parameters require specific applications as opposed to organic pollutants [18].

BOD reflects the degree of organic matter pollution, and it is a measure of the amount oxygen removed by aerobic microorganisms for their metabolic requirement during decomposition of organic materials. Nitrogen and phosphorus are considered as primary drivers of nutrient pollution, and they occur in organic and inorganic forms. Nitrogen in water is usually measured as total nitrogen, ammonium ion, nitrate, nitrite, and total Kjeldahl nitrogen (sum of organic nitrogen and ammonium ion) or as a combination of these parameters to estimate organic or inorganic nitrogen concentrations [25]. Phosphorus in water is usually measured as total phosphorus which is the sum of organic and inorganic forms of phosphorus and includes orthophosphate (PO_4^{3-}), polyphosphates, and organic phosphates [1]. For microbial contamination, indicator organisms are used to detect the presence of pathogens (disease causing organisms). Microorganisms mostly considered are those of fecal origin, and coliform bacteria are most often used to indicate the presence of fecal pollution [26]. Suspended solids are constituents that remain in solid state in water and often occur as part of sediments carried in the water. Measurement of suspended solids is essential as sediments are responsible for contaminant transport in water. Metals can exist as dissolved, colloidal, or suspended forms in water, and their toxicity depends on the degree of oxidation of the metal ion together with the forms in which it occurs [1]. Metals mostly considered with high priority in water pollution are arsenic (As), cadmium (Cd), copper (Cu), chromium (Cr), lead (Pb), mercury (Hg), nickel (Ni), and zinc (Zn) [23]. Nevertheless, selection of any pollutant or combination of pollutants for water quality improvement will depend on the objectives for which the wetland is designed. Based on the river water quality characterization, appropriate equations can be used to determine the required area and organic loading rates of the wetland system.

3.3 Wetland design inflow estimation

The amount of water flow per unit time that passes through a wetland system is one of the important parameters required in the design of a constructed river diversion wetland. Flow rate of water is an important hydrological parameter required to facilitate sizing of a constructed wetland [4]. Even though flow into a wetland can be continuous or intermittent, it however passes through the system at low velocities. There are different approaches employed to determine the quantity of inflow (volumetric inflow rate) into a wetland, depending on the wetland type, treatment objectives, and incoming water to be treated.

For wastewater treatment wetlands, inflow is mostly based on wastewater concentration and generation rates [27]. Mass loading charts with reference to the required level of pollutant removal are mostly used in the United States, while in Europe estimation is based on wastewater generation volume and pollutant concentration [27, 28]. For stormwater constructed wetlands, a range of hydrologic methods are applied to estimate design flows. Typical approaches include the use of routing in response to a storm event like the average recurrence interval (ARI) flow criterion, level-pool routing, and estimation of peak runoff flow rate using curve number (CN) model and rational method [29–31]. The ARI is applied in Australia and level-pool in Malaysia, and the CN is mostly used in the United States. While all these methods are mainly applied to stormwater treatment wetlands, they are however used with reference to specific available data and scenarios in these countries [29]. Moreover, not all wetlands are designed for treatment of maximum expected peak flows; otherwise the vegetation are likely to be damaged due to high flows, and the wetland system would need to be extremely large or the outflow water quality requirement considerably relaxed. Furthermore, the CN model has been examined to be inaccurate due to inherent limitation associated with inconsistency of the fixed ratio (λ) between initial abstraction (I_a) and soil maximum potential retention (S) in the model [32–34]. The rational method was found to be more suitable only for estimating runoff for relatively small catchment that is preferably less than 50 ha [31]. Besides, paucity of site-specific data especially in Africa can make the use of these methods difficult and inaccurate.

For river diversion wetlands, a specific method for estimating design inflow has not been fully established [7]. However, more recently, [8] evaluated the performance of a river diversion wetland for improving quality of river water using relations that can be used to estimate design inflow for a similar wetland system. These relations are presented below.

$$\alpha = \frac{A_w}{A_{rw}} \tag{1}$$

$$\omega = \frac{Q_{w-i}}{Q_{rd}} \tag{2}$$

where α = wetland/river catchment area ratio; ω = wetland/river flow diversion ratio; A_{rw} = river catchment area (ha, m^2); Q_{rd} = average flow volume /discharge in river (m^3, m^3 per unit time); Q_{w-i} = inflow rate (m^3/d); A_w = proposed area of wetland (m^2) based on available space.

Application of the above equations requires estimation of average flow volume of a river. However, flow rates vary over time because of normal variability in precipitation patterns, and a key factor governing hydrological regime of rivers is their discharge variability [35]. Therefore, to determine river flow or discharge regimes, historical flow data are required, including possible seasonality trend of the flows, pattern of past flows (low, moderate, and high flows), and stream gauge information close to the wetland site location [4, 36]. Flow data are important to facilitate understanding of fluctuations in the amount of flowing water in the river and to support development of a rating curve for the river where it is not available. The rating curve has been an important tool widely used for routing purposes in hydrology to estimate discharge in natural rivers [37]. It is a graphical representation that gives relationship between flow regimes and stage heights or water levels of a river at a given site and over a period of time [35, 38]. However, very few rivers have absolutely stable flow characteristics, and thus the rating curve may require revision over time and under unsteady conditions. A comprehensive review of the various equations developed by several authors for correcting unsteady to steady flow condition was presented by [38].

3.3.1 Using the rating curve for estimation of river flow regime

Another crucial aspect of wetland design is the estimation of average river flow regimes. The river flow regimes are required to:

- Guide in determining the amount of water per unit time that can be diverted into the wetland system without compromising the flow needed for survival of the river ecosystem.

- Aid the design and estimation of inflow regime(s) for which the wetland system will be operated since the goal of the wetland is to improve quality of river water.

Therefore, obtaining or developing appropriate rating curve may be necessary to facilitate characterization of flow regimes of the river. Based on the rating curve, the river flows can be classified into low, moderate, and high flows. **Figure 2** shows a typical river rating curve with flows classified into three regimes as indicated. For example, based on the rating curve (**Figure 2**), three flow regimes (0.29 m^3/s, 1.97 m^3/s, and 3.96 m^3/s) (marked with dotted red lines) were selected corresponding to low, moderate, and high flows of the river, respectively. The classification of the flow regimes into low, moderate, and high flows was based on their computed flow velocities as presented in **Table 1**.

For flood or peak flow control wetlands, high flows are often considered for the design, while for water quality improvement, moderate to low flows are mostly the target. Where high flow is to be used for design of river diversion wetland intended for water quality improvement, it may be necessary to include a retention basin in the design to slow down flow energy and allow for gradual release into the system.

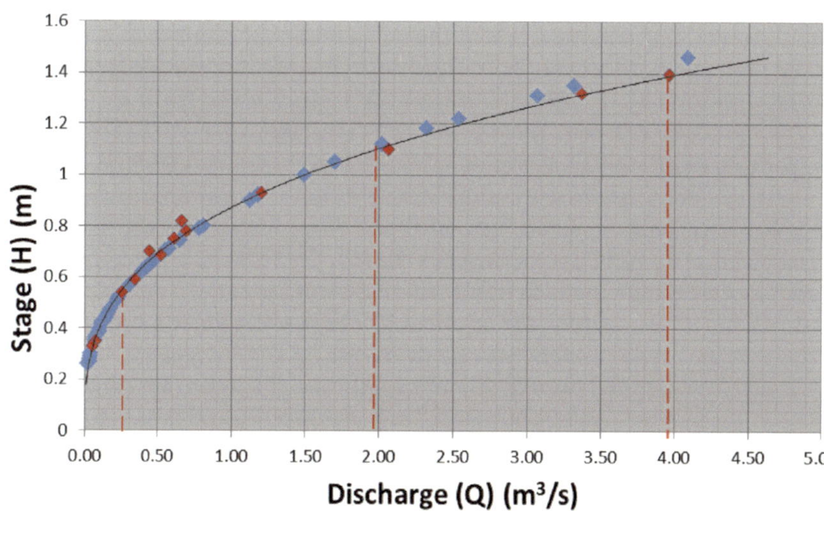

Figure 2.
Typical river rating curve with flows classified into three regimes.

Flow regimes (m^3/s)	Mean cross-sectional area of river gauging section (m^2)	Velocity (m/s)	Velocity* groups (m/s)	Classification
0.29	3.34	0.09	<0.10	Low flow
1.97	3.34	0.59	0.10–0.60	Moderate flow
3.96	3.34	1.19	≥0.70	High flow

**Values of velocity groups adapted from [39].*

Table 1.
Classification of the flow regimes.

Since the river diversion wetland under discussion is intended to be designed for water quality improvement, only a portion of the river flow regimes is required to be diverted into the wetland system per unit time. Thus, to determine the quantity of the design inflow rate of the wetland, Eq. (3) derived from Eqs. (1) and (2) can be used together with the average river flow regime(s).

$$Q_{w-i} = \frac{A_w Q_{rd}}{A_{rc}} \tag{3}$$

where all parameters remain the same as previously defined in Eqs. (1) and (2).

The wetland can be designed to operate with the three river flow regimes (low, moderate, and high) to take into account seasonal flow variability or a single flow regime depending on the objective and availability of space within the site.

4. Computing hydrodynamic parameters of the wetland

4.1 Wetland design equations

The design of constructed wetlands is generally based on empirical equations using zero- or first-order plug flow kinetics as basis for predicting pollutant removal and improving water quality [13]. With zero-order kinetics, the reaction rate does not change with concentration but varies with temperature [4], while first-order kinetics simply implies that the rate of removal of a particular pollutant is directly proportional to the remaining concentration of the pollutant at any point within the wetland [40]. Plug flow means that every portion of flow entering into the wetland takes almost the same amount of time to pass through it which is rarely the case [41]. The kinetic equations also considered FWS wetlands as attached growth biological reactors similar to those found in conventional wastewater treatment systems [23]. Generally, two types of equations are popular that use two different approaches in the design of FWS wetlands based on "rule-of-thumb" (no account for the many complex reactions that occur in a constructed wetland). There is the volume-based or zero-order kinetic equation which uses hydraulic retention time to optimize pollutant removal [42, 43]. The second is the area-based or first-order kinetic equation where the entire wetland area is used to provide the desired pollutant treatment [44]. The key difference between the two equations is in the use of kinetic rate constants. Volume-based equation assumes horizontal or linear kinetics and uses volumetric and temperature-dependent rate constant, with calculations being based on available volume of the wetland and average water temperature. The area-based equation assumes vertical or areal kinetics and uses rate constants which are independent of temperature but related to the wetland surface area. The volume-based equation was developed by [43], and the equations are presented below:

$$\frac{C_e}{C_i} = e^{-K_T t} \tag{4}$$

$$K_T = K_R \theta_R^{(T_w - T_R)} \tag{5}$$

where C_e = outflow pollutant concentration (mg/l); C_i = inflow pollutant concentration (mg/l); t = nominal hydraulic residence or retention time (d); K_T = reaction rate constant for BOD at T_w (/d); K_R = rate constant at T_R (/day); θ_R = temperature coefficient for rate constant; T_R = reference temperature (°C); T_w = ambient or water temperature (°C).

The area-based model equation was developed by [44], and the equations are presented below:

$$\frac{C_e - C^*}{C_i - C^*} = e^{-\frac{K_l}{h_l}} \tag{6}$$

$$h_l = \frac{Q_{w-i}}{A_w} \tag{7}$$

where C^* = background pollutant concentration (mg/l); K_l = reaction rate constant for phosphorus and fecal coliform (m/d); h_l = hydraulic loading rate (m/d); Q_{w-i} = inflow rate (m^3/d); A_w = proposed area of wetland (m^2) based on available space and other parameters as defined in Eq. (1).

The volume-based model was developed based on those parameters that are removed primarily by biological processes such as biochemical oxygen demand (BOD), ammonia (NH_4), and nitrate (NO_3). The areal equation considered more parameters and in addition includes total suspended solids (TSS), total phosphorus (TP), total nitrogen (TN), and fecal coliform (FC). According to [45], while the [43] method provides a relatively conservative area estimate, [44] approach may require considerable land space, depending on the pollutant concentration limit. Furthermore, the [45] model appears to be less sensitive to different climatic conditions as temperature changes are only considered significant for nitrogen removal [46]. However, temperature plays an important role in constructed wetland systems as it enhances higher biological activity and productivity which may lead to better performance of the systems [47, 48]. For this reason, the use of these models may lead to wide variations in performance due to effect of changes in climatic conditions. Additionally, many authors have developed more complex models like the Monod-type and mechanistic compartmental models [49, 50]. However, the [43, 44] models appear to be more straightforward and can be applied with ease by wetland designers [13]. Data limitation on operational performance of constructed wetlands prevented the development of equations which can clearly describe the kinetics of known wetland processes [23]. Thus, optimal design of constructed wetland systems has not yet been determined. However, in order to take advantage of [43, 44] models and ease complexity of computation, [24] presented a simplified approach for the design and sizing of FWS constructed wetlands using the two equations. The approach was based on performance criteria for the removal of four water quality parameters that included BOD, nitrogen, phosphorus, and coliform bacteria. According to [24], rates of BOD and nitrogen removal are principally temperature dependent and therefore utilized equations proposed by [43] model for removal of these parameters. On the other hand, the reduction of phosphorus and coliform bacteria was assumed to be governed by physical processes which are less temperature-dependent, and thus [44] equations were used. In addition, [24, 51] proposed the following relationships for nominal hydraulic retention time and removal of total nitrogen (TN), respectively.

$$t = \frac{V_{w-n}}{Q_{w-i}} = \frac{A_w y_w \varnothing}{Q_{w-i}} = \frac{y_w \varnothing}{h_l} \tag{8}$$

$$\frac{C_e}{C_i} = e^{-K_{TN}t} + e^{K_{TD}t} - e^{K_{TN}t}e^{K_{TD}t} \tag{9}$$

where V_{w-n} = wetland nominal volume (m^3); y_w = theoretical or nominal depth of wetland water flow (m); $\varnothing$ = porosity (percent, expressed as decimal fraction); K_{TD} = reaction rate constant for denitrification (/d); K_{TN} = reaction rate constant for nitrification (/d) and other parameters as defined in Eqs. (1), (3), and (4).

The authors recommended that the above equations can be used together with those presented by [43, 44] to determine the hydrodynamic and size parameters of a new FWS flow constructed wetland, depending on the target pollutant or combination of pollutants (BOD, nitrogen, phosphorus, and coliform bacteria) required to be removed from the wastewater. As indicated by [52], the approach presented by [24] is useful in the design of a new FWS constructed wetland and for performance evaluation of existing ones.

4.2 Using a combination of equations for river diversion wetlands

The use of a combination of wetland design equations proposed by [24, 51] was found to be useful for determination of river diversion wetlands' hydrodynamic parameters. These parameters include nominal hydraulic retention time and hydraulic loading rate.

4.2.1 Hydraulic retention time (HRT)

Determination of nominal hydraulic retention time is important for design guide and estimating possible pollutant removal ability of the wetland system. Thus, the nominal HRT for a river diversion wetland can be estimated based on the kinetic equations governing the removal of basic water quality parameters (BOD, nitrogen,

Parameter	Empirical equations	Equation no.	Source
BOD (mg/l)	$\frac{C_e}{C_i} = e^{-K_T t}$	(1)	[41]
	$K_T = 0.678\,(1.06)^{T_w-20}$	(11)	[12]
TN (mg/l)	$\frac{C_e}{C_i} = e^{-K_{TN}t} + e^{K_{TD}t} - e^{K_{TN}t}e^{K_{TD}t}$	(9)	[12]
	$K_{TN} = 0.2187(1.048)^{T_w-20}$	(12)	
	$K_{TD} = (1.048)^{T_w-20}$	(13)	
TP (mg/l)	$\frac{C_e}{C_i} = e^{\frac{K_l}{h_l}}$	(10)	[22]
	$t = \frac{y\varnothing}{h_l}$	(14)	[13]
FC (CFU/100 ml)	$\frac{C_e}{C_i} = e^{\frac{K_l}{h_l}}$	(10)	[22]
	$t = \frac{y\varnothing}{h_l}$	(14)	[13]

Note: K_l = reaction rate constant for TP = 0.0273 m/d and FC = 0.3 m/d); $\varnothing$ = porosity or space available for water flow through vegetation (0.65–0.75); T_w= water or ambient temperature.

Table 2.
Parameters and equations for computing design HRT.

phosphorus, and coliform bacteria), often used for sizing of constructed wetlands. **Table 2** shows the parameters and kinetic equation used for determining the nominal HRT.

4.2.2 Hydraulic loading rate (HLR)

The HLR of the wetlands system can be computed using Eq. (7) by [44]. The determination of the HLR is essential to guide in the design and can assist to avoid overloading the system. Thus, the design may confirm the organic loading rate is within the wetland limit; an equation developed by [51] can be used to compare the K_T value with the loading rate. The equation is presented as:

$$K_T \leq \frac{-10\ \mathrm{In}\left({}^{C_e}/_{C_i}\right)}{C_i y_w \varnothing} \tag{10}$$

where all parameters remain the same as defined in Eqs. (1), (2), and (4).

5. Wetland sizing and configuration

Sizing is an important component of wetland design and vital for pollutant removal processes to take place. Most of the design recommendations provided certain approaches to wetland sizing to maximize removal of pollutants. For wastewater treatment wetlands, population equivalent (PE) is mostly employed for the determination of design wetland area. The required surface area is usually expressed as unit area per population equivalent (m^2/PE). For example, 5–10 m^2/PE was recommended for FWS, while for SSF it ranges between 2 and 5 m^2/PE depending on the type (HSSF, VSF, and hybrids) [27]. For stormwater wetlands, the typical approach is to consider relative percentage of the contributing catchment area or connected impervious area, and 1–5% of the contributing watershed was recommended as actual sizing criterion [4]. For full-scale river diversion wetlands, a minimum of 2–7% of the total catchment area was recommended as wetland area [20]. However, such sizing criteria pose challenges of overestimation and do not account for any performance consideration [53]. Therefore, such prescribed wetland sizing criteria may be unrealistic due to space limitation and cost. Nevertheless, an approach derived based on empirical determination of actual area required for pollutant removal with reference to hydraulic loading rate as presented by [24] appears to be more realistic for estimating actual area of river diversion wetlands intended for water quality improvement. Thus, the actual area required for such a wetland system can be determined using Eq. (16) which was derived from Eq. (8) by [24].

$$A_{wc} = \frac{Q_{w-i}\, t}{y_w \varnothing} \tag{11}$$

where A_{wc} = actual area of the wetland (m^2) and other parameters as defined in Eq. (8).

For ease of operational control (flow control and water level adjustment) and increased removal efficiency, multiple wetland units often referred to as cells may be used where possible than a single unit wetland. This is particularly more applicable to design of off-stream river diversion wetland. Multiple cells have the advantages of providing greater flexibility in design and operation and enhancing the performance of the system by decreasing the potential for short-circuiting.

Wetland cell size depends primarily on water quality treatment needs and cost considerations.

The actual area of the wetland is then computed using Eq. (16). Based on the computed values, the actual area of the wetland is thus selected as the maximum of areas obtained for each of the target pollutants (BOD, nitrogen, phosphorus, and coliform bacteria).

Wetland system configuration is an important element in the design of river diversion constructed wetland technology. After determining an appropriate wetland size, it is necessary to define the system configuration or layout by choosing an appropriate aspect ratio. Aspect ratio represents length (L) to width (W) ratio (L/W) of the wetland. It was suggested that choosing a good aspect ratio can assist to minimize short-circuiting and maximize flow distribution within the wetland system for biological activities [54]. Aspect ratio of as low as 1:1 was recommended for SSF [55], while length to width ratio of between 3:1 and 5:1 was recommended for FWS from an optimal point of view by [23]. However, based on findings by [56], 10:1 was recommended for FWS for good hydraulic efficiency. For water quality improvement, a river diversion wetland should be designed to operate with the most efficient aspect ratio.

Wetland bed slopes are also critical to maintain a uniform water depth throughout the wetland system and facilitate drainage. In order to minimize short-circuiting, a uniform bed slope from inlet to outlet is recommended. Thus, the bed slope for SSF should be 2% or less, while that for FWS should be 0.5% or less [14]. A river diversion wetland can also be designed to operate with similar bed slope as recommended for FWS since they are related in mode of operation.

6. Water conveying system: inlet and outlet structures of the wetland

This aspect of the wetland design focused on selecting or designing a water conveying system, inlet and outlet control structures that can facilitate flow and distribute inflow and drain outflow water from the wetland effectively. Depending on the type of river diversion wetland, flow diversion structure may be designed to consist of either a pipe or channel system and should function to provide a controlled flow of water to the wetland. However, it is necessary to be explicit about flow capacity at the time of design so that appropriate sizing of flow diversion structure can be made. Generally, the design flow conveyance structure is based on hydraulic; therefore the reader is referred to hydraulic books for detailed information.

In order to ensure that the inflow water is uniformly distributed across the entire wetland area, multiple entry openings or gates should be considered rather than single to deliver the range of design flow regimes required. Flow control structures should be used to control inflow rate and maintain water levels. Control valves or weirs or a combination can be used depending on the type of inlet structured selected. Since the wetland system is for water quality improvement, high incoming water velocities should be discouraged. Therefore, energy dissipation system may be required for the incoming water to provide protection for the wetland inlet. The inlet openings should be designed large enough to avoid obstruction. Inlet zones should provide access for sampling and flow monitoring.

Wetland outlet design is essential in avoiding possible dead zones and controlling water level and for monitoring flow and water quality. Depending on the size of the wetland, a combination of outlets (primary and secondary) or multiple outlets consisting of hydraulic control structures can be considered to collect and discharge treated water for the range of design flow regimes and maintain required water storage level. The purpose of the primary outlets is for water quality control, while the secondary is to act as a spillway and control flows in excess of the maximum design flow regime. Different types of control structures are available that can be used to control water level within the wetland. These may include number of individual pipes that fit together in a combination to obtain the desired water level, drop structures, or weirs. The design requirements of drop control structures and weirs can be found in hydraulic books. The outlet or water level control structure should be able to completely dewater the wetland when needed and allow for changes to be made easily.

7. Conclusion

The management and restoration of water bodies like rivers should go beyond protection through the use of regulations. It should also make the most of opportunities that arise from using ecosystem properties to enhance self-purification capacity of rivers for water quality improvement. A key consideration is the use of constructed river diversion wetlands.

This chapter provided guidance on the design of a river diversion constructed wetland aimed at improving quality of river water. The use of a combination of empirical equations was presented to guide in the estimation of the actual wetland area rather than relying on an assumed rate. The design approach using these equations may present a promising method for the design of river diversion wetlands. Furthermore, this novel approach may be useful to wetland experts as some of the procedures adopted are not popular in wetland studies. This may provide opportunity for wetland designers to document approaches that have been found promising and come up with suitable design criteria for constructed river diversion wetlands.

Acknowledgements

The authors wish to express appreciation for the funding support provided by the Regional Water and Environmental Sanitation Centre, Kumasi (RWESCK), at the Kwame Nkrumah University of Science and Technology (KNUST), Kumasi, which is funded by the Government of Ghana and the World Bank under the Africa Centres of Excellence Project. Also, the authors wish to thank the National Water Resources Institute (NWRI), Nigeria, for providing opportunity and support for the study that led to the generation of knowledge and information presented in this chapter. We declare that the views expressed in this chapter are those of the authors and do not necessarily reflect those of the World Bank, Ghana Government, KNUST, and NWRI.

Conflict of interest

None declared.

Author details

Sani Dauda Ahmed[1,2]*, Sampson Kwaku Agodzo[1] and Kwaku Amaning Adjei[1]

1 Regional Water and Environmental Sanitation Centre Kumasi (RWESCK), Kwame Nkrumah University of Science and Technology, Kumasi, Ghana

2 National Water Resources Institute, Kaduna, Nigeria

*Address all correspondence to: sanidaud@yahoo.com

References

[1] Carr GM, Neary JP. Water Quality for Ecosystem and Human Health. 2nd ed. Ontario, Canada: United Nations Environment Programme (UNEP)-Global Environment Monitoring Systems (GEMS)/Water Programme; 2008

[2] Wang HT, Omosa IB, Keller AA, Li FT. Ecosystem protection, integrated management and infrastructure are vital for improving water quality in Africa. Environmental Science and Technology. 2012;**46**(9):4699-4700

[3] Meybeck M. Global analysis of river systems: From earth system controls to anthropocene syndromes. Philosophical Transactions of the Royal Society of London B: Biological Science. 2003; **358**(1440):1935-1955

[4] ITRC (Interstate Technology and Regulatory Council). Technical and Regulatory Guidance Document for Constructed Treatment Wetlands. USA: Interstate Technology and Regulatory Council (ITRC); 2003. Available from: www.itrcweb.org

[5] Mao X, Yuan D, He L, Wei X, Chen Q, Bian L, et al. How long, narrowly constructed wetlands purify irrigation return water: A case study of Ulansuhai Lake, China. Advances in Meteorology. 2015;**2015**:1-6. DOI: 10.1155/2015/142036

[6] Borges AKP, Tauk-Tornisielo SM, Domingos RN, Angelis DF. Performance of the constructed wetland system for the treatment of water from the Corumbatai River. Brazilian Archives of Biology and Technology. 2008;**51**(6): 1279-1286

[7] Fink DF, Mitsch WJ. Hydrology and nutrient biogeochemistry in a created river diversion oxbow wetland. Ecological Engineering. 2007;**30**:93-102

[8] Jia Z, Tang S, Luo W, Hai Y. Water quality improvement through five constructed serial wetland cells and its implication on nonpoint-source pollution control. Hydrological Sciences Journal. 2016;**61**(16):2946-2956. DOI: 10.1080/02626667.2016.1171323

[9] Zalewski M. Ecohydrology—The use of ecological and hydrological processes for sustainable management of water resources. Hydrological Sciences Journal. 2002;**47**:825-834

[10] Bodin H. Wastewater treatment in constructed wetlands: Effects of vegetation, hydraulics and data analysis methods [PhD thesis]. Sweden: Linkoping UniversityAccession No. 1509; 2013

[11] Konnerup D, Koottatep T, Brix H. Treatment of domestic wastewater in tropical subsurface flow constructed wetlands planted with *Canna and Heliconia*. Ecological Engineering. 2009; **35**(2):248-257

[12] Mangangka IR. Role of hydraulic factors in constructed wetland and bioretention basin treatment performance [PhD thesis]. Australia: Queensland University of Technology; 2013

[13] Rousseau DPL, Vanrolleghemb PA, De Pauwa N. Model-based design of horizontal subsurface flow constructed treatment wetlands: A review. Water Research. 2004;**38**:1484-1493

[14] Ellis JB, Shutes RBE, Revitt DM. Guidance Manual for Constructed Wetlands. Research Development Technical Report No. P2-159/TR2. Bristol, UK: Environment Agency; 2003

[15] Vymazal J. Constructed wetlands for wastewater treatment. Water. 2010;**2**: 530-549. DOI: 10.3390/w2030530

[16] Kadlec RH, Wallace SD. Treatment Wetlands. 2nd ed. Boca Raton, New York, USA: CRC Press; 2009

[17] Vymazal J. Emergent plants used in free water surface constructed wetlands: A review. Ecological Engineering. 2013; **61**(Part B):582-592. 1-11. DOI: 10.1016/j.ecoleng.2013.06.023

[18] Westerhoff P, Sharif F, Halden R, Herckes P, Krajmalnik-Brown R. Constructed wetlands for treatment of organic and engineered nanomaterial contaminants of emerging concerns. In: Web Report No: 4334. Denver, USA: Water Research Foundation (WRF); 2014

[19] Mitsch WJ, Day W. Restoration of wetlands in the Mississippi–Ohio–Missouri (MOM) river basin: Experience and needed research. Ecological Engineering. 2006;**26**:55-69. DOI: 10.1016/j.ecoleng.2005.09.005

[20] Verhoeven JTA, Arheimer B, Yin C, Hefting MM. Regional and global concerns over wetlands and water quality. Trends in Ecology and Evolution. 2006;**21**(2):96-103

[21] Meybeck M, Kimstach V, Helmer R. Strategies for water quality assessment. In: Chapman D, editor. Water Quality Assessments—A Guide to Use of Biota, Sediments and Water in Environmental Monitoring. 2nd ed. Cambridge, Great Britain: UNESCO/WHO/UNEP University Press; 1996

[22] Kimwaga RJ, Mwegoha WJS, Mahenge A, Nyomora AM, Lugali LG. Factors for success and failures of constructed wetlands in sanitation service chain. In: Dissemination of the Sustainable Wastewater Technology of Constructed Wetlands in Tanzania (VLIR Research Project). 2013

[23] USEPA. (U.S. Environmental Protection Agency) Constructed Wetlands for Treatment of Municipal Wastewater Manual (EPA 625/R-99/010). Cincinnati, OH: U.S. EPA; 2000

[24] Economopoulou MA, Tsihrintzis VA. Design methodology of free water surface constructed wetlands. Water Resources Management. 2004;**18**: 541-565

[25] Chapman D, Kimstach V. Selection of water quality variables. In: Chapman D, editor. Water Quality Assessments—A Guide to Use of Biota, Sediments and Water in Environmental Monitoring. 2nd ed. Cambridge, Great Britain: UNESCO/WHO/UNEP University Press; 1996

[26] Ashbolt NJ, Grabow WOK, Snozzi M. 13: Indicators of microbial water quality. In: Fretwell L, Bartram J, editors. Water Quality: Guidelines, Standards and Health (World Health Organization (WHO)). London, UK: IWA Publishing; 2001

[27] Miller J. Constructed Wetlands Technology Assessment and Design Guidance (Manual). USA: Iowa Department of Natural Resources (IDNR); 2007

[28] Gikas GD, Tsihrintzis VA. Municipal wastewater treatment using constructed wetlands. Water Utility Journal. 2014;**8**:57-65

[29] HTCKL (Humid Tropics Hydrology and Water Resources Centre). Technical Report on Constructed Wetland. The Regional Humid Tropics Hydrology and Water Resources Centre for Southern Asia and the Pacific. Malaysia, Kuala Lumpur: HTCKL; 2014

[30] Millhollon EP, Rodrigue PB, Rabb JL, Martin DF, Anderson RA, Dans DR. Designing a constructed wetland for the detention of agricultural runoff for water quality improvement. Journal of Environmental Quality. 2009;**38**: 2458-2467. DOI: 10.2134/jeq2008.0526

[31] WSUD (Water Sensitive Urban Design). Constructed wetlands (Engineering Guidelines) for Brisbane City. Australia, Brisbane City; Brisbane City Councils; 2005

[32] Ajmal M, Khan TA, Kim T. A CN-based ensembled hydrological model for enhanced watershed runoff prediction. Water. 2016;**8**(20):1-17. DOI: 10.3390/w8010020

[33] Duran-Barroso P, Gonzalez J, Valdes JB. Sources of uncertainty in the NRCS model: Recognition and solution. Hydrological Processes. 2017;**31**(22):1-9. DOI: 10.1002/hyp.11305

[34] Yuan Y, Nie W, McCutcheon SC, Taguas EV. Initial abstraction and curve numbers for semiarid watersheds in southeastern Arizona. Hydrological Processes. 2012;**28**(3):1-11. DOI: 10.1002/hyp9592

[35] WMO (World Meteorological Organization). Manual on Stream Gauging: Volume II—Computation of Discharge (WMO-No. 1044). Geneva, Switzerland: World Meteorological Organization (WMO); 2010

[36] Lenhart C, Gordon B, Gamble J, Current D, Ross N, Herring L, et al. Design and hydrologic performance of a tile drainage treatment wetland in Minnesota, USA. Water. 2016;**8**(549): 1-19. DOI: 10.3390/w8120549

[37] Guven A, Aytek A, Azamathulla H. A practical approach to formulate stage–discharge relationship in natural rivers. Neural Computation and Application. 2012;**23**(3-4):873-880. DOI: 10.1007/s00521-012-1011-5

[38] Dottori F, Martina MLV, Todini E. A dynamic rating curve approach to indirect discharge measurement. Hydrology and Earth System Sciences Discussions. 2009;**6**:859-896

[39] Biggs BJF. Hydraulic habitat of plants in streams. Regulated River Resource and Management. 1996;**12**: 131-144

[40] Popov V, Ahmed S, Qureshi MI. Simulation of constructed wetland performance using the advection diffusion equation. WIT Transactions on Ecology and the Environment. 2008; **109**(IV):363-371. DOI: 10.2495/WM080381

[41] Stairs DB. Flow characteristics of constructed wetlands: Tracer studies of the hydraulic regime [MSc thesis]. USA: Oregon State University; 1993. pp. 81

[42] Crites RW, Tchobanoglous G. Small and Decentralized Wastewater Management Systems. New York, NY: McGraw Hill Co.; 1998

[43] Reed SC, Crites RW, Middlebrooks EJ. Natural Systems for Waste Management and Treatment. 2nd ed. New York, USA: McGraw-Hill, Inc.; 1995

[44] Kadlec RH, Knight RL. Treatment Wetlands. Boca Raton, USA: Lewis Publishers; 1996

[45] Kadlec RH. The inadequacy of first-order treatment wetland models. Ecological Engineering. 2000;**15**:105-119

[46] Kayombo S, Mbwette TSA, Katima JHY, Ladegaard N, Jorgensen SE. Waste stabilization ponds and constructed wetlands design manual. Tanzania, Dar Es Salaam: UNEP-IETC with the Danish International Development Agency (Danida); 2005

[47] Kadlec RH, Reddy KR. Temperature effects in treatment wetlands. Water Environment Research. 2001;**73**(5): 543-557

[48] Kivaisi AK. The potential for constructed wetlands for wastewater treatment and reuse in developing countries: A review. Ecological Engineering. 2001;**16**:545-560

[49] Mitchell C, McNevin D. Alternative analysis of BOD removal in subsurface flow constructed wetlands employing

Monod kinetics. Water Research. 2001; **35**(5):1295-1303

[50] Wynn TM, Liehr SK. Development of a constructed subsurface-flow wetland simulation model. Ecological Engineering. 2001;**16**:519-536

[51] Economopoulou MA, Tsihrintzis VA. Design methodology and area sensitivity analysis of horizontal subsurface flow constructed wetlands. Water Resources Management. 2003;**17**(2):147-174

[52] Sarah KL. Natural treatment and onsite processes. Water Environmental Research. 2005;**77**(6):1389-1424

[53] Moreno-Mateos D, Pedrocchi C, Comin FA. Effects of wetland construction on water quality in a semi-arid catchment degraded by intensive agricultural use. Ecological Engineering. 2010;**36**:631-639

[54] Bendoricchio G, Cin LD, Persson J. Guidelines for free water surface wetland design. EcoSys Bd. 2000;**8**: 51-91

[55] Steiner GR, Freeman RJ. Configuration and substrate design considerations for constructed wetlands wastewater treatment. In: Hammer DA, editor. Constructed Wetlands for Wastewater Treatment. Chelsea, USA: Lewis Publishers; 1989

[56] Persson J, Somes NLG, Wong THF. Hydraulics efficiency of constructed wetlands and ponds. Water Science and Technology. 1999;**40**:291-300

Chapter 5

The Tourism Impacts of Lake Erie Hazardous Algal Blooms

Matthew Bingham and Jason Kinnell

Abstract

Nutrient loading and warming waters can lead to hazardous algal blooms (HABs). Policymakers require cost-effective valuation tools to help understand impacts and prioritize adaptation measures. This chapter evaluates the tourism impacts of HABs in Western Lake Erie based on HABs that occurred in 2011 and 2014, both through a unique temporal and spatial specification of HAB severity as well as input/output analysis and decomposition of trips and profitability.

Keywords: hazardous algal blooms, HABs, nutrient loading, socioeconomic, benefits transfer, Lake Erie, input/output, tourism

1. Introduction

Attractive inland waters such as western Lake Erie can provide significant tourism services [1]. Hazardous algal blooms (HABs) that tend to result from warm water and nutrient loading result in murky and unpleasant water (**Figure 1**), potentially interrupting the $12.9 billion tourism industry in the region and putting

Figure 1.
Maumee Bay State Park, Ohio, 2013 HAB. Source: Ref. [2].

at risk up to 119,491 jobs in the sector [3]. Tourism losses can further disrupt local economic activity as tourists provide important demand for residential and hotel rentals, as well as other local expenditures such as restaurants and shops. As a result, profitability impacts to tourism can influence residential and commercial property values. Consequently, the economic implications of HABs in western Lake Erie are far-reaching and complex. In 2014 the authors of this chapter analyzed the socio-economic impacts of HABs in western Lake Erie. This chapter summarizes an evaluation of HAB effects to tourism that was part of a larger study of impacts to regional economic welfare including effects on property values, tourism, and recreation [1].

2. Ecological services and economic benefits

The study employs a forward-looking perspective in order to identify the economic benefits of reductions in future HABs. The economic methods employed are based on willingness to pay for ecological services. Reductions in these services impact economic values. That results in changes to behaviors and activities, ultimately affecting economic benefits. This process is depicted in **Figure 2**. This study examines the impact of reduction in quality of ecological services, which leads to changes in behaviors and activities, ultimately creating economic costs. This process is depicted in **Figure 2**.

As **Figure 2** depicts, there are many potentially interrelated ecological services. As a result, changes in these services affect behaviors and activities that are also interrelated. For example, lower boating quality could affect local boating trips, curbing tourism, which reduces commercial activity and profits as well as both commercial and residential property value.

Conceptually, this study relies on the assumption that individuals express their desires through their choices and trade-offs. Economic gains and losses are measured through consumer surplus—the difference between the amount someone is willing to pay for a good or service and the market value of that same item. For example, if someone is willing to pay $7 for an item with a market price of $4, then that item has a consumer surplus of $3.

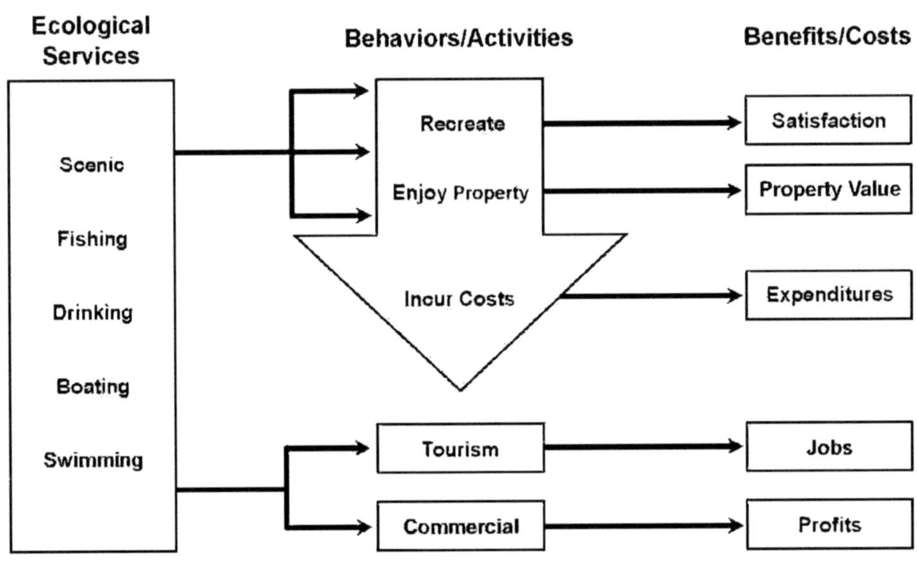

Figure 2.
Economic relationships.

DOI: http://dx.doi.org/10.5772/intechopen.93625

3. HAB scenarios studied in this effort

HABs of varying levels of severity are likely to recur in Lake Erie. Their size and location are difficult to predict, but mitigation may allow for the avoidance of potentially large and far-reaching economic effects. Consequently, when considering the immediate (i.e., within-year) effects, this study uses past HABs to predict the economic effects that would accompany reductions in future HABs.

This study focuses on the most damaging recent HABs in 2011 and 2014, and the consequent service reductions for those years. While information about beach closures is available, there are no data specifically analyzing reductions in tourism, or quantitative analyses of the impacts of these two HABs While visual data showing reductions in ecological service (such as contaminated shorelines or clogged marines) are readily available, a lack of quantitative or written analysis hinders precise analysis of the date, location, and severity of past HABs.

Given this limitation, this study uses news reports and satellite images to create a scale of HAB severity [4]. Since most overhead images of Lake Erie's algal blooms are not precisely dated, the study relies on date-stamped satellite images from NOAA such as that depicted below (**Figure 3**).

For several years up to 2012, NOAA posted Medium-Spectral Resolution Imaging Spectrometer (MERIS) imagery of Lake Erie. Since then, NOAA has posted images of Lake Erie HABs from the Moderate Resolution Imaging Spectro-Radiometer (MODIS) on the AQUA satellite. Both MERIS and MODIS imagery are dated at least weekly [6]. An example satellite view is depicted above. This study uses a scale ranging from 0 to 1 to quantify HAB severity in a given area of Lake Erie.

This study uses the finest degree possible of both temporal specificity—weekly analysis—and spatial specificity—county-level for mainland shorelines in addition to three island groupings. Severity ratings by week and month were developed for 2011 and 2014 from July through October. **Table 1** below analyzes July of 2011.

This information was incorporated into the evaluation of effects to tourism.

Figure 3.
Satellite view of Lake Erie HAB. Source: Ref. [5].

Location	2011 July weeks			
	1st	2nd	3rd	4th
Essex mainland	0	0	0.25	0
Pelee Island	0	0	0	0
Wayne (southern tip)	0	0	0	0
Monroe	0	0	0.50	0.50
Lucas	0	0	0.50	0.25
Ottawa mainland	0	0	0	0
Bass Islands	0	0	0	0
Sandusky	0	0	0.50	0.25
Erie mainland	0	0	0	0
Kelleys Island, Erie County	0	0	0	0

Table 1.
Severity rating for HABs in the Western Basin of Lake Erie, July 2011. Sources: [6–9].

4. Tourism and commerce

Since tourism, business demand, and commercial property values are all closely related, by affecting tourism HABs can in turn negatively impact all three economic sectors in areas close to western Lake Erie. For example, a well-publicized HAB event would almost certainly reduce tourism, in turn lowering revenue for businesses such as local restaurants, hotels, and charter boat operators. As these businesses lose revenue, they would likely purchase fewer supplies, affecting other businesses upstream in the supply chain. Finally, since these businesses would be expected to purchase less labor due to lower demand, either by hiring less or through layoffs, the local economy suffers as a result of lost local wages.

Ultimately, these sorts of effects would be reflected in business balance sheets as reduced revenues and profitability. Additionally, since affected businesses' values are most likely tied to their assets and the real estate they occupy (for example, a marina is not easily converted to some other use), on-going balance sheet effects would ultimately lead to reductions in commercial real estate values.

There are many challenges to understanding the implications of changes in tourism from HABs. The clearest challenge obstructing a precise analysis of these impacts is a lack of data either on the amount of tourism at risk or the specific impact of HABs on tourism. For example, while county-level data exists for total expenditures on tourism, this includes tourism which would not be interrupted by HABs or other discouraging factors.

An additional challenge relates to the distinction between economic benefits (willingness to pay) and economic impacts (expenditures), and the measurement of the economic benefits that arise from economic impacts (profits). For example, consider a restaurant owner who loses $10,000 in revenue because of a HAB. The owner's willingness to pay to recover that revenue; is (roughly speaking) the lost profit on that revenue. This is more difficult to identify than lost revenue. Understanding the negative effects of HABs upstream in a supply chain requires knowing what expenditures were foregone, which depends on the operation's variable cost situation with respect to employees (salaried or not) already purchased foodstuffs (perishable or not) and utilities. To address this issue, the study identifies expenditure changes and then characterizes benefits associated with those changes.

An additional issue is that changes in tourism may represent changed rather than lost trips. A tourist who does not go to the western basin because of HABs might instead go to the central basin, or somewhere else. As a result, changes in demand in one area have an opposite effect in other areas. To address this, we limit the geographical scope of the study to a region affected by HABs.

Finally, because commercial property values tend to be linked to business profitability, evaluating both risks double-counting. This study focuses on business profitability.

The remainder of this chapter presents the detailed methods and results. Counties studied are United States counties depicted below (**Figure 4**).

Due to differences in available data, slightly different methods are applied for Ohio, and Michigan.

4.1 Ohio tourism

As different sorts of information are available by region, varying approaches are applied. This sub-section explores potential effects in Lucas, Ottawa, Sandusky, and Erie counties. The approach relies on estimates of expenditures per trip. Expenditure and trip data in Ohio are collected from [3, 10] which indicate $110 per Ohio day visitor in 2013. This is 57.4% of total visitor spending and 80% of total Ohio visitors. Some 33% are from Toledo and Cleveland.

Spending from overnighters in 2013 was estimated at $335 per day—42.6% of total Ohio visitor spending. These visitors were 20% of total visitors. Of these, 20% are with relatives and friends. Average of nights per trip was 3.2 nights per trip and that of members per party was 3.4. Eighteen percent of these visitors went to a beach at a lake. Consumers spend the most on transportation, as well as food and beverage, since both day and overnight visitors spend money in these categories. Lodging only accounts for 11% of spending, while retail and recreation expenditures are almost one-third of Ohio visitor spending.

These expenditure rates can be subdivided based on trip type and expenditures. For example, day visitors spend $110 per visitor with none of that being for air travel or lodging. Overnight visitors' costs vary depending on if visitors stay with friends/family or in commercial lodging. For the purposes of this study, we presume overnight visitors who stay with friends and family do not spend money on lodging and overnight visitors who stay with friends and family spend an average of $244. Those who stay in commercial lodging places spend about 10% more on food and beverages than overnight visitors who stay with friends and family. On average this is $358 per day for each overnight visitor who pays for lodging.

Figure 4.
Counties studied.

Per-day expenditures vary by type of visit. In order to capture the full effect of changes in tourism using available tourism information, the effect of consumer expenditures must be extrapolated in terms of their implications for expenditures in other parts of the supply chain. To do so, we apply a mathematical-economic technique called input/output analysis [11]. Input/output analysis can be used to assess the effects of direct changes in expenditures through indirect impacts which arise in supplying industries and induced impacts which result from changes in local employment impacts to local expenditures.

Impacts are estimated using IMPLAN [12] with equations and data from ZIP codes on the shoreline of Lake Erie in Lucas County, Ohio. IMPLAN contains detailed input-output information on more than 500 economic sectors at the national, state, county, and ZIP code level.

Expenditures are apportioned over these sectors at the rate that they appear in the IMPLAN data and then simulations are conducted using IMPLAN. The sum of per-trip indirect and induced effects is a fraction of direct effects.

The approach for estimating tourist trips and dollars at risk in Ohio begins with estimates of by county tourism economic impacts in 2013. These are available from [3].

These are converted into a composite trip. Modeling in IMPLAN indicates the economic impact of an average tourist day is $210. This approach provides estimates of tourist trips from outside Ohio's western basin shoreline counties, before narrowing the scope to trips that would potentially be affected by HABs, i.e. trips related to Lake Erie which occur when HABs are present in Lake Erie. Based on [10] which indicates that hotel stays are evenly distributed over the year and that 18% of tourist trip to Ohio are visits to lakeside beaches, the late summer and early fall account for 12% of annual days. Tourist days that are at risk from HABs are calculated as 2.16% of trips ($0.18 \times 0.12 = 0.0216$). The percentage of shoreline trips is not available, and 10% is specified. This process results in estimates of Ohio tourist dollars at risk that range from $66 million to $305 million.

Based on a percent of diverted trips of 5 and 10% Ohio tourist dollar losses on the low-end total approximately $3 million. On the high end, they exceed $30 million. Considering these estimates of lost revenue, it remains to consider the benefits associated with these. Changes in profit are the best-available representation of the benefit (willingness to pay) for changes in revenue. Profit is the difference between costs and revenues.

The authors of [13] report restaurants earn from 4 to 6% median income on revenue before taxes. According to calculations derived from [14] losing a marginal customer could impact restaurant profit anywhere from 5 to 68%, so long as labor and operating costs remain constant. This implies high-end lost profits of $20.79 million and low-end estimates of $165,000.

4.2 Michigan tourism

Wayne and Monroe counties are in Michigan and adjacent to Lake Erie. Because only a small portion of Wayne County is exposed and impacts there are minimal Wayne County was not evaluated. Using methods similar to those that were applied for Ohio the number and types of trips that would sum to 14 million visitors to Monroe County were evaluated [15, 16].

These 14 million visitors indicate a total of 74,288 trips at risk from the presence of HABs. This further breaks down to 24,728 day trips, 25,276 unpaid overnight trips, 16,355 at a hotel, and 7930 at a bed and breakfast that are at-risk. This is associated with $18.2 million. With indirect and induced effects included a total of $24.78 million in tourism economic impact is at risk in Monroe County.

Using methods as described for Ohio, there are high-end lost profits of $1.685 million and low-end estimates of $124,000.

5. Conclusions

Tourism is a dynamic activity that can be easily affected by negative events such as HABs. Ohio tourist dollars at risk from HABs range from $66 million to $305 million. Associated high-end lost profits are $21 million but could be under $1 million. In Michigan, about $25 million in tourism economic impact was judged to be at risk, which was associated with lost profits of $1.7 million on the high end. Deriving these results from available data requires numerous assumptions, and result in large ranges of uncertainty.

Acknowledgements

Underlying efforts were funded by the International Joint Commission. The authors are grateful for assistance from Frank Lupi and Sanjiv Sinha.

Conflict of interest

There are no conflicts of interest.

Author details

Matthew Bingham* and Jason Kinnell
Veritas Economic Consulting, Cary, North Carolina, United States

*Address all correspondence to: matthew.bingham@veritaseconomics.com

References

[1] Bingham M, Sinha S, Lupi F. Economic Benefits of Reducing Harmful Algal Blooms in Lake Erie. Submitted to the International Joint Commission. Ann Arbor, Michigan: Environmental Consulting & Technology, Inc.; 2015. p. 66

[2] U.S. Geological Survey. Lake Erie Waters at the Beach with Buoy, Public Domain Image [Internet]. 2014. Available from: https://www.usgs.gov/media/images/lake-erie-waters-beach-buoy [Accessed: 12 February 2020]

[3] Tourism Economics. The economic impact of tourism in Ohio. In: Presented at the TourismOhio Advisory Board's 2014 Symposium on the Future of Tourism in Ohio. Wayne, Pennsylvania: Tourism Economics; 2014

[4] Ohio Department of Health, Ohio Environmental Protection Agency, and Ohio Department of Natural Resources. State of Ohio Harmful Algal Bloom Response Strategy for Recreational Waters [Internet]. 2012. Available from: http://www.epa.ohio.gov/portals/35/hab/HABResponseStrategy.pdf [Accessed: 30 April 2015]

[5] U.S. Geological Survey. Lake Erie Algal Bloom, Public Domain Image [Internet]. 2014. Available from: https://www.usgs.gov/media/images/lake-erie-algal-bloom [Accessed: 12 February 2020]

[6] National Oceanic and Atmospheric Administration Great Lakes Environmental Research Laboratory. Harmful Algal Blooms in Lake Erie—Experimental HAB Bulletin Archive [Internet]. 2015. Available from: http://www.glerl.noaa.gov/res/waterQuality/lakeErieHABArchive/ [Accessed: 04 May 2015]

[7] International Joint Commission Working Committee 2, Land Use and Management. Final Report [Internet]. 1993. Available from: http://www.ijc.org/files/publications/ID1243.pdf [Accessed: 13 May 2015]

[8] Michalak A, Anderson E, Beletsky D, Boland S, Bosch N, Bridgeman T, et al. Record-setting algal bloom in Lake Erie caused by agricultural and meteorological trends consistent with expected future conditions. Proceedings of the National Academy of Sciences of the United States of America. 2013;**110**(16):6448-6452

[9] Shuchman R, Sayers M, Raymer Z, Grimm A. Multi-Satellite Harmful Algal Bloom Observation Summary 1997-2014 Western Basin Lake Erie [Internet]. 2015. Available from: http://www.glerl.noaa.gov/res/waterQuality/docs/Table_Update_Jan2015.pdf [Accessed: 13 May 2015]

[10] Longwoods International. TourismOhio. In: Presented at the TourismOhio Advisory Board's 2014 Symposium on the Future of Tourism in Ohio; 26 June 2014; Columbus, Ohio [Internet]. 2014. Available from: http://discoverohio.com/admin/new/Uploads/MeetingMinutes/3bce4ace-8fbc-4144-bddb-a231a162780fLongwoods%20I.pdf [Accessed: 30 March 2015]

[11] Leontief W. Input-Output Economics. New York: Oxford University Press; 1986

[12] IMPLAN Group, LLC. IMPLAN system (Lucas County, Ohio 2013 Data). Huntersville, North Carolina: IMPLAN Group, LLC; 2014

[13] National Restaurant Association. Restaurant operations report [Internet]. 2015. Available from: http://www.restaurant.org/News-Research/Research/Operations-Report [Accessed: 01 May 2015]

[14] Locsin A. The Average Profit Margin for a Restaurant [Internet]. 2015. Available from: http://smallbusiness.chron.com/average-profit-margin-restaurant-13477.html [Accessed: 19 May 2015]

[15] Monroe County Planning Department and Commission. Monroe County, Michigan: 2013 Comprehensive Economic Development Strategy Second Draft. Monroe, Michigan: Monroe County Planning Department and Commission; 2013. Available from: https://www.co.monroe.mi.us/docs/13_CEDS_2ndDraft_complete.pdf [Accessed: 30 April 2015]

[16] D.K. Shifflet & Associates Ltd. Year-End 2010 Visitor Profile: An Inside Look at the Leisure Travel Market in Michigan. McLean, Virginia: D.K. Shifflet & Associates Ltd. p. 2011

Chapter 6

Constructed Wetlands in Wastewater Treatment and Challenges of Emerging Resistant Genes Filtration and Reloading

Donde Oscar Omondi and Atalitsa Caren Navalia

Abstract

A wetland is a unique and distinct ecosystem that is flooded by water, either permanently or seasonally, where oxygen-free processes prevail, and the primary distinctive factor of wetlands from other landforms or water bodies is the occurrence of adaptive vegetation of aquatic plants, characteristic to the unique hydric soil. A constructed wetland is an artificial shallow basin filled with substrate, usually soil or gravel, and planted with vegetation that has tolerance to saturated conditions. As much as the use of constructed wetland has been recommended in the treatment of various forms of wastewater, the system efficiency is a factor of very many natural and artificial factors, with the emerging pollutants and contaminants such as resistant genes being the most complicated contaminants to eliminate through the system. Indeed, the emerging pollutants in forms of antibiotic resistant genes (ARGs) have remained prevalent in aquatic environments such as wetlands that receive ARG-loaded sewage. Therefore, this chapter covers a discussion on constructed wetlands in wastewater treatment and challenges of emerging contaminants, such as resistant genes filtration and reloading mechanisms, and provides recommendation for the proper handling and removal of such pollutants from the wetlands' functional system.

Keywords: antibacterial resistant genes, constructed wetlands, emerging pollutants, wastewater treatment, wetlands

1. Introduction

Wetland is a unique and distinct ecosystem that is flooded by water, either permanently or seasonally, where oxygen-free processes prevail, and the primary distinctive factor of wetlands from other landforms or water bodies is the occurrence of adaptive vegetation of aquatic plants, characteristic to the unique hydric soil [1, 2]. The modified form of wetland is termed "constructed wetland." Constructed wetlands for water treatment are complex, integrated systems of water, plants, animals, microorganisms, and the environment [3, 4]. Wetlands play a number of functions, including water purification, water storage, processing and recycling of carbon and other micro and macro nutrients, stabilization of shorelines, and support of plants and animals. While wetlands are generally reliable,

IntechOpen

self-adjusting systems, an understanding of how natural wetlands are structured and how they function greatly increases the likelihood of successfully constructing a wetland treatment system [5, 6].

The cleansing of water has always occurred through natural processes as the water flows through rivers, lakes, streams, and wetlands, and in the last several decades, systems have been constructed to use some of these processes for water quality improvement [7]. Wetlands are now highly preferred as systems for improving the quality of point and nonpoint sources of water pollution, including stormwater runoff, domestic wastewater, agricultural wastewater, as well as coal mine drainage [4]. To enhance sustainability in wastewater management, the use of constructed wetlands has been applied in the treatment of different forms of wastes. Artificially created wetlands have been successful in the treatment of petroleum refinery wastes, wastes from sugar factory, leachates from landfills and composts, wastes from aquaculture systems, wastes from pulp and paper mills, and wastes that emanate from slaughter houses, textile mills, and plants that process sea food. Under the management of these wastes, the constructed wetlands can serve as the sole treatment or may be part of an integrated wastewater treatment system [8].

Antimicrobial resistance (AMR) is defined as the ability of a microbe to resist the effects of medication that was once successful and efficient in treating the microbe [9]. The term antibiotic resistance (ABR) is a subset of AMR, as it applies only to bacteria becoming resistant to antibiotics. The AR phenotypes can arise within a microorganism through the lateral and horizontal gene transfers and mutation. The mutations of the chromosomal DNA alter the existing bacterial proteins, through transformation, resulting in the creation of mosaic proteins and/or as a result of the transfer and acquisition of new genetic material between bacteria of the same or different species or genera [10]. The emerging pollutants in forms of antibiotic resistance genes (ARGs) have remained prevalent in aquatic environments such as wetlands that receive ARG-loaded sewage [11].

As much as the use of constructed wetland has been recommended in the treatment of various forms of wastewater, the system efficiency is a factor of very many natural and artificial factors, with the emerging pollutants and contaminants such as resistant genes being the most complicated contaminants to eliminate through the system [11, 12]. Moreover, some studies have reported constructed wetlands as reservoirs to various forms of resistant genes, which trap them and release them to other aquatic systems, hence contributing to their higher concentration in streams, rivers, or lakes [13]. Numerous suggestions have been provided to improve wetland's functional effect, efficiency, and predictability and provide a proper ecosystem management [6, 7]. This chapter covers a discussion on the constructed wetlands in wastewater treatment and the challenges of emerging related contaminants, such as resistant genes, and provides recommendation for the proper handling and removal of such wastes from the wetland's functional system.

2. Wetlands as functional systems

Wetlands are ecotones/transitional areas between land and water, with indistinct boundaries between the wetland area and uplands or deep water [14]. The definition expansion of the term wetland covers a broad range of systems that range from marshes, bogs, swamps, wet meadows, tidal wetlands, floodplains, and ribbon (riparian) zones along stream channels. However, all wetlands, whether they are natural or artificial, freshwater, or salty, pose a single characteristic or numerous characteristics, and they occur within the surface or near-surface water, whether they are permanently or temporarily submerged under water [15]. In most wetlands, hydrologic conditions are such that the substrate is saturated long enough during the growing season, a mechanism that creates oxygen-poor conditions in the substrate, limiting the vegetation to those species that are adapted to low-oxygen environments [16].

Wetlands provide a number of functions and benefits. Wetland functions are inherent processes occurring in wetlands; wetland values are the attributes of wetlands that society perceives as beneficial [17]. The wetland hydrology is generally one of slow flows with either shallow waters or saturated substrates, which allows sediments and other pollutants, including emerging contaminants to settle as the water passes through the wetland system. The occurrence of slow flows provides prolonged contact times between the water and the surfaces within the wetland [15]. The wetland treatment mechanisms are anchored on the complex mass of organic and inorganic materials, with diverse opportunities for gas/water interchanges, which foster a diverse community of microorganisms that break down or transform a wide variety of substances [7]. Within the wetland's ecosystems, there are dense growths of vascular plants adapted to saturated conditions, which slow the water, create microenvironments within the water column, and provide attachment sites for the microbial communities as well as other contaminants. The litter that accumulates as plants die back in the fall creates additional material and exchange sites and provides a source of carbon, nitrogen, and phosphorous to fuel microbial processes [18].

Even though, not all wetlands can perform all functions and values, majority of them provide several benefits. When subjected to appropriate ecological management without any threats, majority of wetlands can provide the following:

I. Water quality services

II. Flood storage services under excessive precipitation and the desynchronization of storm

III. Nutrients and other materials cycling services

IV. Habitat for fish and wildlife

V. Services for passive recreation, such as bird watching and photography

VI. Services for active recreation, such as hunting education and research

VII. Services for esthetics and landscape enhance merit

2.1 Constructed wetlands

A constructed wetland is an artificial shallow basin filled with substrate, usually soil or gravel, and planted with vegetation that has tolerance to saturated conditions. Water is then directed into the system from one end and flows over the surface (surface flow) or through the substrate (subsurface flow) and gets discharged from the other end at the lower point through a weir or other structure, which controls the depth of the water in the wetland [11]. Several forms of constructed wetlands have been introduced, including surface flow wetlands, subsurface flow wetlands, and hybrid systems that integrate surface and subsurface flow wetland types [6, 19]. Constructed wetland systems can also be combined with conventional treatment technologies to provide higher treatment efficiency [8]. The choice of constructed wetland types depends on the existing environmental conditions and how appropriate they are for domestic wastewater, agricultural wastewater, coal mine drainage, and stormwater [6].

Constructed wetlands have been widely used in the treatment of primary or secondary domestic sewage effluents, and others have been used to treat domestic wastewater and have also been modeled to handle high organic loads associated with agriculture or domestic wastewater [5]. A large number of constructed wetlands have also been built to treat drainage from active and abandoned coal mines [20]. The constructed wetland technology has recently been used in the control and management of stormwater flows, and its application in reducing the impacts by stormwater floods within urban areas is expanding globally [21]. The constructed wetland technology is not only preferred in stormwater flow control but also in the treatment of wastewater, and its preference is based on its low cost, low energy requirement, and need for minimal operational attention and skills. Due to its numerous merits and high sustainability potential, there is an increasing extensive research on its practical application to expand the knowledge on its operation and to provide more insight on its appropriate design, performance, operation, and maintenance for optimum environmental benefits. Even though the constructed wetlands are sturdy and effective systems, their performance depends on the periodic improvements to handle emerging contaminants such as antibiotic and antibacterial resistant genes, and for them to remain effective, they must be carefully designed, constructed, operated, and maintained [11, 12].

2.2 Components of constructed wetland

Constructed wetland is a system that puts together different units that work together to ensure that its intended purpose is achieved. Constructed wetland systems entail a properly designed and constructed basin that holds water, a substrate that provides filtration pathways, habitat/growth media for the needed organisms, and also communities of microbes and aquatic invertebrates, which in most cases develop naturally. Most importantly, constructed wetlands also hold vascular plants whose nature depends on the intended purification role and efficiency. The efficiency of the constructed wetlands in waste treatment depends on the interaction and maintenance of these components [22].

In a constructed wetland system, natural geochemical and biological processes within a wetland realm are involved in the treatment of metals, explosives, and other contaminants that exist within the water. Normally, there are three primary components in a constructed wetland. Constructed wetland has an impermeable layer (generally clay). It also has a gravel layer that acts as a substrate needed for the provision of nutrients and support to the root zone. It also has an above-surface vegetation zone [16]. The impermeable layer within the constructed wetland system prevents infiltration of wastes down into underground aquifers. The gravel layer and root zone comprise of a layer where water flows and bioremediation and denitrification occur. The above-ground vegetative layer contains the well-adopted plant material. Within the wetlands, both the aerobic and anaerobic processes occur, and these can be divided into separate cells [5, 16]. Groundwater can be made to flow through pumping or naturally by gravity through the wetland. Within the anaerobic cells, plants and other natural microbes are involved in the degradation of the contaminant. The aerobic cell performs the work of further improving the water quality through continued exposure to the plants and the movement of water between cell compartments. The use of straw, manure, or compost with little or no soil substrate has been beneficial in the wetlands constructed primarily for the removal of metals. However, for wetlands constructed to treat explosives-contaminated water, certain plant species are used to enhance the degradation through a process termed phytoremediation [23].

2.2.1 Water

Wetlands are formed on substrates that are fully or partially submerged in water, where a relatively impermeable subsurface layer prevents the surface water from seeping into the ground [1, 2]. These conditions can be created with few modifications to form a constructed wetland. A constructed wetland can be built almost anywhere in the landscape by shaping the land surface to collect the surface water and by sealing the basin to retain the water [7]. Hydrology that enhances the linking of all the functions in a wetland system stands as the most important design factor to be considered in constructed wetlands, as it is often the primary factor in the success or failure of most constructed wetlands. Therefore, planning and putting up of constructed wetlands require the contribution of a qualified hydrologist to ensure that all the hydrological requirements and conditions are taken care of [24]. Even though the hydrology of most constructed wetlands is very much similar to the other surface and near-surface water, it does differ in several important respects. Small changes in hydrology can have fairly significant effects on a wetland's functionality and its treatment effectiveness and efficiency. Indeed, due to the large surface area of the water and its shallow depth, a wetland system interacts strongly with the atmosphere through rainfall and evapotranspiration. This (the combined loss of water by evaporation from the water surface and loss through transpiration by plants) and the density of the vegetation of a wetland strongly affect the constructed wetlands' hydrology. This can be experienced through the obstruction of water flow paths as the water finds its sinuous way through the network of stems, leaves, roots, and rhizomes, and it can also occur through the blockage of exposure to wind and sun [7, 24, 25]. Water always acts as a vehicle for delivering the pollutants to the system and also for discharging the untapped pollutants away from the system [24].

2.2.2 Substrate

Substrates for constructed wetlands can come in the form of sediment or litter. Substrates used to construct wetlands include soil, sand, gravel, rock, and organic materials such as compost [26]. Due to low water velocities and high productivity typical of wetlands, the sediment and litter accumulation occurs within the wetlands. The substrates, sediments, and litter have numerous functions that are beneficial to the efficiency of the constructed wetlands. They provide support to many of the living organisms in wetlands, and the substrate permeability also affects the movement of water through the wetland and provides numerous chemical and biological processes, many of which are microbial in nature and also enhance the transformation of pollutants within the substrates. The substrates also provide storage for many contaminants, and the accumulation of litter increases the amount of organic matter in the wetland, which provides sites for material exchange and microbial attachment. Through this process, carbon source is realized as well as the energy source that drives some of the important biological reactions in wetlands.

Flooding of the constructed wetlands with water has a contribution in its functional mechanism. The physical and chemical characteristics of soils and other substrates are altered when they are wholly or partially under water. For example, under saturated substrate, the water replaces the atmospheric gases within the pore spaces and the microbial-driven metabolism results in the consumption of the available oxygen. Therefore, since oxygen is consumed more rapidly than it can be replaced by diffusion from the atmosphere, the substrates change to anoxic condition (without oxygen). Such conditions become significant in the removal of pollutants such as nitrogen and metals. However, substrates can also act as reservoirs for most contaminants, with high concentration of emerging contaminants such as resistant genes being detected in the constructed wetland substrates [27, 28].

2.2.3 Vegetation

Constructed wetlands can work with both the vascular plants (the higher plants) and nonvascular plants (algae), and the photosynthesis process by algae increases the dissolved oxygen content of the water which in turn affects nutrients and metals [18, 29]. Constructed wetlands also attract large organisms such as birds which can feed on contaminants. Additionally, they form attachment surfaces for other protozoans and other microorganisms such as zooplanktons, phytoplanktons, and bacterioplanktons which also aid in the elimination of pollutants and contaminants [30, 31]. Vegetation acts as the main trapping and retention points for most contaminants. Studies have continued to detect a high concentration of emerging contaminants such as resistant genes within the root systems of most constructed wetland vegetations [11, 32].

2.2.4 Other life-forms

Constructed wetlands' performance is also a factor of other life-forms. Organisms within the wetlands include microorganisms and other larger animals. The regulation functions by the microorganisms and their metabolism processes are the fundamental functions of the wetlands systems [33]. The microorganisms are varied in species and possess the required adaptions to drive the functions of the wetland systems. The known significant microorganisms include bacteria, yeasts, fungi, protozoa, and rind algae. The biomass generated from these microbes (microbial biomass) forms a major useful sink for organic carbon and many nutrients. Additionally, the microbial activities also transform a great number of organic and inorganic substances into innocuous or insoluble substances as well as alter the reduction/oxidation (redox) conditions of the substrate, and thus not only affect the processing capacity of the wetland but also enhance the recycling of nutrients. Some microbial transformation processes are aerobic as they require free oxygen to occur, while others are anaerobic as they occur under the absence of free oxygen. However, most of the bacterial species are also facultative anaerobes in nature. These groups are capable of functioning under the constructed wetland conditions of either aerobic or anaerobic in response to changing environmental conditions [6, 34].

The level of water within a constructed system is crucial to the microbial activities, and microbial populations undergo adjustments to changes in the water delivered to them. Populations of microbes can rapidly expand under the condition of suitable energy-containing materials. However, when environmental conditions become unsuitable, many microorganisms become dormant and can remain dormant for years [35]. The microbial community of a constructed wetland can be affected by toxic substances, such as pesticides and heavy metals, and care must be taken to prevent such chemicals from being introduced at damaging concentrations. The biodiversity with the constructed wetlands is rich, and this is based on the favorable habitat that the system provides to different forms of organisms, which range from animals to plants, including invertebrates and vertebrates. The invertebrate animals, which include insects and worms, contribute to the treatment process by actively fragmenting detritus and consuming organic matter [36].

Additionally, the larvae of many insects are also aquatic and they undertake the consumption of a significant amount of material during their larval stages, which may last for several years in most insect species. The invertebrates also perform a number of ecological roles; for example, dragonfly nymphs have been confirmed to be important predators of mosquito larvae which results in biocontrol of malaria in most waterlogged areas. Despite invertebrates being the most important animals as far as water quality improvement is concerned, constructed wetlands also harbor a variety of amphibians, turtles, birds, and mammals, all of which are important in the systems' ecological balancing [37].

3. Constructed wetlands for wastewater treatment

The mechanisms that are available to improve water quality within a constructed wetland system are numerous and often interrelated. The mechanisms involve the settling of suspended particulate matter; the filtration and chemical precipitation through contact of the water with the substrate and litter; chemical transformation; adsorption and ion exchange on the surfaces of plants, substrate, sediment, and litter; the breakdown and transformation of pollutants by microorganisms and plants uptake; and transformation of nutrients by microorganisms and plants as well as the predation and natural die-off of pathogens [36]. The removal can be undertaken biologically through microbiological degradation through catabolism and anabolism, protozoic predation and digestion, and through plant uptake and storage; chemically through adsorption (ionic and covalent) oxidation, reduction, and UV degradation and physically through filtration and settlement, which filters some materials and degrades others [38–40].

Constructed wetland treatment technology incorporates the principal components of wetland ecosystems that promote degradation and control of contaminants by plants, degradation by microbial activity, and increased sorption, filtering, and precipitation [38–40]. The treatment need dictates the nature of technology required and requires proper selection of designs, such as surface or subsurface flow, single or multiple cells, and parallel or series flow. Putting up of constructed wetland systems are sometimes part of a treatment train that integrates processes in series such as settling ponds, oil/water separators, and physical/chemical treatment methods. The removal mechanisms within the constructed wetlands can act uniquely, sequentially, or simultaneously on each contaminant group or species [3, 4]. For instance, the volatile organic compounds (VOCs) in contaminated groundwater are primarily eliminated through the integrative physical mechanism of diffusion-volatilization. Further to this, mechanisms such as adsorption to suspended matter, photochemical oxidation, and biological degradation may also play a role. Within a constructed wetland treatment system, physical removal mechanisms of contaminants include settling, sedimentation, and volatilization. Gravitational settling is responsible for most of the removal of suspended solids. The most effective treatment wetlands are those that foster these mechanisms.

3.1 Merits and demerits of constructed wetlands in waste handling

The long-term effectiveness of constructed wetlands to contain or treat some contaminants is not well known. Wetland aging may contribute to a decrease in contaminant removal rates over time. However, constructed wetlands are a cost-effective and technically feasible approach to treating wastewater and runoff for several reasons [41].

Constructed wetlands' demerits outweigh the merits. Some of the merits are that they can be less expensive and more affordable to build than other forms of treatments, their cost of operation and maintenance (required supplies and energy) are low, and the operation and maintenance only require periodic and not continuous on-site labor. Furthermore, the constructed wetlands are able to tolerate fluctuations in flow, they sustainably facilitate water recycling and reuse, they provide favorable habitat for many wetland organisms, and the system can be built to fit harmoniously into the landscape. Constructed wetlands have the ability to provide numerous benefits in addition to water quality improvement, such as wildlife habitat that supports tourism and other sporting, and they enhance the esthetic enhancement of open spaces. Therefore, due to all the above economic, ecological, and esthetic benefits, constructed wetlands are environmentally sensitive treatment approaches that are viewed with favor by the general public [42].

The use of constructed wetlands is also subject to limitations that are associated with the use and putting up of the system. Compared to conventional wastewater treatment systems, constructed wetlands generally require larger land areas. Even though wetland treatment may be economical relative to other options, this only applies to where land is available and affordable. The constructed wetland's performance efficiency may be less consistent as compared to the conventional treatment. The treatment efficiency of constructed wetlands may vary; this variation may be seasonal in response to changing environmental conditions, including rainfall and drought or spatial in relation to the existing weather conditions in different places. While the average performance over the year may be acceptable, but due to such fluctuations in performance efficiency, wetland treatment cannot be relied upon if the effluent quality must meet stringent discharge standards at all times. The biological components are always sensitive to toxic chemicals, such as ammonia, and other pesticides that are periodically flushed or surged by the flowing water, and this may temporarily reduce treatment effectiveness and reduce the efficiency. For proper survival and improved efficiency, constructed wetlands also require a minimum amount of water. While wetlands can tolerate temporary drawdowns, they cannot withstand complete drying and some plants in it can also not tolerate complete submergence [1]. The use of constructed wetlands for wastewater treatment and stormwater control is a fairly recent development. There is yet no consensus on the optimal design of wetland systems, nor is there much information on their long-term performance. Furthermore, its ability and potential to eliminate emerging contaminants such as resistant genes have not been fully realized [32].

4. Constructed wetlands and drug-resistant bacteria and related genes

Antibiotic-resistant genes (ARGs) originate from hospitals, wastewater treatment plants effluents and sewage sludge, and animal slurry in farmland. Soils, surface water (e.g., seas and rivers), and sediments are contaminated by these large arrays of antibiotic resistance genes [43]. Resistant genes are the major courses of antibiotic resistance, which is one of the upcoming crucial concerns to global health care with considerable effect in rising morbidity, mortality, and costs associated with major public health problems. Antimicrobial resistance occurs naturally over time, usually through genetic changes. However, the misuse and overuse of antimicrobials is accelerating this process [44]. Horizontal and lateral gene transfers have greatly contributed to the increasing number of drug-resistant pathogens within the environment (**Figure 1**).

Antibiotic resistance has the potential to affect people at any stage of life as well as the health-care, veterinary, and agriculture industries, making it one of the world's most urgent environmental and public health problems. [45]. Its chain of spread spans from contaminated wastewater discharges from the hospitals to the consumption of contaminated food material (**Figure 2**). The occurrence of antibiotic-resistant genes in the environment is considered one of the most urgent threats to modern health care and environmental quality and safety. It is often assumed that the abundance and diversity of known resistance genes are representative also for the non-characterized fraction of the resistome in a given environment [46]. Antibiotic resistance genes are ubiquitous in the environment, which has led to the suggestion that there is a high risk these genes can cause in the spread of the disease [46, 47].

Constructed wetlands, though designed to remove and eliminate pollutants from wastewater, can also be the hot spots for horizontal or vertical gene transfer, enabling the spread of antibiotic resistance genes between different microorganisms. Antibiotic resistance occurs due to changes or mutations in the DNA of the microorganism, or due to the acquisition of antibiotic resistance genes from other microbial species through gene transfer. The transfer of genetic materials between unrelated individuals is termed horizontal gene transfer, while the transfer of genetic materials from parent to their offspring is termed vertical gene transfer [48]. Horizontal gene transfer is the major source of ARGs as well as the emergence of pathogenic forms

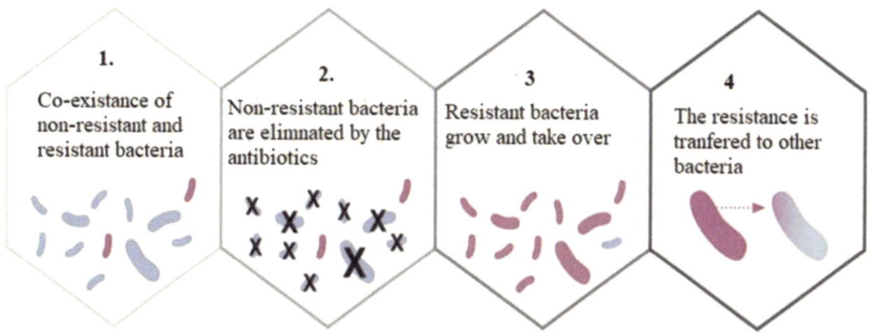

Figure 1.
The transfer of resistant genes between resistant and nonresistant microbes.

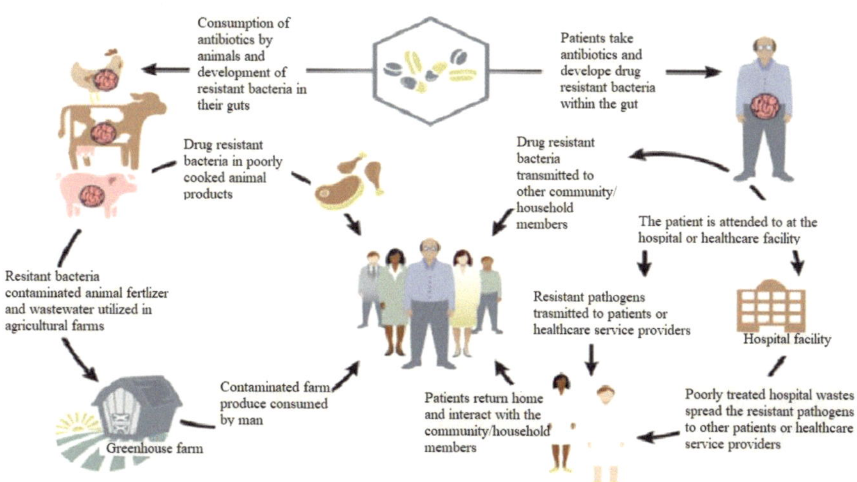

Figure 2.
Resistant genes contamination pathways.

of microorganisms with new virulence [11, 12]. Constructed wetlands being the reservoir for various strains and species of microorganisms may provide the media for such transfers to occur, hence contributing to problems of drug resistance [11]. There is rising concern due to the wide presence of antibiotics in the constructed wetlands, as it not only causes serious toxic effects on organisms but also promotes the spread of antibiotic-resistant genes (ARGs), even with low concentrations in the environment. ARGs being spread through horizontal or vertical gene transfers can also be spread and maintained in microbial populations, even without selection pressure from antibiotics, and wetlands systems provide favorable transfer grounds [12].

The recognition that the environment could serve as a source for resistance genes to human pathogens has spurred interest in investigating the distribution of resistance genes in various environments to better understand the process [10, 32]. Wastewater and wastewater treatment plants such as constructed wetlands can act as reservoirs and environmental suppliers of antibiotic resistance through filtration and load of resistant genes into the aquatic ecosystems [13]. Indeed, wastewater has been confirmed to be the major route by which the antimicrobials, ARBs, and ARGs are introduced into the natural ecosystem from the human settings. Although wastewater treatment plants such as the constructed wetlands significantly reduce the load of bacteria, the final effluents may contain ARBs, sometimes even at higher concentrations than in the raw wastewater [11, 12].

Considerable research has been conducted on the behavior and fate of ARBs and ARGs discharged from different forms of wastewater to soil through the application of animal manure wastewater irrigation and to aquatic environments through wastewater discharge and runoff. The impact of discharging ARGs in treated wastewater to aquatic systems as well as associated ARG amplification and attenuation dynamics has neither been adequately researched nor discussed. Indeed, intracellular and free ARGs in surface and groundwater can propagate through horizontal gene transfer to indigenous pathogenic microbes. Furthermore, these ARBs may eventually reach and colonize humans through multiple pathways resulting in acute infections or long-term silent colonization that can eventually evolve into an infection [13].

4.1 Challenges of emerging resistant genes in constructed wetlands

There are various routes through which the antibiotic-resistant genes can enter the environment. One major route is when the antibiotic-resistant pathogens and associated metabolites are released from hospitals through urine and feces from patients as hospital wastewater. After the release, the effluent physical chemical characteristics and the prevailing environmental factors determine the biodegradation, adsorption, and uptake processes of these drug-resistant pathogens and related genes, eventually shaping the abundance and diversity of the available drug-resistant bacteria. Similarly, antibiotics may be released into the wastewater treatment system via people taking antibiotics from home (**Figure 3**). From the wastewater treatment plants, the antibiotics can load into sludge, which are later dispersed on fields as fertilizer or released as runoff directly into the receiving surface water [49, 50]. Further to this, wastewater can also be treated by releasing it into constructed wetlands. In such cases, the constructed wetlands will be exposed to antibiotic contaminants from the wastewater. Even though the constructed wetland is expected to filter all the contaminants, including the drug-resistant pathogens and related genes, that is always not the case, the receiving effluents may still receive some amount of drug-resistant pathogens and related genes as effluent loads from the constructed wetland system. Additionally, antibiotics are also used therapeutically or as growth promoters in livestock and poultry. Antibiotics and

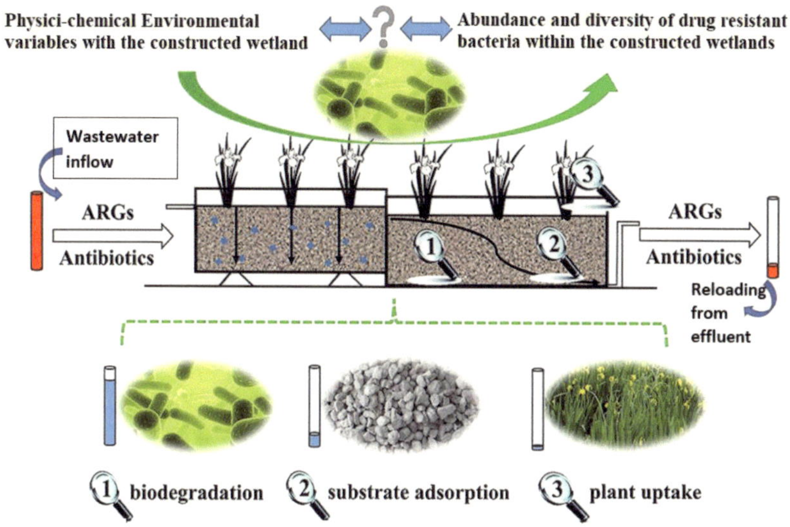

Figure 3.
Contribution of constructed wetland in the ARG removal and reloading.

their metabolites can spread through animal excrements and end up in the treatment systems such as the constructed wetlands, which can eventually release the treated effluents into the fields and groundwater, or in the case of antibiotic use in fish farms, directly into the aquatic environment. It is also worth noting that wherever antibiotics are spread, it is also likely that resistant bacteria follow the same routes of dispersal [51, 52]. Due to these interactions and movements of the drug-resistant pathogens and related genes, there have been increased levels of antibiotics, ARGs, and drug-resistant bacteria within the environment. Furthermore, the environmental bacterial flora which also harbor ARGs and potential ARGs continue to increase within the receiving aquatic environments. Therefore, these types of environments are the likely resistance hot spots where ARGs proliferate and new resistant strains are created by and transferred to other parts of the environment. Due to the increased spread of drug-resistant bacteria and related genes, the routes by which humans come into contact with these bacteria are also increasing. These may include consumption of crops grown by contaminated sludge used as fertilizer, drinking of water drawn from contaminated groundwater or surface water, and frolicking in marine water linked to contaminated surface water. When these resistant bacteria enter humans, they have the opportunity to spread their ARGs to the human microbiome and, through constructed wetlands in wastewater treatment, the cycle repeats [50, 53].

While ARGs in their environmental context may originally have had other primary functions aside from conferring resistance to antibiotics, these genes have now been recruited as resistance genes in pathogenic bacteria. Reuse of treated wastewater is increasingly seen as one of the solutions to tackle the water scarcity problem and to limit the pollution load to surface water. Yet, using reclaimed water for non-potable purposes and particularly to irrigate food crops presents an exposure pathway for antibiotics and antibiotic-resistant bacteria and genes (ARB & G) to enter the human food chain. Wastewater reuse is currently of particular concern as the potential source of selective pressure that elevates the levels of antibiotic resistance in native bacteria [54]. Aquatic ecosystems are considered important matrices for the release, mixing, persistence, and spread of ARBs and ARGs associated with horizontally transferable genetic elements [11, 12]. Presently, existing regulations give little attention to the protection and management of wetlands, making them to increasingly get exposed to resistant gene-loaded human excreta, raw sewage, untreated wastewater, and other pollutants from diverse sources, making natural and constructed wetlands to be the potential reservoirs of ARBs carrying ARGs that might spread to microbes as well as man [55].

5. Conclusion and recommendations

The knowledge on antibiotic resistance in wastewater has continued to expand, but proper management for complete elimination with zero reloading into the environment has not been achieved. Indeed, in the past few years, introduction of high-tech molecular studies has increased the understanding on this study subject. However, there are still numerous gaps on the subject, such as how active are horizontal and lateral gene transfers in wastewater, what are the specific main driving factors to the transfer mechanisms, and what is the role of the wastewater treatment plants in increasing the spread of drug-resistant microbes. Indeed, even though constructed wetlands have been commercially used to control and degrade municipal and industrial wastewater, there is need for caution on how exotic wastes such as explosives and those that harbor resistant genes are handled by these systems. With the growing concerns that environmental concentrations of antibiotics exert a selective pressure on clinically relevant bacteria, for the control of such acute strains, there is need for a major shift toward a more localized management of the water cycle, pioneering low-cost wastewater treatment technologies, and more efficient monitoring strategies based on a limited number of indicators that would facilitate the assessment of the anthropogenic impact on the water cycle. Furthermore, there is need to better understand the dispersion processes and the fate of pathogenic and antibiotic-resistant bacteria in the environment, in order to prevent risks to humans and their environment, while also controlling and reducing as much as possible the anthropogenic bacterial input into the environment.

Acknowledgements

The authors appreciate Egerton University for providing review materials and an opportunity to produce this chapter.

Author details

Donde Oscar Omondi* and Atalitsa Caren Navalia
Department of Environmental Science, Egerton University, Kenya

*Address all correspondence to: oscardonde@yahoo.com

References

[1] Iqbal A, Shang Z. Wetlands as a carbon sink: Insight into the Himalayan Region. In: Shang Z, Degen A, Rafiq M, Squires V, editors. Carbon Management for Promoting Local Livelihood in the Hindu Kush Himalayan (HKH) Region. Cham: Springer; 2020

[2] Keough JR, Thompson TA, Guntenspergen GR, et al. Hydrogeomorphic factors and ecosystem responses in coastal wetlands of the Great Lakes. Wetlands. 1999;**19**:821-834. DOI: 10.1007/BF03161786

[3] Siegrist RL. Treatment using constructed wetlands. In: Decentralized Water Reclamation Engineering. Cham: Springer; 2017

[4] Vymazal J, Greenway M, Tonderski K, Brix H, Mander Ü. Constructed wetlands for wastewater treatment. In: Verhoeven JTA, Beltman B, Bobbink R, Whigham DF, editors. Wetlands and Natural Resource Management. Ecological Studies (Analysis and Synthesis). Vol. 190. Berlin, Heidelberg: Springer; 2006

[5] Abou-Elela SI. Constructed wetlands: The green technology for municipal wastewater treatment and reuse in agriculture. In: Negm A, editor. Unconventional Water Resources and Agriculture in Egypt. The Handbook of Environmental Chemistry. Vol. 75. Cham: Springer; 2017

[6] Donde OO. Wastewater management techniques: A review of advancement on the appropriate wastewater treatment principles for sustainability. Environmental Management and Sustainable Development. 2017;**6**(1): 40-58. DOI: 10.5296/emsd.v6i1.10137

[7] Thorslund J, Jarsjo J, Jaramillo F, Jawitz JW, Manzoni S, Basu NB, et al. Wetlands as large-scale nature-based solutions: Status and challenges for research, engineering and management. Ecological Engineering. 2017;**108**(Part B):489-497. DOI: 10.1016/j.ecoleng.2017.07.012

[8] Van-Biervliet O, McInnes RJ, Lewis-Phillips J, et al. Can an integrated constructed wetland in Norfolk reduce nutrient concentrations and promote in situ bird species richness? Wetlands. 2020. DOI: 10.1007/s13157-019-01247-7

[9] Wi T, Lahra MM, Ndowa F, Bala M, Dillon J-AR, Ramon-Pardo P, et al. Antimicrobial resistance in *Neisseria gonorrhoeae*: Global surveillance and a call for international collaborative action. PLoS Medicine. 2017;**14**(7):e1002344. DOI: 10.1371/journal.pmed.1002344

[10] Ochman H, Lawrence JG, Groisman EA. Lateral gene transfer and the nature of bacterial innovation. Nature. 2000;**405**:299-304. DOI: 10.1038/35012500

[11] Lamori JG, Xue J, Rachmadi AT, et al. Removal of fecal indicator bacteria and antibiotic resistant genes in constructed wetlands. Environmental Science and Pollution Research. 2019;**26**:10188-10197. DOI: 10.1007/s11356-019-04468-9

[12] Chen J, Liu Y, Su H, et al. Removal of antibiotics and antibiotic resistance genes in rural wastewater by an integrated constructed wetland. Environmental Science and Pollution Research. 2015;**22**:1794-1803. DOI: 10.1007/s11356-014-2800-4

[13] Stavros C, Andreas K, Konstantina T, Iosifina G, Danae V, Nicolas K. Evaluation of a constructed wetland for wastewater treatment: Addressing emerging organic contaminants and antibiotic resistant bacteria. New Biotechnology.

2019;**52**:94-103. DOI: 10.1016/j.nbt.2019.05.006

[14] Cózar A, Bergamino N, Mazzuoli S, et al. Relationships between wetland ecotones and inshore water quality in the Ugandan coast of Lake Victoria. Wetlands Ecology and Management. 2007;**15**:499-507. DOI: 10.1007/s11273-007-9046-6

[15] Mięsiak-Wójcik K, Turczyński M, Sposób J. Diverse sediment permeability and implications for groundwater exchange in closed Lake-Wetland catchments (West Polesie, East Poland). Wetlands. 2018;**38**:779-792. DOI: 10.1007/s13157-018-1027-4

[16] Magee T, Kentula M. Response of wetland plant species to hydrologic conditions. Wetlands Ecology and Management. 2005;**13**:163-181. DOI: 10.1007/s11273-004-6258-x

[17] McCartney M, Rebelo LM, Sellamuttu S. Wetlands, livelihoods and human health. In: Finlayson C, Horwitz P, Weinstein P, editors. Wetlands and Human Health. Wetlands: Ecology, Conservation and Management. Vol. 5. Dordrecht: Springer; 2015

[18] Ge X, Cao X, Song X, Wang Y, Si Z, Zhao Y, et al. Bioenergy generation and simultaneous nitrate and phosphorus removal in a pyrite-based constructed wetland-microbial fuel cell. Bioresource Technology. 2020;**296**:122350. DOI: 10.1016/j.biortech.2019.122350

[19] Vymazal J. Removal of enteric bacteria in constructed treatment wetlands with emergent macrophytes: A review. Journal of Environmental Science and Health. Part A, Toxic/Hazardous Substances & Environmental Engineering. 2005;**40**(6-7):1355-1367

[20] Pat-Espadas AM, Loredo Portales R, Amabilis-Sosa LE, Gómez G, Vidal G. Review of constructed wetlands for acid mine drainage treatment. Watermark. 2018;**10**. DOI: 10.3390/w10111685

[21] Mangangka IR, Liu A, Goonetilleke A, Egodawatta P. Storm water treatment. In: Enhancing the Storm Water Treatment Performance of Constructed Wetlands and Bioretention Basins. Springer Briefs in Water Science and Technology. Singapore: Springer; 2016

[22] Hedges PD, Fermor PM, Dušek J. The hydrological sustainability of constructed wetlands for wastewater treatment. In: Vymazal J, editor. Wastewater Treatment, Plant Dynamics and Management in Constructed and Natural Wetlands. Dordrecht: Springer; 2008

[23] Herath I, Vithanage M. Phytoremediation in constructed wetlands. In: Ansari A, Gill S, Gill R, Lanza G, Newman L, editors. Phytoremediation. Cham: Springer; 2015

[24] Fitzgerald SK. The role of constructed wetlands in creating water sensitive cities. In: Nagabhatla N, Metcalfe C, editors. Multifunctional Wetlands. Environmental Contamination Remediation and Management. Cham: Springer; 2018

[25] Donde OO, Xiao B. Understanding wastewater treatment mechanisms: A review on detection, removal and purification efficiencies of faecal bacteria indicators across constructed wetlands. Environmental Reviews. 2017;**25**(4):444-451. DOI: 10.1139/er-2017-0017

[26] Wang HX, Xu JL, Sheng LX, Liu XJ. A review of research on substrate materials for constructed wetlands. Materials Science Forum. 2018;**913**:917-929

[27] Vymazal J. The use of hybrid constructed wetlands for wastewater

treatment with special attention to nitrogen removal: A review of a recent development. Water Research. 2013;**47**(14):4795-4811. DOI: 10.1016/j.watres.2013.05.029

[28] Wu H, Zhang J, Ngo HH, Guo W, Hu Z, Liang S, et al. A review on the sustainability of constructed wetlands for wastewater treatment: Design and operation. Bioresource Technology. 2015;**175**:594-601. DOI: 10.1016/j.biortech.2014.10.068

[29] Wu X, Mitsch WJ. Spatial and temporal patterns of algae in newly constructed freshwater wetlands. Wetlands. 1998;**18**:9-20. DOI: 10.1007/BF03161438

[30] Eivers RS, Duggan IC, Hamilton DP, et al. Constructed treatment wetlands provide habitat for zooplankton communities in agricultural peat lake catchments. Wetlands. 2018;**38**:95-108. DOI: 10.1007/s13157-017-0959-4

[31] Montemezzani V, Duggan IC, Hogg ID, Craggs RJ. Assessment of potential zooplankton control treatments for wastewater treatment high rate algal ponds. Algal Research. 2017;**24**(Part A):40-63

[32] Donde OO et al. A novel integrative performance evaluation of constructed wetland covering removal of viable bacterial cells and related pathogenic, virulent and multi-drug resistant genes from wastewater systems. Water Process Engineering. 2020;**33**(101060). DOI: 10.1016/j.jwpe.2019.101060

[33] Sánchez O. Constructed wetlands revisited: Microbial diversity in the –omics era. Microbial Ecology. 2017;**73**:722-733. DOI: 10.1007/s00248-016-0881-y

[34] Kuschk P et al. The status of research on constructed wetlands. In: Vitale K, editor. Environmental and Food Safety and Security for South-East Europe and Ukraine. NATO Science for Peace and Security Series C: Environmental Security. Dordrecht: Springer; 2012

[35] Fisher J, Acreman MC. Wetland nutrient removal: A review of the evidence. Hydrology and Earth System Sciences. 2004;**8**:673-685. DOI: 10.5194/hess-8-673-2004

[36] Ruhí A, Fairchild GW, Spieles DJ, Becerra-Jurado G, Moreno-Mateos D. Invertebrates in created and restored wetlands. In: Batzer D, Boix D, editors. Invertebrates in Freshwater Wetlands. Cham: Springer; 2016

[37] Harris LD. The nature of cumulative impacts on biotic diversity of wetland vertebrates. Environmental Management. 1998;**12**:675-693. DOI: 10.1007/BF01867545

[38] Töre GY, Meriç S, Lofrano G, De Feo G. Removal of trace pollutants from wastewater in constructed wetlands. In: Lofrano G, editor. Emerging Compounds Removal from Wastewater. Springer Briefs in Molecular Science. Dordrecht: Springer; 2012

[39] Vo H, Bui X, Nguyen T, et al. Insights of the removal mechanisms of pharmaceutical and personal care products in constructed wetlands. Current Pollution Report. 2018;**4**:93-103. DOI: 10.1007/s40726-018-0086-8

[40] Vymazal J. Removal of nutrients in various types of constructed wetlands. Science of the Total Environment. 2007;**380**(1-3):48-65. DOI: 10.1016/j.scitotenv.2006.09.014

[41] Donde OO, Cuicui T, Yingying T, Bangding X. Efficacy of macrophyte dominated wastewater enclosure as post-treatment alternative in domestic wastewater quality polishing for eradication of faecal pathogenic bacteria pollution. Process Safety and Environment Protection.

2018;**114**:192-205. DOI: 10.1016/j.psep.2017.12.023

[42] Mishra S, Chowdhary P, Bharagava RN. Conventional methods for the removal of industrial pollutants, their merits and demerits. In: Bharagava R, Chowdhary P, editors. Emerging and Eco-Friendly Approaches for Waste Management. Singapore: Springer; 2019

[43] Chen H, Jing L, Yao Z, Meng F, Teng Y. Prevalence, source and risk of antibiotic resistance genes in the sediments of Lake Tai (China) deciphered by metagenomic assembly: A comparison with other global lakes. Environment International. 2019;**127**:267-275. DOI: 10.1016/j.envint.2019.03.048

[44] Zhao W, Wang B, Yu G. Antibiotic resistance genes in China: Occurrence, risk, and correlation among different parameters. Environmental Science and Pollution Research. 2018, 2018;**25**(22):21467-21482

[45] Friedman ND, Temkin E, Carmeli Y. The negative impact of antibiotic resistance. Clinical Microbiology and Infection. 2016;**22**(5):416-422. DOI: 10.1016/j.cmi.2015.12.002

[46] Kümmerer K. Resistance in the environment. Journal of Antimicrobial Chemotherapy. 2004;**54**(2):311-320. DOI: 10.1093/jac/dkh325

[47] Radhouani H, Silva N, Poeta P, Torres C, Correia S, Igrejas G. Potential impact of antimicrobial resistance in wildlife, environment and human health. Frontiers in Microbiology. 2014;**5**:23. DOI: 10.3389/fmicb.2014.00023.

[48] Maryury BJ, Calero-Cáceres W, Muniesa M. Transfer of antibiotic-resistance genes via phage-related mobile elements. Plasmid. 2015;**79**:1-7. DOI: 10.1016/j.plasmid.2015.01.001

[49] Gekenidis MT, Qi W, Hummerjohann J, Zbinden R, Walsh F, et al. Antibiotic-resistant indicator bacteria in irrigation water: High prevalence of extended-spectrum beta-lactamase (ESBL)-producing *Escherichia coli*. PLoS One. 2018;**13**(11):e0207857. DOI: 10.1371/journal.pone.0207857

[50] Jun C, Deng WJ, Liu YS, Hu LH, He LY, Zhao JL, et al. Fate and removal of antibiotics and antibiotic resistance genes in hybrid constructed wetlands. Environmental Pollution. 2019;**249**:894-903. DOI: 10.1016/j.envpol.2019.03.111

[51] Czekalski N, Berthold T, Caucci S, Egli A, Bürgmann H. Increased levels of multiresistant bacteria and resistance genes after wastewater treatment and their dissemination into Lake Geneva, Switzerland. Frontiers in Microbiology. 2012, 2012;**3**:106

[52] Adefisoye MA, Okoh AI. Id entification and antimicrobial resistance prevalence of pathogenic *Escherichia coli* strains from treated wastewater effluents in Eastern Cape, South Africa. Microbiology Open. 2016;**5**(1):143-151

[53] Bai X, Ma X, Xu F, Li J, Zhang H, Xiao X. The drinking water treatment process as a potential source of affecting the bacterial antibiotic resistance. Science of the Total Environment. 2015;**533**:24-31

[54] Fahrenfeld N, Ma Y, O'Brien M, Pruden A. Reclaimed water as a reservoir of antibiotic resistance genes: Distribution system and irrigation implications. Frontiers in Microbiology. 2013;**4**:130

[55] Rizzo L, Manaia C, Merlin C, Schwartz T, Dagot C, Ploy M, et al. Urban wastewater treatment plants as hotspots for antibiotic resistant bacteria and genes spread into the environment: A review. Science of the Total Environment. 2013, 447:345-360

Section 3

Remote Sensing and Modelling of Inland Waters

Chapter 7

Modeling Thermal Stratification Effects in Lakes and Reservoirs

Scott A. Wells

Abstract

A brief overview of characteristics of stratified water bodies is followed by an in-depth analysis of the governing equations for modeling hydrodynamics and water quality. Equations are presented for continuity or the fluid mass balance; x-momentum, y-momentum, and z-momentum equations; mass constituent balance equation; the heat balance equation for temperature; and the equation of state (relating density to temperature and concentration of dissolved and suspended solids). Additional equations and simplifications such as the water surface equation and changes to the pressure gradient term are shown. Many of the assumptions that are made in water quality models are discussed and shown. Typical water quality source-sink terms for temperature, dissolved oxygen, algae, and nutrients are listed. A summary of some typical water quality models for lakes and reservoirs is shown. Two case studies showing how models can predict temperature and dissolved oxygen dynamics in stratified reservoirs are shown. The brief summary looks at ways to improve water quality and hydrodynamic models of lakes and reservoirs.

Keywords: water quality modeling, hydrodynamic modeling, temperature modeling, reservoir modeling, dissolved oxygen modeling, reservoir, lake, stratification

1. Characteristics of lakes and reservoirs

Lakes and reservoirs are bodies of water that often serve multiple beneficial uses, such as water supply for municipal and agricultural use, recreation use, fishery enhancement, flood control, and power generation. Their physical, biological and chemical characteristics determine to a large extent how those beneficial uses are met. Survey texts, such as Wetzel [1] and Hutchinson [2], describe the important limnological processes that affect lake and reservoir water quality. An overview of reservoir dynamics and water quality is well-summarized in Martin et al. [3].

Lakes are different from man-made reservoirs where outlet (and perhaps inlet) hydraulic structures regulate the flow rates and often internal hydrodynamics of the reservoir. Not only does this flow regulation affect the reservoir temperature stratification, but also in consequence affects its water quality. An important distinction between rivers and lakes/reservoirs is the cycle of stratification that can occur throughout the year since most rivers are well-mixed vertically.

In some river systems though, stratification can occur if there are natural pools. For example, in the Chehalis River basin in Washington, USA, the Chehalis River is usually well-mixed except in pools of slow-moving water. This is shown where a large area of the Chehalis river has little to no channel slope and exhibits lake-like characteristics in **Figure 1**.

Stratification in turn is related to the density of water as a function of temperature and dissolved substances. The progression of stratification during a summer period is shown in **Figure 2** in a mountain lake during a summer period where the upper well-mixed layer, the epilimnion, is separated from the lower layer, the hypolimnion, by the strong density (temperature) gradient. **Figure 3** shows the typical inverse stratification in the wintertime. Oftentimes, ice formation on the surface can impede gas transfer and create winter-time oxygen deficits even though there is reduced biological activity as a result of the cold temperatures.

The progression of summer stratification can also influence the progression of dissolved oxygen depletion (see **Figure 4** for Tenkiller Reservoir, OK, USA). This

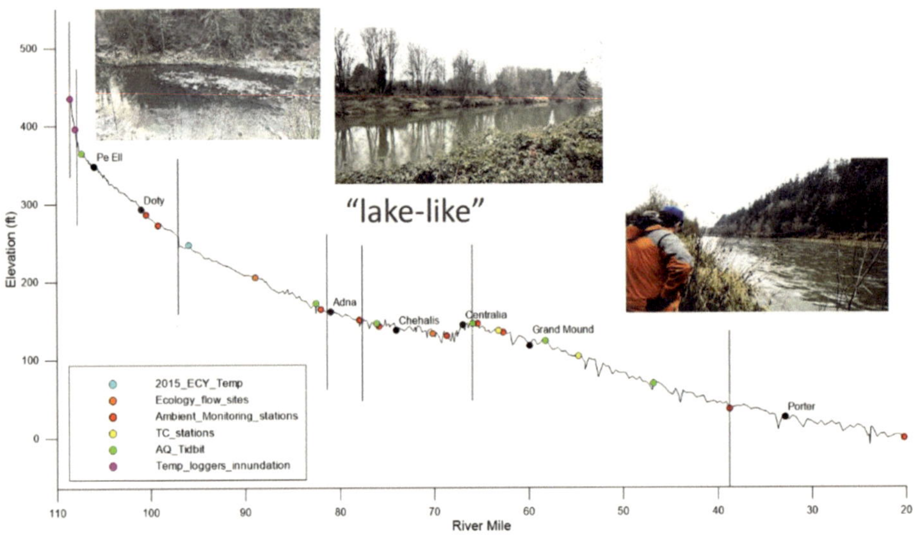

Figure 1.
Elevation drop along the Chehalis River, WA, USA, showing a section that is lake-like where summer stratification occurs. Sampling sites (multi-colored dots) are also shown.

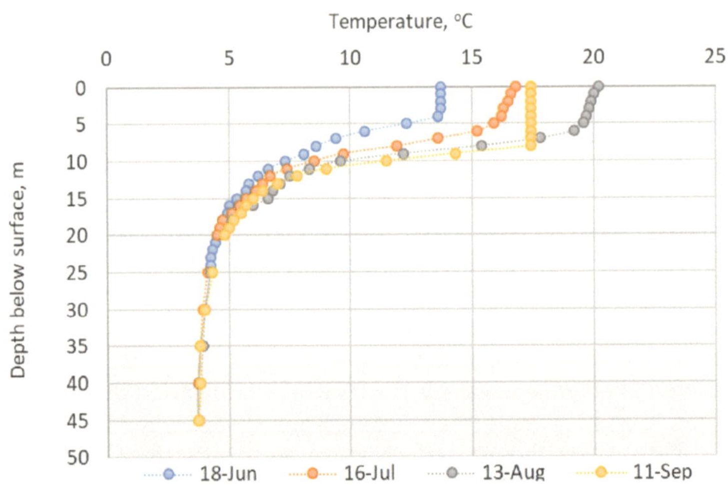

Figure 2.
Progression of stratification in summer of Bull Run Lake, OR, USA, during 1997.

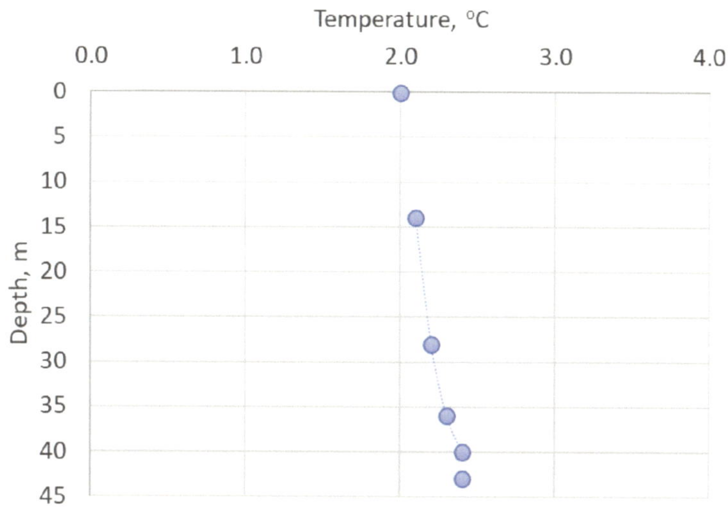

Figure 3.
Bull Run Lake, OR, USA, temperature profile on January 19, 1993.

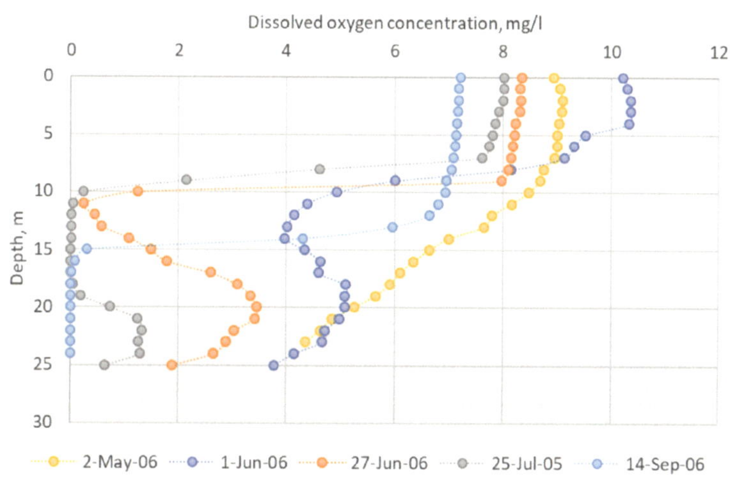

Figure 4.
Tenkiller reservoir dissolved oxygen profiles in 2006 showing progression of summer oxygen depletion.

seasonal depletion in **Figure 4** includes both the metalimnetic minimum (caused by hydrodynamic interflow of low-dissolved oxygen water at the base of the epilimnion) and the hypolimnetic depletion as a result of sediment oxygen demand.

Also, as a result of internal seiching, wind dynamics, surface cooling, and solar radiation input, the vertical profiles for water quality parameters can vary during the day. For example, Hemlock Lake temperature and dissolved oxygen vertical profiles are shown in **Figures 5** and **6**, respectively, for the morning (9 am) and early afternoon (1 pm). Variation of 1–2°C and 4–5 mg/l dissolved oxygen concentrations were noted over the 4-hour time difference between profiles.

Showing the effect of diurnal wind on seiching dynamics, **Figure 7** shows a temperature buoy at a depth of 15 m in Chester Morse Lake, WA, USA, where variations of 2–3°C can be common diurnally as wind-induced seiching occurs.

In order to describe these changes in water quality in a lake or reservoir, the next section describes the mathematical framework for modeling lakes and reservoirs.

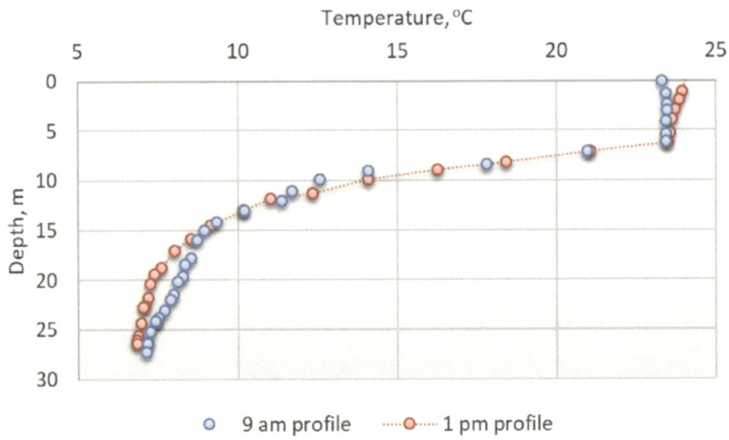

Figure 5.
Hemlock Lake, NY, USA temperature profile July 13, 2013 at 9 am and 1 pm.

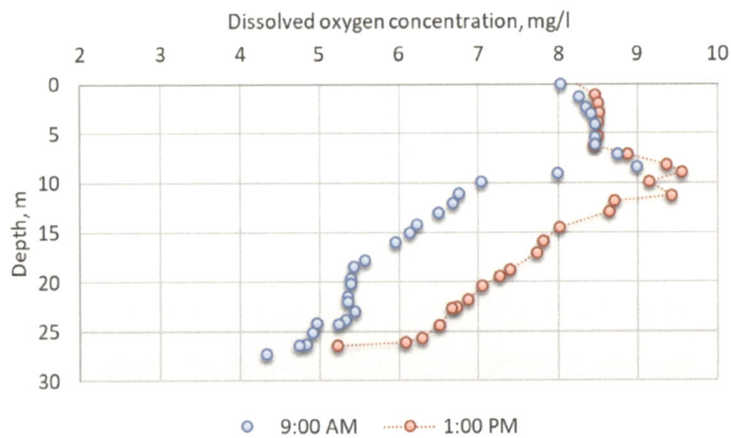

Figure 6.
Hemlock Lake, NY, USA dissolved oxygen profile July 13, 2013 at 9 am and 1 pm.

2. Governing equations for lake and reservoir water quality modeling

The basic governing equations for hydrodynamics and water quality were discussed by Wells et al. [4] and summarized and simplified here. The hydrodynamic governing equations include conservation of water mass and momentum. The water quality governing equations include conservation of constituent mass and heat including processes such as advection, turbulent diffusion, molecular diffusion (and dispersion if there is spatial averaging). An equation of state is used to relate the water density to salinity, temperature, and suspended solids that can affect fluid momentum.

2.1 Governing equations for mass, momentum, constituent mass and heat conservation

The equations for fluid motion are based on mass and momentum conservation. The development of the governing equations is based on a control volume of homogeneous properties. The conservation of fluid mass is the change in fluid mass within the control volume equaling the sum of mass inflows to the control volume and the sum of mass outflows from the control volume. The conservation of

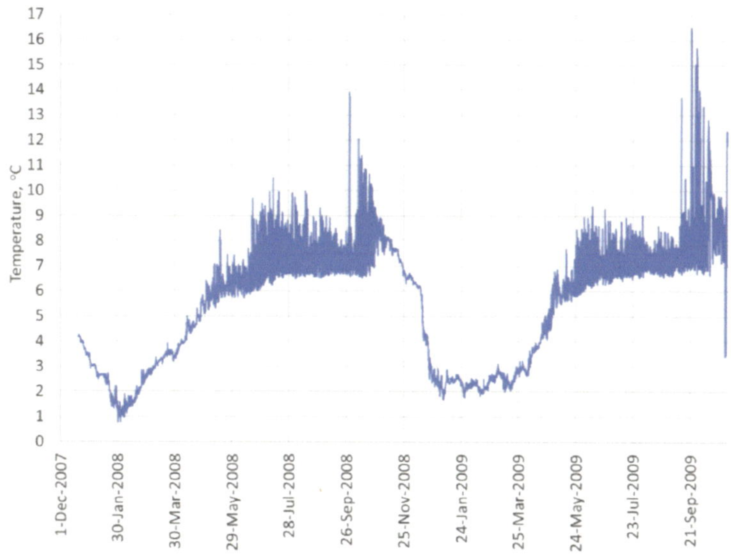

Figure 7.
Internal seiching as evident in temperature dynamics at a depth of 15 m in Chester Morse Lake, WA, USA. Variations of 2°C occur at a diurnal time scale are evident during the later spring and summer as a result of wind seiching and closeness to vertical temperature gradient.

momentum is based on evaluating the sum of forces acting on a control volume in x, y, or z (for a Cartesian system) and equating these to the acceleration of a control volume as shown in **Figure 8**. Mathematically, conservation of momentum is described as $\sum \vec{F} = m\vec{a}$, where $\vec{F}$: vector forces acting on control volume, m: mass within control volume, $\vec{a}$: acceleration of fluid within control volume.

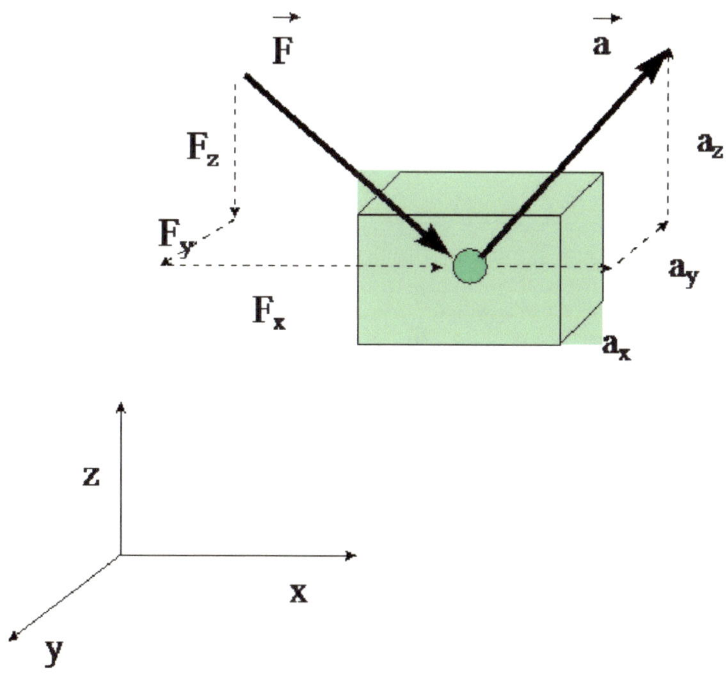

Figure 8.
Example of a force acting on a control volume resulting in the acceleration of the fluid within the control volume.

The general coordinate system used in the development of the governing equations is shown in **Figure 9**. The rotation of the coordinate system can result in significant horizontal accelerations of fluids. This is usually restricted to large water bodies such as large lakes (such as the Great Lakes in the USA) and oceanic systems. The body force that causes horizontal accelerations because of the spinning coordinate system is termed the Coriolis force.

The continuity (or conservation of fluid mass) and the conservation of momentum equations for a rotating coordinate system [5–7] are the governing equations used to determine the velocity field and water level.

The final form of the governing equations is obtained by making the following assumptions:

- the fluid is incompressible, where $\frac{\Delta\rho}{\rho} << 1$ where ρ is the fluid density and Δρ is the change in density,
- the centripetal acceleration is a correction to gravitational acceleration,
- the Boussinesq approximation (which is related to the incompressibility assumption) is applied to all terms in the momentum equation except those dealing with density gradient induced accelerations, i.e. $\frac{1}{\rho} = \frac{1}{\rho_o + \Delta\rho} \approx \frac{1}{\rho_o}$ where $\rho = \rho_o + \Delta\rho$, ρ_o is a base value,
- all velocities and pressure are turbulent time averages, i.e., $u = \overline{u} + \acute{u}$, where $\overline{u} = \frac{1}{T}\int_t^{t+T} u dt$ and $\acute{u}$ is the temporal fluctuation of u about the mean, and similarly for the velocity in the y-direction, $v = \overline{v} + \acute{v}$, the velocity in the z direction $w = \overline{w} + \acute{w}$, and the pressure, $p = \overline{p} + \acute{p}$

The governing equations become after time averaging and simplifying:

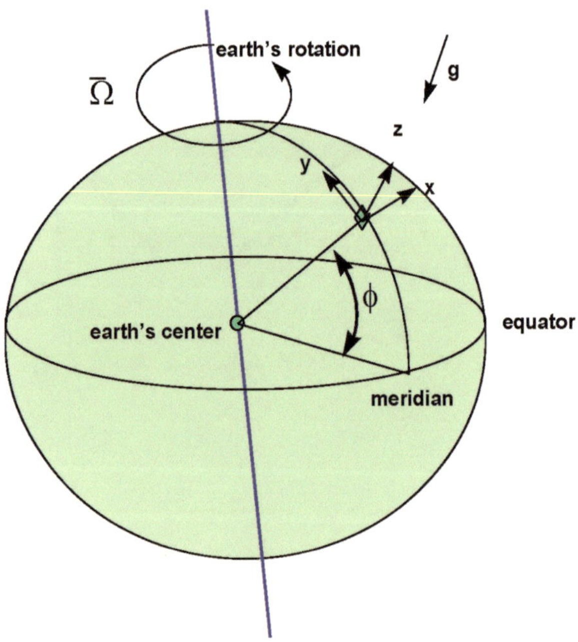

Figure 9.
Definition sketch of coordinate system for governing equations where x is oriented east, y is oriented north, and z is oriented upward opposite gravity, Ω is the angular velocity of the earth spinning on its axis and ϕ is the latitude.

2.1.1 Continuity

$$\frac{\partial \overline{u}}{\partial x}+\frac{\partial \overline{v}}{\partial y}+\frac{\partial \overline{w}}{\partial z}=0 \tag{1}$$

where $\overline{u}$: temporal mean velocity in the x-direction, $\overline{v}$: temporal mean velocity in the y-direction, $\overline{w}$: temporal mean velocity in the z-direction. The continuity equation is usually also integrated vertically to provide the water surface equation, such that $\frac{\partial \overline{\eta}}{\partial t}=\frac{\partial}{\partial x}\int_{\eta}^{h}\overline{u}dz+\frac{\partial}{\partial y}\int_{\eta}^{h}\overline{v}dz-\int_{\eta}^{h}qdz$ where q is removal from or inflow to a model cell in units of flow rate per unit length, z = h is the location of the bottom referenced to a datum, and z=η is the water surface level referenced to a datum. This equation is used to solve for the water surface elevation.

2.1.2 X-momentum equation

$$\underbrace{\frac{\partial \overline{u}}{\partial t}}_{\text{unsteady acceleration}}+\underbrace{\overline{u}\frac{\partial \overline{u}}{\partial x}+\overline{v}\frac{\partial \overline{u}}{\partial y}+\overline{w}\frac{\partial \overline{u}}{\partial z}}_{\text{convective acceleration}}-\underbrace{2\Omega_z\overline{v}}_{\text{Coriolis acceleration}}$$
$$=-\underbrace{\frac{1}{\rho}\frac{\partial \overline{p}}{\partial x}}_{\text{pressure gradient}}+\underbrace{\frac{\mu}{\rho}\left(\frac{\partial^2 \overline{u}}{\partial x^2}+\frac{\partial^2 \overline{u}}{\partial y^2}+\frac{\partial^2 \overline{u}}{\partial z^2}\right)}_{\text{viscous stresses}}+\underbrace{\frac{1}{\rho}\left(\frac{\partial \tau_{xx}}{\partial x}+\frac{\partial \tau_{xy}}{\partial y}+\frac{\partial \tau_{xz}}{\partial z}\right)}_{\text{turbulent stresses}} \tag{2}$$

where: $\tau_{xx}=\rho\overline{u'u'}$ where τ_{xx} is the turbulent shear stress acting in x direction on the x-face of control volume, $\tau_{xy}=\rho\overline{u'v'}$ where τ_{xy} is the turbulent shear stress acting in x direction on the y-face of control volume, $\tau_{xz}=\rho\overline{u'w'}$where τ_{xz} is the turbulent shear stress acting in x direction on the z-face of control volume, μ = dynamic viscosity, Ω = component of Coriolis acceleration where: $\Omega_z=\Omega_E \sin\varphi$, $\Omega_y=\Omega_E\cos\varphi$, ϕ = latitude, Ω_E = earth's rotation rate, and assuming $2\Omega_y\overline{w}$ is negligible. In general, the molecular viscous stresses are negligible except at boundaries. Analogous to laminar shear stress, the turbulent shear stresses are often parameterized as $\tau_{xx}=\mu_{turbulent-xx}\frac{\partial \overline{u}}{\partial x}=\rho\overline{u'u'}$, $\tau_{xx}=\mu_{turbulent-xy}\frac{\partial \overline{u}}{\partial y}=\rho\overline{u'v'}$, $\tau_{xz}=\mu_{turbulent-xz}\frac{\partial \overline{u}}{\partial z}=\rho\overline{u'w'}$ where the term $\mu_{turbulent}$ is the turbulent eddy viscosity analogous to molecular viscosity. The pressure is usually decomposed into the following terms: $\overline{p}=\overline{p}_a+g\int_{\eta}^{z}\overline{\rho}dz$ where p_a is the atmospheric pressure on the water surface and g is the acceleration due to gravity. The pressure gradient in the x-momentum then becomes after simplification $-\frac{1}{\rho}\frac{\partial \overline{p}}{\partial x}=-\frac{1}{\rho}\frac{\partial \overline{p}_a}{\partial x}+g\frac{\partial \overline{\eta}}{\partial x}-\frac{g}{\rho}\int_{\eta}^{z}\frac{\partial \overline{\rho}}{\partial x}dz$.

2.1.3 Y-momentum equation

$$\frac{\partial \overline{v}}{\partial t}+\overline{u}\frac{\partial \overline{v}}{\partial x}+\overline{v}\frac{\partial \overline{v}}{\partial y}+\overline{w}\frac{\partial \overline{v}}{\partial z}+2\Omega_z\overline{u}=-\frac{1}{\rho}\frac{\partial \overline{p}}{\partial y}+\frac{\mu}{\rho}\left(\frac{\partial^2 \overline{v}}{\partial x^2}+\frac{\partial^2 \overline{v}}{\partial y^2}+\frac{\partial^2 \overline{v}}{\partial z^2}\right)+\frac{1}{\rho}\left(\frac{\partial \tau_{yx}}{\partial x}+\frac{\partial \tau_{yy}}{\partial y}+\frac{\partial \tau_{yz}}{\partial z}\right) \tag{3}$$

where: $\tau_{yx}=\rho\overline{v'u'}$ where τ_{yx} is the turbulent shear stress acting in y direction on the x-face of control volume, $\tau_{yy}=\rho\overline{v'v'}$ where τ_{yy} is the turbulent shear stress acting in y direction on the y-face of control volume, $\tau_{yz}=\rho\overline{v'w'}$ where τ_{yz} is the turbulent shear stress acting in y direction on the z-face of control volume, and assuming $-2\Omega_x\overline{w}$ is negligible. Analogous to laminar shear stress, the turbulent shear stresses are often parameterized as $\tau_{yx} = \mu_{turbulent-yx}\frac{\partial\overline{v}}{\partial x} = \rho\overline{v'u'}$, $\tau_{yy} = \mu_{turbulent-yy}\frac{\partial\overline{v}}{\partial y} = \rho\overline{v'v'}$, $\tau_{yz} = \mu_{turbulent-xz}\frac{\partial\overline{v}}{\partial z} = \rho\overline{v'w'}$. The pressure is usually decomposed into the following terms: $\overline{p} = \overline{p}_a + g\int_\eta^z \overline{\rho}dz$, and the pressure gradient in the y-momentum then becomes after simplification $-\frac{1}{\rho}\frac{\partial\overline{p}}{\partial y} = -\frac{1}{\rho}\frac{\partial\overline{p}_a}{\partial y} + g\frac{\partial\overline{\eta}}{\partial y} - \frac{g}{\rho}\int_\eta^z \frac{\partial\overline{\rho}}{\partial y}dz$.

2.1.4 Z-momentum equation

$$\frac{\partial\overline{w}}{\partial t} + \overline{u}\frac{\partial\overline{w}}{\partial x} + \overline{v}\frac{\partial\overline{w}}{\partial y} + \overline{w}\frac{\partial\overline{w}}{\partial z} = -g - \frac{1}{\rho}\frac{\partial\overline{p}}{\partial z} + \frac{\mu}{\rho}\left(\frac{\partial^2\overline{w}}{\partial x^2} + \frac{\partial^2\overline{w}}{\partial y^2} + \frac{\partial^2\overline{w}}{\partial z^2}\right) + \frac{1}{\rho}\left(\frac{\partial\tau_{zx}}{\partial x} + \frac{\partial\tau_{zy}}{\partial y} + \frac{\partial\tau_{zz}}{\partial z}\right) \quad (4)$$

where: $\tau_{zx} = \rho\overline{w'u'}$ where τ_{zx} is the turbulent shear stress acting in z direction on the x-face of control volume, $\tau_{zy} = \rho\overline{w'v'}$ where τ_{zy} is the turbulent shear stress acting in z direction on the y-face of control volume, $\tau_{zz} = \rho\overline{w'w'}$ where τ_{zz} is the turbulent shear stress acting in z direction on the z-face of control volume, and neglecting the Coriolis terms $-2\Omega_y\overline{u} + 2\Omega_x\overline{v}$. Analogous to laminar shear stress, the turbulent shear stresses are often parameterized as $\tau_{zx} = \mu_{turbulent-zx}\frac{\partial\overline{w}}{\partial x} = \rho\overline{w'u'}$, $\tau_{zy} = \mu_{turbulent-zy}\frac{\partial\overline{w}}{\partial y} = \rho\overline{w'v'}$, $\tau_{zz} = \mu_{turbulent-zz}\frac{\partial\overline{w}}{\partial z} = \rho\overline{w'w'}$. In cases where vertical accelerations are much less than horizontal accelerations, this equation can be reduced to the hydrostatic equation, i.e., $\frac{1}{\rho}\frac{\partial\overline{p}}{\partial z} = -g$.

2.2 Conservation of constituent mass and heat: the ADVECTIVE diffusion equation

The conservation of constituent mass in a control volume is a sum of all the fluxes (advective and diffusive) into and out from the control volume plus sources and sinks (chemistry, biology, physics, withdrawals, inputs) within the control volume. Summing up the fluxes in each direction, assuming that the fluid is incompressible and that the molecular diffusivity, D, is homogeneous and isotropic, the advective diffusion equation becomes

$$\underbrace{\frac{\partial c}{\partial t}}_{\substack{\text{unsteady change}\\ \text{in concentration}}} + \underbrace{u\frac{\partial c}{\partial x} + v\frac{\partial c}{\partial y} + w\frac{\partial c}{\partial z}}_{\text{advective mass transport}} = \underbrace{D\left[\left(\frac{\partial^2 c}{\partial x^2}\right) + \left(\frac{\partial^2 c}{\partial y^2}\right) + \left(\frac{\partial^2 c}{\partial z^2}\right)\right]}_{\text{diffusive mass transport}} + \underbrace{S}_{\text{sources/sinks}} \quad (5)$$

where c is the concentration [M/L^{-3}], S is the sources and sinks of reactions occurring in the control volume, or the reaction rate [ML^{-3} T^{-1}].

This equation is a 3-D, unsteady equation that applies to all flow conditions: laminar and turbulent. Since we cannot determine the instantaneous velocity field, the x-y-and z momentum equations were time averaged and hence were only able to practically predict the temporal mean velocity. Similarly, we time average the conservation of mass/heat equation using time averages of the velocity field.

The instantaneous velocity and concentration are decomposed into a mean and an unsteady component. Similar to the velocity field shown earlier, for concentration, c, this becomes $c = \overline{c} + c'$ where $\overline{c} = \frac{1}{T}\int_t^{t+T} c dt$ and $c'$is the fluctuation about the mean.

Substituting the time average and fluctuating components of concentration and velocities into the 3D governing equation and time averaging we obtain:

$$\underbrace{\frac{\partial \overline{c}}{\partial t}}_{\substack{\text{unsteady change}\\ \text{in concentration}}} + \underbrace{\overline{u}\frac{\partial \overline{c}}{\partial x} + \overline{v}\frac{\partial \overline{c}}{\partial y} + \overline{w}\frac{\partial \overline{c}}{\partial z}}_{\text{mean advective mass transport}} = \underbrace{D\left[\left(\frac{\partial^2 \overline{c}}{\partial x^2}\right) + \left(\frac{\partial^2 \overline{c}}{\partial y^2}\right) + \left(\frac{\partial^2 \overline{c}}{\partial z^2}\right)\right]}_{\text{molecular diffusive mass transport}}$$
$$+ \underbrace{\left[\frac{\partial}{\partial x}\left(E_x\frac{\partial \overline{c}}{\partial x}\right) + \frac{\partial}{\partial y}\left(E_y\frac{\partial \overline{c}}{\partial y}\right) + \frac{\partial}{\partial z}\left(E_z\frac{\partial \overline{c}}{\partial z}\right)\right]}_{\text{turbulent diffusive mass transport}} + \underbrace{\overline{S}}_{\text{sources/sinks}} \quad (6)$$

where the turbulent mass fluxes in x, y and z were assumed to be defined as a gradient, diffusion-type process, such as $\left(\overline{u'c'}\right) = -E_x\frac{\partial \overline{c}}{\partial x}$, $\left(\overline{v'c'}\right) = -E_y\frac{\partial \overline{c}}{\partial y}$, $\left(\overline{w'c'}\right) = -E_z\frac{\partial \overline{c}}{\partial z}$, E_x is the turbulent mass diffusivity in x $[L^2/T]$, E_y is the turbulent mass diffusivity in y$[L^2/T]$, E_z is the turbulent mass diffusivity in z $[L^2/T]$. The new terms in the governing equation represent mass transport by turbulent eddies. As the intensity of turbulence increases, turbulent mass transport increases.

In turbulent fluids, E_x, E_y, and $E_z >> D$, and D can be neglected (except at boundaries or density interfaces where turbulent intensity may approach zero). The turbulent diffusion coefficients can be thought of as the product of the velocity scale of turbulence and the length scale of that turbulence. These coefficients are related to the turbulent eddy viscosity. In general, these turbulent diffusion coefficients are non-isotropic and non-homogeneous.

Spatial averaging of this equation leads to the introduction of "dispersion" coefficients which account for the transport of mass as a result of spatial irregularities in the velocity field.

These equations are also valid for heat transport and temperature modeling by substituting the concentration of heat, $\rho c_p T$, where T is temperature, c_p is the coefficient of specific heat at constant pressure and ρ is the density, such that the governing equation for temperature, T, becomes after simplification

$$\underbrace{\frac{\partial \overline{T}}{\partial t}}_{\substack{\text{unsteady change}\\ \text{in temperature}}} + \underbrace{\overline{u}\frac{\partial \overline{T}}{\partial x} + \overline{v}\frac{\partial \overline{T}}{\partial y} + \overline{w}\frac{\partial \overline{T}}{\partial z}}_{\text{mean advective heat transport}} = \underbrace{D_T\left[\left(\frac{\partial^2 \overline{T}}{\partial x^2}\right) + \left(\frac{\partial^2 \overline{T}}{\partial y^2}\right) + \left(\frac{\partial^2 \overline{T}}{\partial z^2}\right)\right]}_{\text{molecular diffusive heat transport}}$$
$$+ \underbrace{\left[\frac{\partial}{\partial x}\left(E_x\frac{\partial \overline{T}}{\partial x}\right) + \frac{\partial}{\partial y}\left(E_y\frac{\partial \overline{T}}{\partial y}\right) + \frac{\partial}{\partial z}\left(E_z\frac{\partial \overline{T}}{\partial z}\right)\right]}_{\text{turbulent diffusive heat transport}} + \underbrace{\frac{\overline{S}}{\rho c_p}}_{\text{heat flux}} \quad (7)$$

where D_T is the molecular thermal conductivity for heat and E_x, E_y, and E_z are the heat and mass turbulent eddy diffusivities assuming they are of the same order of magnitude.

2.3 Equation of state

Since density is an important variable for the momentum equation to account for density-driven flows, the computation of density is accomplished through an equation of state where density is computed from dissolved and suspended solids concentrations $(c_{dissolved\ solids}, c_{suspended\ solids})$ and temperature, T, such as

$$\overline{\rho} = f\left(\overline{T}, \overline{c_{dissolved\ solids}}, \overline{c_{suspended\ solids}}\right) \tag{8}$$

Typical equations of state for fresh and saltwater have been published by Gill [8] and Ford and Johnson [9].

2.4 Solution of governing equations

There are six equations (continuity or conservation of fluid mass, conservation of momentum in x, y and z, and conservation of constituent mass or heat, equation of state) that we are solving for six unknowns: turbulent time average concentration (or temperature), velocities in x, y, and z, density and turbulent time average pressure (or water surface), i.e. $\overline{c}$ *or* $\overline{T}, \overline{u}, \overline{v}, \overline{w}, \overline{\rho}$, *and* $\overline{\eta}$ or $\overline{p}$. The mathematical solution is dependent on specifying the following: (1) turbulent shear stresses or Reynolds stresses by specification of the turbulent eddy viscosities, (2) turbulent mass (heat) fluxes by specification of E_x, E_y and E_z, (3) initial and boundary conditions, (4) dynamic molecular viscosity and molecular diffusivity for computations at interfaces or boundaries (otherwise, they are usually neglected since all natural water bodies are highly turbulent), and (5) the Coriolis acceleration (if 2D horizontal or 3D for large water bodies).

Determination of the turbulent eddy viscosities and eddy diffusivities is often based on what are termed closure models that are based on the turbulent Schmidt number (Sc = ratio of turbulent viscosity to turbulent diffusivity of mass) and the turbulent Prandtl number (Pr = ratio of turbulent viscosity to turbulent conductivity of heat). Most experimental evidence suggests that the turbulent Sc and Pr numbers are close to unity for turbulent flows and that turbulent Sc or Pr numbers vary only little between flows. Even though many models use a constant value of these ratios such that mass and heat transfer turbulent coefficients are approximately equal, buoyancy affects that value [10–12].

Determination of turbulent eddy viscosities have been based on multiple approaches: (1) eddy viscosity models as a function of water stability [13–16], (2) Mixing length models [17, 18], (3) One equation models for turbulent kinetic energy [19], (4) Two-equation k-ε models for turbulent kinetic energy and dissipation [11] and (5) Reynolds stress and algebraic stress models [11]. In many models, once the turbulent eddy viscosity is known, then the turbulent diffusion coefficients are computed from $E \sim \frac{\mu_{turbulent}}{\rho}$ where the approximation is based on typical Sc or Pr numbers. Many water quality and temperature models for lakes and reservoirs use some form of a k-ε turbulence model [20].

Vertical boundary conditions for the hydrodynamic model usually involve a surface shear stress condition for the wind and a bottom shear stress condition for frictional resistance based on a specified friction coefficient (for example, Chezy or Manning's). Vertical boundary conditions for temperature and water quality constituents are assumed to be known fluxes at the surface and bottom.

Horizontal boundary conditions for mass or heat include mass or heat fluxes as a result of advection and for hydrodynamics include water level (or head) or flow conditions. The flow conditions in outlets to stratified reservoirs can be complicated because of local vertical accelerations in the vicinity of the outlet. In many models, the vertical acceleration of a fluid parcel is assumed to be much less than the horizontal accelerations and hence the vertical momentum equation simplifies to the hydrostatic equation. In order to model the complicated outlet hydraulics in a reservoir, special selective withdrawal algorithms are often used [21, 22]. These allow the computation of flow from multiple vertical layers without having to solve the full-vertical momentum equation.

Typical assumptions of the flow field and water quality model are related to the dimensionality of the system (one, two or three-dimensions), whether the flow field is dynamic or steady-state, and the turbulence closure approximation. Based on the model assumptions, the model grid is developed where the governing equations are satisfied at points (differential equation representation) or over control volumes (integral representation). The resulting equations are then solved using numerical methods.

2.5 Sources-sinks for water quality and temperature

The source-sink term in the mass and heat conservation equation can be either positive or negative and is determined by each water quality state variable. The units of $\overline{S}$ in the mass conservation equation are $[ML^{-3}\,T^{-1}]$ with a typical unit of $g/m^3/s$ and in the heat balance equation the units are $[Energy\ L^{-3}\,T^{-1}]$ with a typical unit of $J/m^3/s$. **Table 1** shows some of the typical source sink terms for several water quality state variables. Details of these can be found in Wells [20] and Chapra [23].

State variable	Typical source-sink term	Description
Temperature	$\overline{S} = -\frac{\partial \varphi}{\partial z}$	φ is the heat flux in units of W/m^2 transmitted through the water body. This is the short-wave solar radiation transmitted through the water and is a function of light extinction. The variable z is assumed to be positive downward.
Salinity or conservative substance	$\overline{S} = 0$	No sources and sinks
Suspended solids	$\overline{S}_{SS} = -w_{SS}\frac{\partial c_{SS}}{\partial z}$	w_{ss} is the settling velocity of particles as a positive velocity, c_{SS} is the concentration of suspended solids of a given size fraction. Often multiple size fractions are modeled independently using Stokes' law for settling velocity, w_{ss}. The variable z is assumed to be positive downward.
CBOD	$\overline{S_{particulate}} = -k_{CBODp}c_{CBODp} - w_{CBOD}\frac{\partial c_{CBODp}}{\partial z}$ $\overline{S_{dissolved}} = -k_{CBODd}c_{CBODd}$	Source/sink terms are shown for dissolved CBOD (c_{CBODd}) and particulate CBOD (c_{CBODp}), k_{CBOD} is a BOD decay rate for dissolved and particulate CBOD, and w_{CBOD} is the settling velocity for particulate BOD. Models of CBOD usually use $CBOD_{ultimate}$. Many models also track the P and N associated with this organic matter. Many models track multiple CBOD groups.
Algae	$\overline{S_{algae}} = \mu_{growth}c_{algae} - \mu_{respiration}c_{algae} - \mu_{excretion}c_{algae} - \mu_{mortality}c_{algae} - w_{algae}\frac{\partial c_{algae}}{\partial z}$	Source sink terms include the algae growth rate μ_{growth} $[T^{-1}]$ (this is a complicated function of light, limiting nutrient and temperature), $\mu_{respiration}$ $[T^{-1}]$ the "dark" respiration rate, $\mu_{excretion}$ $[T^{-1}]$ the rate of excretion or biomass loss,

State variable	Typical source-sink term	Description
		$\mu_{mortality}$ $[T^{-1}]$ the mortality rate (which often can include zooplankton grazing as a separate loss rate based on zooplankton populations and zooplankton food preferences), and w_{algae} the algae settling rate (this also can have complicated expressions especially for cyanobacteria and other species which migrate up and down in the water column). Often models include multiple algae groups. C_{algae} is the concentration of algae.
Ammonia-N	$\overline{S_{ammonia}} = \delta_{aN}\left(-\mu_{growth}c_{algae} + \mu_{respiration}c_{algae}\right) + \delta_{CBODdN}k_{CBODd}c_{CBODd} + \delta_{CBODpN}k_{CBODp}c_{CBODp} + SOD_N\frac{A}{V} - k_{nitr}c_{ammonia}$	The source/sink terms shown include algae uptake and release (where δ_{aN} is the stoichiometric equivalent of algae to ammonia-N, but the N source can be nitrate), organic matter release as particulate and dissolved CBOD decay (where δ_{CBODdN} is the stoichiometric equivalent of c_{BODd} to N and δ_{CBODpN} is the stoichiometric equivalent of c_{CBODp} to N), and sediment oxygen demand release under anoxic conditions (where SOD_N is the rate of N release in mass/area/time and V is the volume of the computational cell and A is the area of the sediment), nitrification decay rate k_{nitr} $[T^{-1}]$, and $c_{ammonia}$ is the total ammonia concentration.
Dissolved oxygen	$\overline{S_{DO}} = \delta_{aO2}\left(\mu_{growth}c_{algae} - \mu_{respiration}c_{algae}\right) - k_{CBODd}c_{CBODd} - k_{CBODp}c_{CBODp} - SOD\frac{A}{V} - \delta_{NO2}k_{nitr}c_{ammonia} + k_{reaeraton}(c_s - c_{DO})$	The source/sink term includes algae production and respiration (where δ_{aO_2} is the stoichiometric equivalent of dissolved oxygen to algae), CBOD particulate and dissolved water-column decay, sediment oxygen demand, nitrification demand (where δ_{NO_2} is the stoichiometric equivalent of dissolved oxygen to N), and reaeration at the surface only (where $k_{reaeration}$ is the reaeration rate in $[T^{-1}]$ which is generally a function of wind speed in lake and reservoirs and c_s is the saturation value of dissolved oxygen). Other models also include terms for metal oxidation, methane oxidation, and oxidation of hydrogen sulfide.
Nitrate-Nitrite-N	$\overline{S_{NOx}} = k_{nitr}c_{ammonia} - \delta_{aNOx}\left(\mu_{growth}c_{algae}\right) - k_{denit}c_{NOx}$	The source/sink terms include algae uptake (where δ_{aNO_x} is the stoichiometric equivalent of algae to nitrate-N since each algal group can have a preference for ammonia or nitrate as a N source), nitrification source, and a denitrification rate

State variable	Typical source-sink term	Description
		under anoxic conditions only (where k_{denit} is the denitrification rate under anoxic conditions). Other models also include terms for diffusion of nitrate into bottom muds. c_{NOX} is the concentration of nitrite and nitrate.
PO_4-P	$\overline{S_{PO4}} = \delta_{aP}\left(-\mu_{growth}c_{algae} + \mu_{respiration}c_{algae}\right) + \delta_{CBODdP}k_{CBODd}c_{CBODd} + \delta_{CBODpP}k_{CBODp}c_{CBODp} + SOD_P\frac{A}{V}$	The source/sink terms shown include algae uptake and release (where δ_{aP} is the stoichiometric equivalent of algae to P), organic matter release as particulate and dissolved CBOD decay (where δ_{CBODdP} is the stoichiometric equivalent of c_{CBODd} to P and δ_{CBODpP} is the stoichiometric equivalent of c_{CBODp} to P), and sediment oxygen demand release under anoxic conditions (where SOD_P is the rate of P release in mass/area/time and V is the volume of the computational cell, c_{algae} is the algae concentration, and A is the area of the sediment). Other models include adsorption of P onto inorganic particles.

Table 1.
Typical source-sink terms for temperature and some eutrophication water quality state variables.

3. Lake and reservoir water quality models

There are many models used to simulate reservoir and lake water quality. A summary of modeling approaches for lakes is shown in Mooij et al. [24] and Janssen et al. [25]. **Table 2** shows a listing of some common lake and reservoir models.

The choice of a correct framework is dependent on several considerations: (1) dimensionality of the lake/reservoir system (even though all water bodies are in essence 3D, 2D and 1D models can often represent the important processes of water quality and temperature gradients), (2) documentation (up-to-date user manual with example problems), (3) ease of use and expertise required (all models require a degree of file manipulation and many include GUI interfaces that often facilitate running the model for new users), (4) established record of successful projects (as documented in papers and conference proceedings and technical reports) and (5) model processes represent important lake/reservoir processes (for example, if macrophyte growth is an important ecological consideration, does the model represent macrophytes).

In many cases, 3D models do not often do better than other model frameworks. One reason may be that the data and parameter uncertainty increase in higher dimensional models [34]. In a comparison of 2D and 3D models, many examples have shown [28, 35, 36] that 2D models often better represent temperature profiles than some 3D models. There may be many reasons for this, but the important message is that more complicated models do not necessarily mean better model predictions. Another issue with 3D models is the excessive computational time compared to lower dimensional models. In one comparison between a 2D and 3D

Model name	Description	Reference
DYRESM and CAEDYM	1D model based on mixed layer dynamics, separate temperature and water quality models	Tanentzap et al. [26]
CE-QUAL-W2	2D longitudinal-vertical, open source, eutrophication model, hydrodynamics and water quality solved together	Wells [20]
CE-QUAL-R1	1D vertical	Environmental Laboratory [27]
W3	3D, hydrodynamics and water quality solved together	Al-Zubaidi and Wells [28]
EFDC and WASP	3D, hydrodynamics and water quality solved separately, both sigma stretch and z coordinate models	Hamrick [29], Tetra Tech [30]
GLM	1D	Hipsey et al. [31]
ELCOM and CAEDYM	3D-mixed layer dynamic model, hydrodynamics and water quality solved separately	Hipsey et al. [32], Hodges and Dallimore [33]

Table 2.
List of common lake and reservoir water quality models.

model, the 3D model took 30× longer than the 2D model. This will vary depending on model configuration and model. This is becoming more of an issue as models are being used for multiple-decade simulations evaluating climate change and long-term changes in model boundary conditions.

4. Typical results of lake and reservoir modeling

Using the CE-QUAL-W2 model [20] as an example, consider an application to Folsom Reservoir, CA, USA, as presented in Martinez et al. [37].

Folsom lake, located near Sacramento California USA, is a deep-storage reservoir that provides municipal water, power generation and cold water for primarily salmonid fish in the lower American River (see **Figure 10**). The reservoir has multiple outlets that allow the operator to choose different water levels for downstream temperature control.

The model was set-up and calibrated to a 10-year period between January 1, 2001 and December 31, 2011. Boundary conditions for flow, meteorological data, and outflow during this period were developed. A very detailed approach for filling in data gaps was undertaken to provide a good set of boundary conditions. Typical model predictions compared to field data are shown for temperature in **Figure 11** in 2002 and 2007 at multiple longitudinal stations in the reservoir. Error statistics for temperature profiles over the 10-year period using about 27,000 data comparisons were an average mean error of 0.004°C, an average absolute mean error (AME, average absolute value of the error) of 0.56°C, and a root mean square (RMS) average error of 0.71°C. The R^2 correlation between modeled and predicted temperature was 0.996.

In other examples of predicting the thermal regime, Cole [38] has shown that typical errors (AME, RMS) for temperature should often be well less than 1°C with a mean error of close to zero with minimal calibration if the boundary condition data are well-specified.

Oftentimes, the success of modeling other water quality state variables is first dependent on obtaining good temperature calibration results. For example, in a

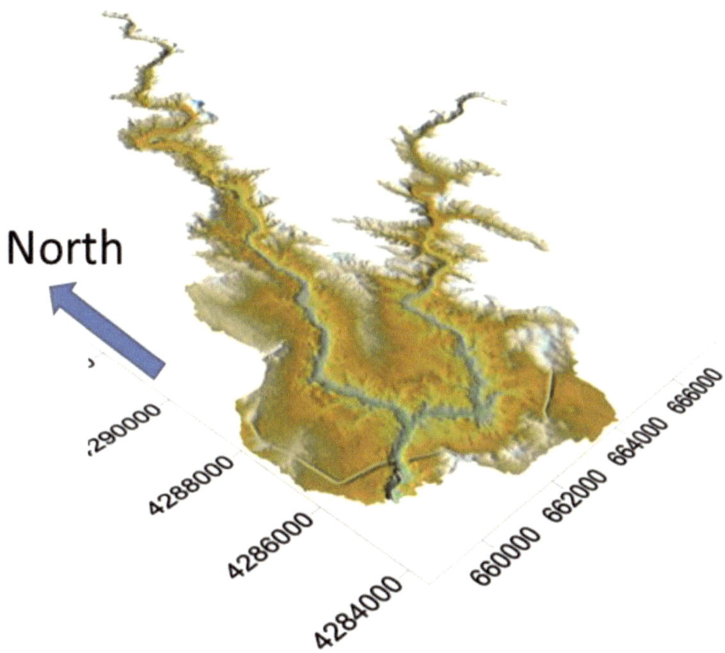

Figure 10.
Folsom reservoir bathymetry showing the north fork and south fork of the American River channels. Axes are labeled in m.

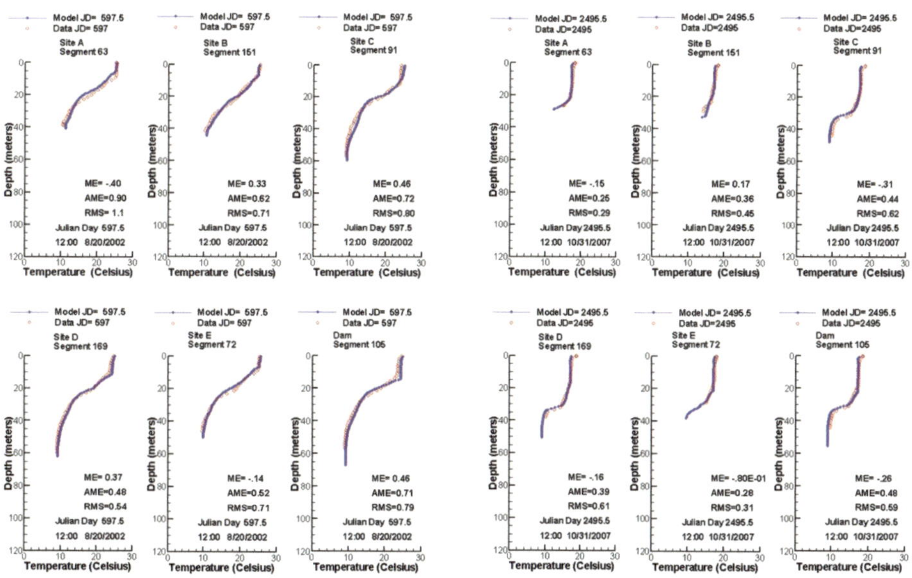

Figure 11.
Folsom reservoir model temperature predictions compared to field data on August 20, 2002 (left) and October 31, 2007 (right) at 6 different stations in Folsom reservoir.

higher elevation pristine lake, Chester Morse Lake, WA, USA, Ceravich and Wells [39] have shown dissolved oxygen profiles mimicking the unusual behavior of the dissolved oxygen profile in a lake with little algae growth as shown in **Figure 12**. Error statistics for dissolved oxygen, which integrates all the water quality processes, were a ME of 0.15 mg/l, a AME of 0.42 mg/l, and a RMS error of 0.49 mg/l for 551 data-model comparisons.

5. Conclusions in hydrodynamic and water quality modeling

The complexity of existing models has often exceeded our capacity in the field to verify model coefficients usually because of cost and time. Deterministic water quality models require an incredible amount of information that is rarely measured. In the CE-QUAL-W2 model, for each algal group the model user must specify approximately 25 values describing rate coefficients for growth, respiration, excretion, mortality, stoichiometry, temperature preferences, N preferences, light saturation limits, and settling velocities. Even though this model has no limit to the number of algal groups one can represent mathematically, in a practical sense modeling living populations and their impact on nutrients, organic matter, pH, temperature, and oxygen is very complex. In the end, the model user tries to balance the known field data with literature values of the coefficients with the goal that if the boundary conditions are well-specified, the model requires little calibration.

If one cannot understand and interpret field data, then it will be challenging for a model to match field measurements. Hence, knowing and understanding the field data as one is setting up the model is important for making sure the model is agreeing with field data trends.

In other cases though, the model is able to discern complex interactions between water quality state variables that may be difficult for the model user to piece together a priori. For example, the unusual dissolved oxygen profiles in the field data and model shown in **Figure 12** is one example where it was unclear the reasons for the unusual vertical profile until the combination of a sharp thermocline, algae growth within the metalimnion, and slow sediment oxygen demand caused the model to match the field data vertical trend.

Water quality models are adding more and more complex algorithms to reproduce admittedly complex phenomena. But this increasing complexity does not necessarily mean a better model or one that better reproduces field data. One example is the use of a complex model of bacterial populations on the Snake River

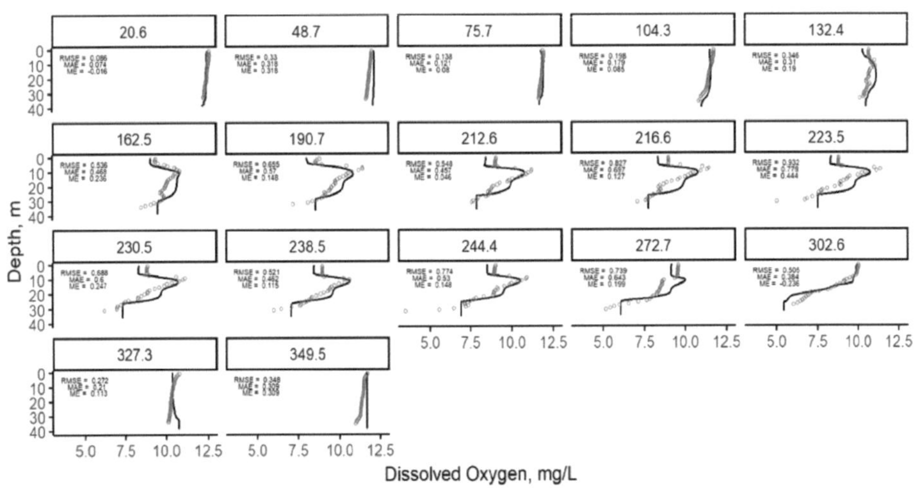

Figure 12.
Predictions (solid lines) and field data (dots) of dissolved oxygen at one sampling site for Chester Morse Lake in 2015. Dates shown are Julian days since January 1, 2015.

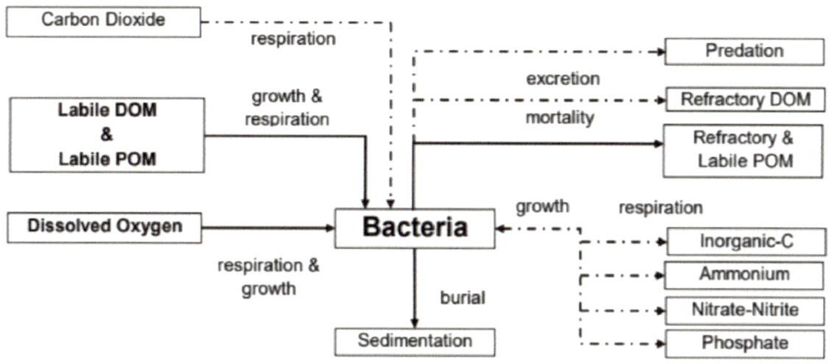

Figure 13.
Bacterial dynamics model compartments in the Snake River from Harrison [40].

in ID/OR, USA, from Harrison [40]. The bacterial populations were modeled based on Reichert et al. [41] as shown in **Figure 13** and compared to a model with only a first order decay rate for organic matter decay (basically neglecting all the complex bacterial dynamics). In predicting the impact of organic matter on dissolved oxygen, the simpler model neglecting bacterial dynamics performed better. This does not mean that complex models may not be useful for research purposes, but more complicated does not mean a better model.

Hence, to improve water quality models, one of the most fruitful areas is working on obtaining better boundary condition data by "smart" filling in of data gaps in time series of field data. This is still a critical component of modeling lakes and reservoirs. In addition, measuring field data on-site for lakes and reservoirs helps tremendously in understanding better the impact of hydrodynamics on water quality.

Author details

Scott A. Wells
Department of Civil and Environmental Engineering, Portland State University, Portland, USA

*Address all correspondence to: wellss@pdx.edu

References

[1] Wetzel R. Limnology Lake and River Ecosystems. NY: Academic Press; 2001

[2] Hutchinson GE. A Treatise on Limnology. New York: John Wiley & Sons; 1967

[3] Martin J, Higgins J, Edinger J, Gordon J. Energy Production and Reservoir Water Quality. Reston, VA: ASCE; 2007

[4] Wells SA, Manson JR, Martin JL. Numerical hydrodynamic and transport models for reservoirs. In: Martin J, Higgins J, Edinger J, Gordon J, editors. Energy Production and Reservoir Water Quality. Reston, VA: ASCE; 2007. Chapter 4

[5] Batchelor GK. An Introduction to Fluid Dynamics. NY: Cambridge University Press; 1967

[6] Sabersky R, Acosta A, Haupmann E. Fluid Flow A First Course in Fluid Mechanics. NY: Macmillan Publishing Co.; 1989

[7] Cushman-Roisin B. Introduction to Geophysical Fluid Dynamics. Englewood Cliffs, NJ: Prentice-Hall; 1994

[8] Gill AE. "Appendix 3, Properties of Seawater", Atmosphere Ocean Dynamics. Vol. 599. New York, NY: Academic Press; 1982. p. 600

[9] Ford DE, Johnson MC. An Assessment of Reservoir Density Currents and Inflow Processes. Technical Rpt. E 83-7. Vicksburg, MS: US Army, Engineer Waterways Experiment Station; 1983

[10] Turner JS. Buoyancy Effects in Fluids. Cambridge: Cambridge University Press; 1973

[11] Rodi W. Turbulence models and their application in hydraulics. In: IAHR. 3rd ed. Rotterdam: A.A. Balkema; 1993

[12] Venayagamoorthy SK, Koseff JR, Ferziger JH, Shih LH. Testing of RANS turbulence models for stratified flows based on DNS data. In: Annual Research Report Briefs. Center for Turbulence Research; 2003. pp. 127-138

[13] Munk WH, Anderson ER. Notes on the theory of the thermocline. Journal of Marine Research. 1948;**1**:276

[14] Mamayev OI. The influence of stratification on vertical turbulent mixing in the sea. Izv. Geophys. Ser. 1958:870-875

[15] Fischer HB, List J, Koh R, Imberger J, Brooks N. Mixing in Inland and Coastal Waters. New York: Academic Press; 1979

[16] Holland PR, Kay A, Botte V. A numerical study of the dynamics of the riverine thermal bar in a deep lake. Environmental Fluid Mechanics. 2001;**1**: 311-332

[17] Nezu I, Nakagawa H. Turbulence in Open-Channel Flows. Rotterdam: A.A. Balkema; 1993

[18] Bloss S, Patterson JC. Modeling turbulent transport in a stratified estuary. ASCE Journal of Hydraulic Engineering. 1988;**114**(9):1115-1133

[19] Martin JL, McCutcheon SC. Hydrodynamics and Transport for Water Quality Modeling. NY: Lewis Publishers; 1999

[20] Wells SA. CE-QUAL-W2: A Two-Dimensional, Laterally Averaged, Hydrodynamic and Water Quality Model, Version 4.2, User Manual Part 2, Hydrodynamic and Water Quality Model Theory. Portland, OR:

Department of Civil and Environmental Engineering, Portland State University; 2019

[21] Brooks NH, Koh RC. Selective withdrawal from density-stratified reservoirs. Journal of Hydraulics Division, ASCE. 1969;**95**(HY4): 1369-1397

[22] Davis JE, Holland JP, Schneider ML, Wilhelms SC. SELECT: A Numerical, One-Dimensional Model for Selective Withdrawal, Instruction Report E-87-2. Waterways Experiment Station, Hydraulics Laboratory, Vicksburg, MS: Corps of Engineers; 1987

[23] Chapra S. Surface Water Quality Modeling. NY: McGraw-Hill; 1997

[24] Mooij WM, Trolle D, Jeppesen E, Arhonditsis G, Belolipetsky PV, Chitamwebwa DBR, et al. Challenges and opportunities for integrating lake ecosystem modelling approaches. Aquatic Ecology. 2010;**44**(3):633-667

[25] Janssen ABG, Arhonditsis GB, Beusen A, Bolding K, Bruce L, Bruggeman J, et al. Exploring, exploiting and evolving diversity of aquatic ecosystem models: A community perspective. Aquatic Ecology. 2015; **49**(4):513-548. DOI: 10.1007/s10452-015-9544-1

[26] Tanentzap AJ, Hamilton DP, Yan ND. Calibrating the dynamic reservoir simulation model (DYRESM) and filling required data gaps for one-dimensional thermal profile predictions in a boreal lake. Limnology and Oceanography: Methods. 2007;**5**: 484-494

[27] Environmental Laboratory. CE-QUAL-R1: A Numerical One-Dimensional Model of Reservoir Water Quality: Users Manual, Technical Report E-82-1. Vicksburg, MS: Corps of Engineers, Waterways Experiment Station, Environmental Laboratory; 1982

[28] Al-Zubaidi HAM, Wells SA. Analytical and field verification of a 3D hydrodynamic and water quality numerical scheme based on the 2D formulation in CE-QUAL-W2. Journal of Hydraulic Research. 2020;**58**(1): 152-171. DOI: 10.1080/00221686.2018.1499051

[29] Hamrick JM. A three-dimensional environmental fluid dynamics computer code: Theoretical and computational aspects. In: Special Report. Vol. **317**. The College of William and Mary, Virginia Institute of Marine Science; 1992. p. 63

[30] Tetra Tech. The Environmental Fluid Dynamics Code User Manual US EPA Version 1.01. Fairfax, VA; 2007

[31] Hipsey MR, Bruce LC, Hamilton DP. GLM—General lake model: Model overview and user information. In: AED Report #26. The University of Western Australia Technical Manual. 2014. p. 22

[32] Hipsey MR, Romero J, Antenucci J, Hamilton D. Computational Aquatic Ecosystem Dynamics Model CAEDYM v2.0 Science Manual. Centre for Water Research, University of Western Australia; 2003

[33] Hodges B, Dallimore C. Estuary and Lake Computer Model Science Manual Code Version 1.5.0. University of Western Australia: Centre for Water Research; 2001

[34] EPA. Guidance on the development, evaluation, and application of environmental models. In: EPA/100/K-09/003. Washington, DC: Office of the Science Advisor; 2009

[35] DeGasperi C. Modeling Lake Sammamish, WA Using a 2-D and 3-D Model, Presentation. Portland State University: Department of Civil Engineering; 2005

[36] Geologic Survey Israel and Tahal. Red Sea Dead Sea Conveyance Study

Program. GSI Report Number: GSI/10/2011. Israel, Jerusalem; 2011

[37] Martinez VI, Wells SA, Addley RC. Meeting temperature requirements for fisheries downstream of Folsom Reservoir, California. In: EWRI, ASCE, Portland, OR: Proceedings World Environmental and Water Resources Congress. 2014. pp. 1081-1092

[38] Cole T. Examples of model applications. In: Wells S, editor. CE-QUAL-W2: A Two-Dimensional, Laterally Averaged, Hydrodynamic and Water Quality Model, Version 4.2, User Manual Part 4, Model Examples. Portland, OR: Department of Civil and Environmental Engineering, Portland State University; 2019

[39] Ceravich A, Wells SA. Water quality and hydrodynamic modeling of Chester Morse Lake. In: Technical Report Prepared for Seattle Public Utilities. WA, Seattle: Department of Civil and Environmental Engineering, Portland State University; 2020

[40] Harrison J. Partitioning snake river organic matter and modeling aerobic oxidation. Brownlee Reservoir [PhD dissertation]. Idaho: Department of Civil Engineering, University of Idaho; 2005

[41] Reichert P, Borchardt D, Henze M, Rauch W, Shanahan P, Somlyody L, et al. River water quality model no. 1. In: IWA Task Group on River Water Quality Modeling, editor. Scientific and Technical Report No. 12. London, UK: IWA Publishing; 2001. p. 131

Chapter 8

Assessment of the CHIRPS-Based Satellite Precipitation Estimates

Franklin Paredes-Trejo, Humberto Alves Barbosa, Tumuluru Venkata Lakshmi Kumar, Manoj Kumar Thakur and Catarina de Oliveira Buriti

Abstract

At present, satellite rainfall products, such as the Climate Hazards Group InfraRed Precipitation with Stations (CHIRPS) product, have become an alternative source of rainfall data for regions where rain gauge stations are sparse, e.g., Northeast Brazil (NEB). In this study, continuous scores (i.e., Pearson's correlation coefficient, R; percentage bias, PBIAS; and unbiased root mean square error, ubRMSE) and categorical scores (i.e., probability of detection, POD; false alarm ratio, FAR; and threat score, TS) were used to assess the CHIRPS rainfall estimates against ground-based observations on a pixel-to-station basis, during 01 January 1981 to 30 June 2019 over NEB. Results showed that CHIRPS exhibits better performance in inland regions (R, PBIAS, and ubRMSE median: 0.51, −3.71%, and 9.20 mm/day; POD, FAR, and TS median: 0.59, 0.44, and 0.40, respectively) than near the coast (R, PBIAS, and ubRMSE median: 0.36, −5.66%, and 12.43 mm/day; POD, FAR, and TS median: 0.32, 0.42, and 0.26, respectively). It shows better performance in the wettest months (i.e., DJF) than in the driest months (i.e., JJA) and is sensitive to both the warm-top stratiform cloud systems and the sub-cloud evaporation processes. Overall, the CHIRPS rainfall data set could be used for some operational purposes in NEB.

Keywords: CHIRPS, Northeast Brazil, satellite rainfall, rainfall, remote sensing, rain gauge, ground-based validation

1. Introduction

Rainfall is a key component of the global water cycle and is essential for a wide range of applications such as crop modeling, hydrometeorology, water resource management, flood and drought monitoring, and climatological applications [1–3]. Accurate and consistent rainfall estimates are also of remarkable importance for the drought-prone regions, such as the semiarid region of Northeast Brazil (NEB), which is at high risk of food insecurity due to the occurrence of prolonged droughts whose impacts affect adversely their water resources and crop production [4–6].

Nowadays, the measurement of precipitation is based on rain gauge stations, meteorological radars, and satellite retrievals [7, 8]. Rainfall data from ground stations provide high accuracy [9], but they are limited in spatial coverage [10]. Meteorological radars suffer from reduced data quality owing to signal blockage or distortion [11]. Satellites can be used for sensing large regions with a high temporal and spatial resolution, though satellite retrieval approaches are prone to biases and systematic errors [12]. Consequently, satellite-based rainfall estimates must be validated against rain gauge data in order to assess their uncertainties before being used [13, 14].

In NEB, despite the efforts of the state climate agencies (e.g., National Center for Monitoring and Early Warning of Natural Disasters, CEMADEN; National Institute of Meteorology, INMET; Meteorology and Hydrologic Resources Foundation of Ceara, FUNCEME; Superintendence for the Development of the Northeast, SUDENE; and National Water Agency, ANA), most of the rain gauge networks currently available are inadequate to produce reliable rainfall analysis, because of their scarce spatial coverage, high proportion of missing data, and short-length records [15]. To overcome these limitations, there is a wide variety of satellite-based rainfall products, such as the Climate Hazards Group InfraRed Precipitation with Stations (CHIRPS).

CHIRPS is a quasi-global rainfall data set with relatively high spatial resolution (°0.05 × °0.05) and long-term temporal coverage (from 1981 to near real time), whose processing chain blends satellite and gauge rainfall estimates [16]. Since early 2014, CHIRPS rainfall estimations are disseminated with different temporal scales (monthly, 10-day, 5-day, and daily time steps) by the University of California at Santa Barbara (UCSB). It has been subjected to various assessments worldwide by comparing to gauge measurements. According to these studies, the CHIRPS rainfall data set performs relatively well at both a regional and global scale, mainly in terms of bias and the Pearson's correlation coefficient when compared to other state-of-the-art satellite rainfall products [1, 8, 17–21].

Unlike other natural regions, very few studies have been carried out to validate CHIRPS rainfall estimates in NEB. Overall, CHIRPS achieves better results during the rainy season (i.e., March to May), but its ability for the rain detection is poor [22]. Moreover, CHIRPS displays a rainfall pattern similar to the rain gauge data in the south-southeast subregion of the NEB, even though some performance scores are lower than the ones derived from the Tropical Rainfall Measuring Mission (TRMM) Multi-satellite Precipitation Analysis (TMPA) 3B42V7 product, particularly from 2012 to 2014 [23]. Interestingly, CHIRPS provides performance better in terms of rain amount than the Multi-Source Weighted-Ensemble Precipitation (MSWEP), SM2RAIN-CCI (Climate Change Initiative), and Climate Prediction Center Morphing Technique (CMORPH) rainfall products over the Cerrado biome of NEB [24]. These findings are promising for operational applications in NEB (e.g., remote drought monitoring). Nevertheless, to our knowledge, a study investigating the performance of the CHIRPS rainfall data set by using new available ground-based observations is still absent.

The purpose of this study is to evaluate the quality of the CHIRPS rainfall estimates in NEB by considering the newest in situ data from the INMET meteorological stations, which is used as a benchmark rainfall data set over a 39-year period (1981–2019).

2. Materials and methods

2.1 Study area

The study was carried out in NEB (~8,515,759 km^2), which is located between 5.2° N–33.7° S and 34.7°–48.7° W [25]. In this region, the annual precipitation decreases from the east and northeast coast (>1500 mm/year) to inland dry regions

(<500 mm/year) [22], due to the impact of the orography [26] and the influence of different meteorological systems, such as the intertropical convergence zone (ITCZ), squall lines (SL), easterly wave disturbances (EWD), upper tropospheric cyclonic vortices (UTCV), frontal systems (FS), mesoscale convective complexes (MCC), and the South Atlantic convergence zone (SACZ) [27]. The rainy season occurs at different times of the year: April to June in the eastern coast of the NEB; November to January in the southern part of the NEB; and March to May in the semiarid northwestern part of the NEB [27]. This region includes two main river basins, namely, the basins of the São Francisco River (where the Sobradinho reservoir is located) and the Parnaíba River. It also contains the Amazonia, Cerrado, Atlantic Forest, and Caatinga biomes, which are strongly related to the spatial distribution of rainfall regimes [6, 15].

2.2 Rainfall data sets

Daily rain gauge observations from rain gauge stations were provided by the INMET (www.inmet.gov.br). The higher values than daily mean ± 3.5 standard deviations (method for detection of outliers) were coded as missing data [20]. The daily rainfall time series with more than 25% missing data per month were omitted [22]. A number of 27 stations were selected with these criterions (temporal coverage: January 1981 to June 2019). It is worth mentioning that 77%, 62%, and 42% of these stations were used in the blending process of CHIRPS during 1981–1998, 1999–2013, and 2014–2019, respectively (see https://bit.ly/2ZZFAvA); therefore, this sample is not a completely independent data set [13]. As depicted in **Figure 1**, most stations are located in the northwest NEB or near the coast.

CHIRPS rainfall estimates were obtained from the UCSB-Climate Hazards Group (CHG) webpage (https://www.chc.ucsb.edu/data; version 2 released in February 2015) at a daily time scale and spatial resolution of 0.05°, starting 1 January 1981 to 30 June 2019. This rainfall product uses a three-step development process. First, infrared precipitation (IRP) pentad (5-day) rainfall estimates are created from satellite data using cold cloud durations (CCD) lower than 235 K as a threshold value and calibrated in relation to the TRMM 3B42-based precipitation pentads by local regression. Then, the IRP pentads are divided by its long-term IRP mean values to present a percent of normal. Second, the percent of normal IRP pentad is multiplied by the corresponding Climate Hazards Precipitation

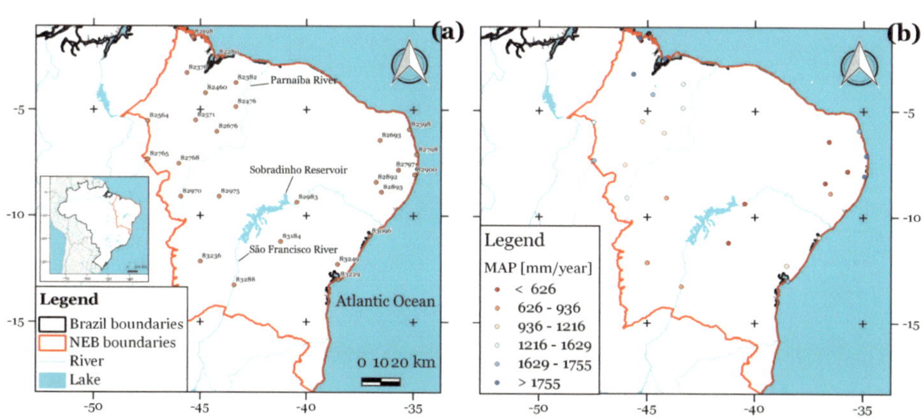

Figure 1.
Geographical location of the study area showing (a) selected stations. The numbers indicate the World Meteorological Organization (WMO) serial of each station; (b) annual mean precipitation for selected stations from 1 January 1981 to 30 June 2019.

Climatology (CHPclim) pentad to generate an unbiased rainfall estimate, with units of millimeters per pentad, called the CHG IR Precipitation (CHIRP). Third, pentadal CHIRP values are disaggregated to daily precipitation estimates based on daily NOAA Climate Forecast System (CFS) fields rescaled to 0.05° resolution. Finally, CHIRPS is produced through blending stations with the CHIRP data sets via a modified inverse distance-weighted algorithm [8]. For more details about the CHIRPS data set, the reader is referred to Funk et al. [16].

2.3 Auxiliary data sets

The land cover, annual rainfall, elevation, and type of climate were used as auxiliary information. The land cover was derived from the Land Cover-Climate Change Initiative (LC-CCI) product [28] (available online at http://maps.elie.ucl.ac.be). The average annual rainfall was estimated from the selected stations. The gauge elevation was obtained from the metadata information at each station. The slope and aspect of the terrain were derived from the Shuttle Radar Topographic Mission (SRTM) (available online at https://earthexplorer.usgs.gov). The type of climate was extracted from the Köppen-Geiger climate classification developed by Beck et al. [29] (available online at https://bit.ly/2Zt90Bu).

2.4 Methodology

The methodology applied in this study is summarized in **Figure 2**. The CHIRPS rainfall data set was chosen because of its low latency (about 3 weeks), high spatial resolution (0.05° × 0.05°), daily temporal resolution, and long-term temporal coverage (1981 to near real time), respectively, so it is potentially suitable for operational purposes in NEB. Firstly, the CHIRPS product was clipped using a shapefile of NEB as a mask. Then, CHIRPS rainfall estimates were extracted using the nearest neighbor (NN) method to generate a paired rainfall data from 1 January 1981 to 30 June 2019 (i.e., the common temporal coverage). The rationale behind the choice of

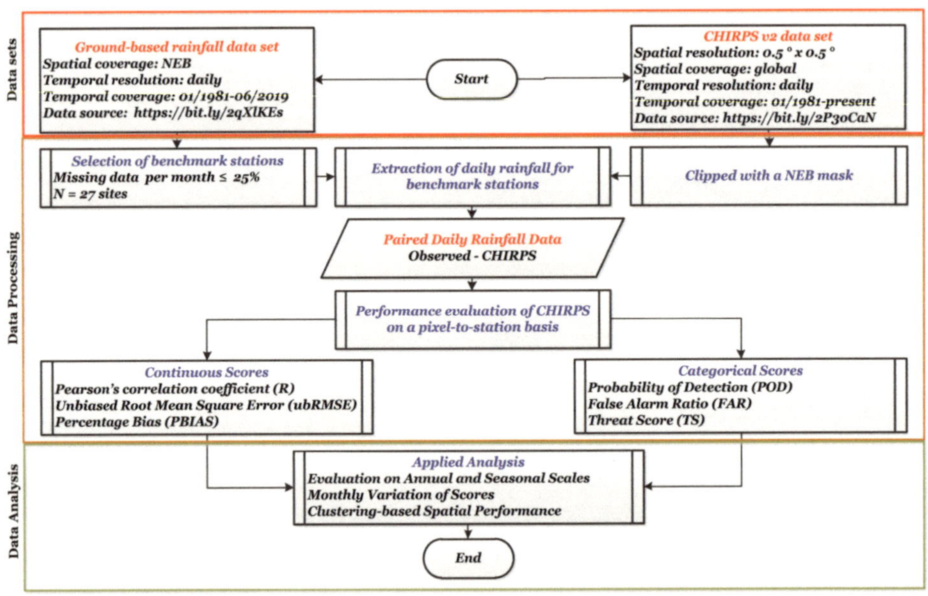

Figure 2.
Simplified flowchart of the methodology used in this study.

the NN method instead of gridded ground-based rainfall data (e.g., via spatial interpolation) is related to the fact that the latter would involve large uncertainties given the lack of a high-density rain gauge network to reproduce adequately the rainfall gradients in NEB [22]. Secondly, an intercomparison of both rainfall data sets was carried out in order to explore the performance of the CHIRPS product at the monthly, seasonal, and annual time scales during the common temporal coverage. Consequently, several metrics on a pixel-to-station basis were computed. The Pearson's correlation coefficient (R), unbiased root mean square error (ubRMSE), and percentage bias (PBIAS) were used as continuous scores. R measures the linear relationship strength between estimations and observations, while ubRMSE and B scores measure how the value of estimates differs from the observed values [20]. To examine the rain detection capability of the CHIRPS product, the probability of detection (POD), false alarm ratio (FAR), and threat score (TS) were used as categorical scores. POD and FAR indicate the fraction of the observed events that were correctly forecasted and the fraction of the predicted events did not occur, respectively. TS is the fraction between hits to all CHIRPS-based events. The categorical scores were derived from a contingency table using a rainfall threshold of 1 mm/day to discriminate between rain and no-rain event [29] (see **Table 1**). This rainfall threshold was chosen due to its previous use in semiarid regions [22, 23, 30]. Finally, in order to investigate the influence of the rainfall station spatial distribution on the performance scores, a cluster analysis based on the k-medoid algorithm was applied using the score values of all stations as cases. This unsupervised classification technique was implemented because it is not sensitive to outliers and reduces noise [31]. The equations, ranges, and optimal values of the performance scores are outlined in **Table 2**.

	Gauge ≥ threshold	**Gauge < threshold**
CHIRPS ≥ threshold	A	B
CHIRPS < threshold	C	D

Table 1.
Contingency table to estimate categorical scores. A, number of hits; B, number of false alarms; C, number of misses; D, number of correct negatives; threshold, rainfall threshold (1 mm/day).

Name	**Formula**	**Range**	**Perfect score**
Pearson's correlation coefficient	$R = \frac{\sum(G-\overline{G})(S-\overline{S})}{\sqrt{\sum(G-\overline{G})^2}\sqrt{\sum(S-\overline{S})^2}}$	[−1, 1]	1
Root mean square error	$RMSE = \sqrt{\frac{1}{N}(S-G)^2}$	[0, ∞)	0
Percentage bias	$B = 100\frac{\sum(S-G)}{N}$	(−∞, ∞)	0
Unbiased root mean square error	$ubRMSE = \sqrt{RMSE^2 - (B/100)^2}$	[0, ∞)	0
Probability of detection	$POD = \frac{A}{A+C}$	[0, 1]	1
False alarm ratio	$FAR = \frac{B}{A+B}$	[0, 1]	0
Threat score	$TS = \frac{A}{A+B+C}$	[0, 1]	1

Table 2.
Formulas of continuous and categorical scores. G, gauge-based rainfall measurement (mm/day); S, CHIRPS-based rainfall estimate (mm/day); $\overline{G}$ *and* $\overline{S}$*, average for G and S, respectively (mm/day); N, number of data pairs; A, B, and C for POD, FAR, and TS, as per **Table 1**.*

3. Results

For clarity, this section is split into three parts: (1) evaluation on annual and seasonal scales; (2) monthly variation of scores; and (3) clustering-based spatial performance.

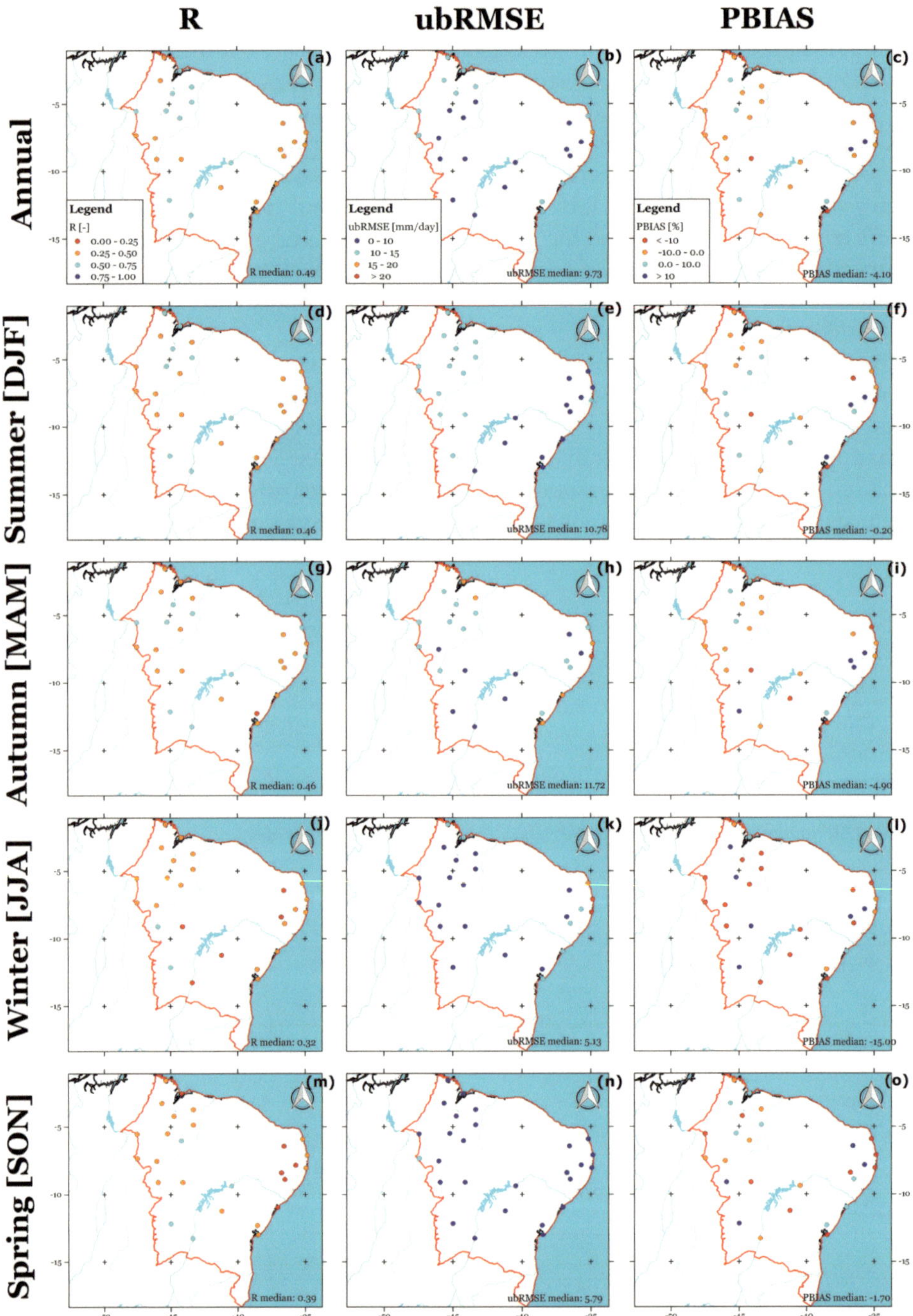

Figure 3.
Spatial distribution of R, ubRMSE, and PBIAS derived from the CHIRPS rainfall estimates against ground observations for (a–c) annual; (d–f) summer; (g–i) autumn; (j–l) winter; and (m–o) spring. The median value of each score is reported.

3.1 Evaluation on annual and seasonal scales

Figure 3 shows the spatial distribution of the continuous scores obtained after the pixel-to-station comparison of the CHIRPS rainfall estimates against the gauge-based data set during the study period. The seasons were defined as summer (Dec-Jan-Feb), autumn (Mar-Apr-May), winter (Jun-Jul-Aug), and spring

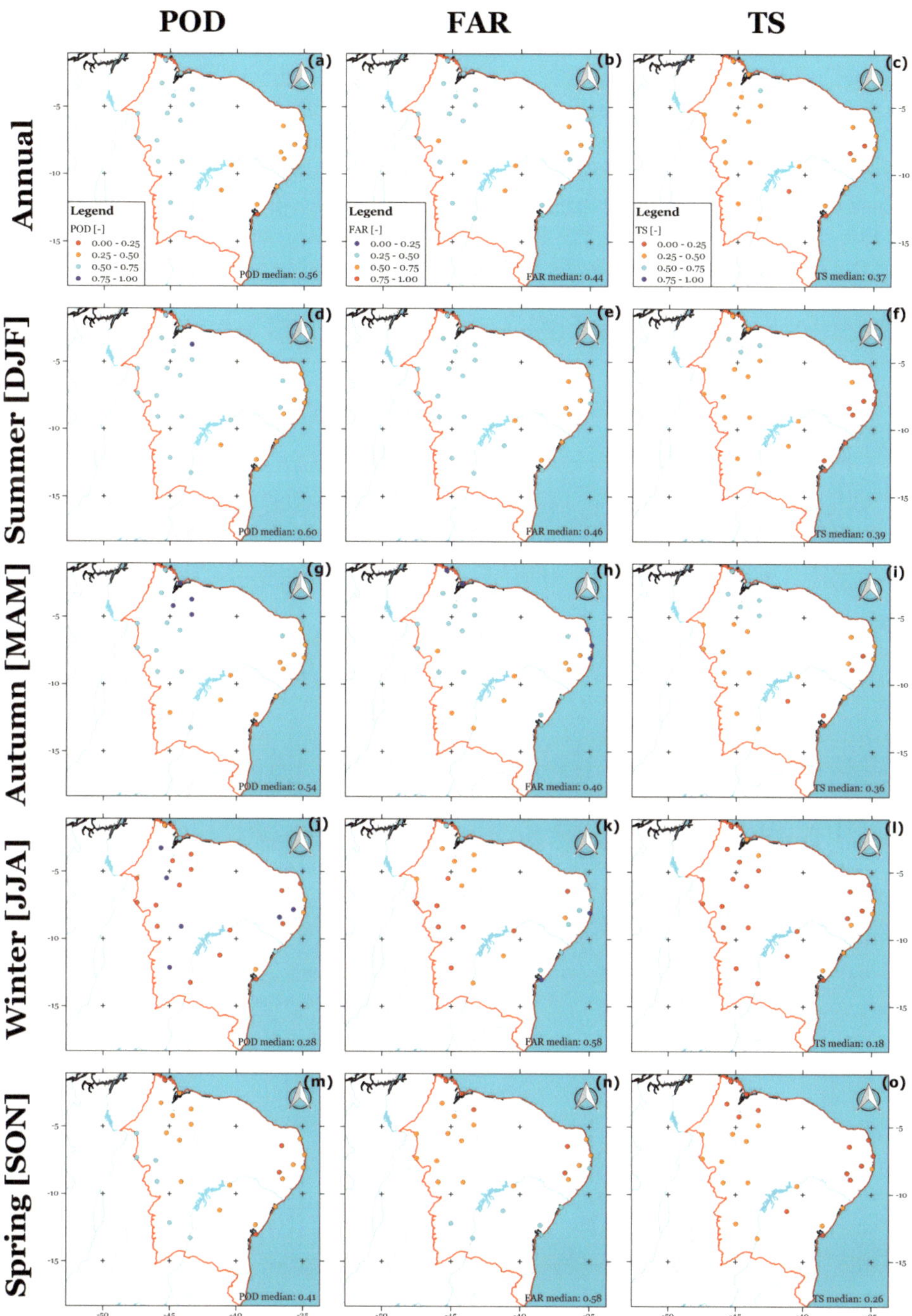

Figure 4.
Spatial distribution of POD, FAR, and TS derived from the CHIRPS rainfall estimates against ground observations for (a–c) annual; (d–f) summer; (g–i) autumn; (j–l) winter; and (m–o) spring. The median value of each score is reported.

(Sep-Oct-Nov) because the NEB is located in the southern hemisphere. The R, ubRMSE, and PBIAS median values listed in each subpanel were obtained by averaging these values from all stations via median to minimize the effects of extreme values. The CHIRPS product showed relatively good agreement with observations in terms of R, ubRMSE, and PBIAS at annual time scale (R median: 0.49; ubRMSE median: 9.73 mm/day; PBIAS, −4.10%), particularly in the northwest NEB ($R > 0.50$, ubRMSE and PBIAS near zero). Interestingly, the R median value begins to decrease from above 0.46 in summer to 0.32 in winter, but it rebounds and increases to values above 0.39 in spring. The ubRMSE values showed a similar pattern, with the higher ubRMSE values in summer and autumn (ubRMSE > 10 mm/day) and lower values in winter and spring (ubRMSE < 6 mm/day). The comparison revealed also that CHIRPS tends to underestimate the amount of rainfall in the course of a year (PBIAS annual median: −4.10%), especially during the transition from summer to winter (PBIAS median from −0.20% to −15.00%).

For the annual time scale, the POD, FAR, and TS mean values were 0.56, 0.44, and 0.37, respectively (**Figure 4**), indicating an acceptable rain detection ability in terms of POD, even though with a medium probability of false alarms in the central NEB. Similar to R and ubRMSE (**Figure 2**), the higher POD and TS values occurred

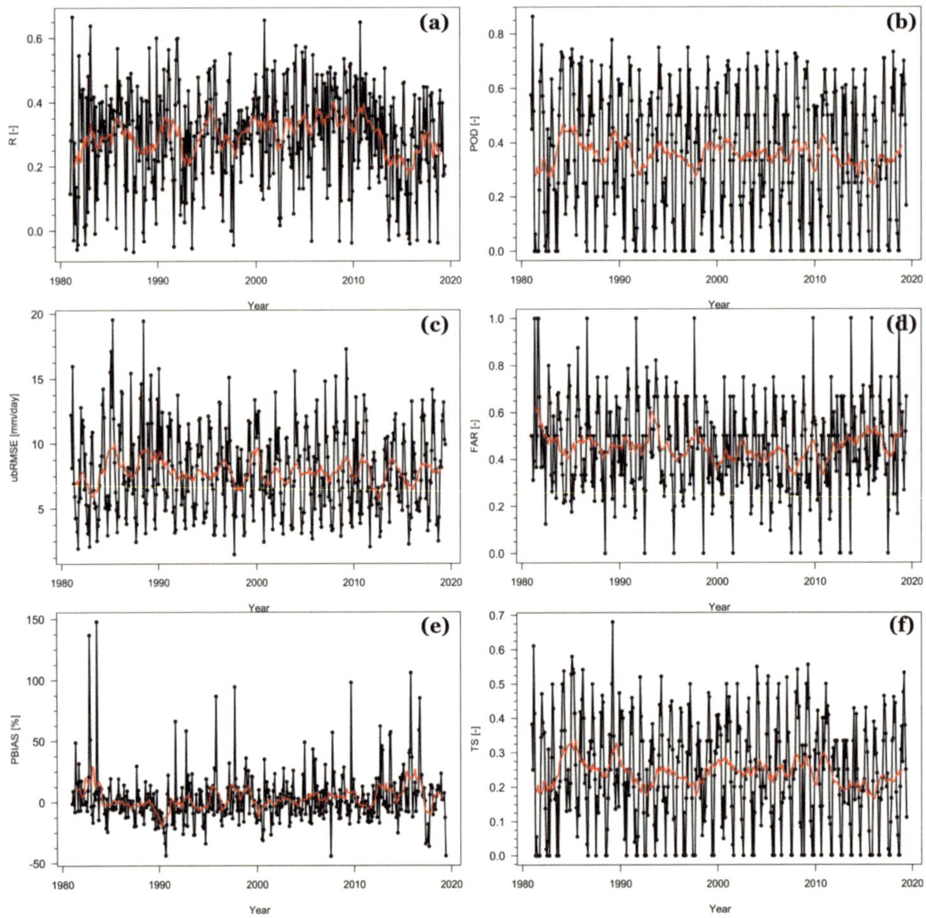

Figure 5.
Monthly time series for (a) R (dimensionless); (b) POD (dimensionless); (c) ubRMSE (mm/day); (d) FAR (dimensionless); (e) PBIAS (%); and (f) TS (dimensionless) derived from the CHIRPS rainfall estimates against ground observations (black line) for all NEB during the period 1981–2019. The red line depicts a 12-month moving average.

in summer and autumn (POD median > 0.50; TS median > 0.30), while lower values were observed in winter and spring. As expected, the FAR exhibited an inverse response to POD throughout the year (i.e., FAR median > 0.55 in winter and spring with lower values in summer and autumn).

3.2 Monthly variation of scores

Figure 5 shows the median of the scores for all stations, months, and years. The median values of R, ubRMSE, and PBIAS ranged between −0.06 and 0.66, 1.48 mm/day and 19.54 mm/day, and −44.50% and 147.80%, respectively. The lowest R values were observed in August (R median: 0.16) and the highest R values in March (R median: 0.41). According to the PBIAS time series, CHIRPS tends to underestimate (overestimate) the amount of rainfall between May and August (September and April), which is consistent with the findings from **Figure 3**. A moderate linear relationship between the monthly averaged values of PBIAS and ubRMSE was also found (R = −0.35, p-value <0.05), suggesting that PBIAS tends to increase when ubRMSE decreases. Furthermore, R, ubRMSE, and PBIAS did not exhibit a long-term trend (not shown for brevity), even though they showed high values for the coefficient of variation (i.e., 51.86%, 41.82%, and 675.49%, respectively).

The temporal variation of POD, FAR, and TR is shown in **Figure 5**. They varied from 0.00 to 0.86, from 0.00 to 1.00, and from 0.00 to 0.68, respectively. The highest POD and TR values were observed in February and March and the lowest in July and August. This means that CHIRPS shows better performance during the rainy season in terms of detection of rain events, which is in line with those inferences obtained from **Figure 4**. Moreover, the lowest FAR values were observed in July and August, indicating a minimum rate of false alarms during the driest

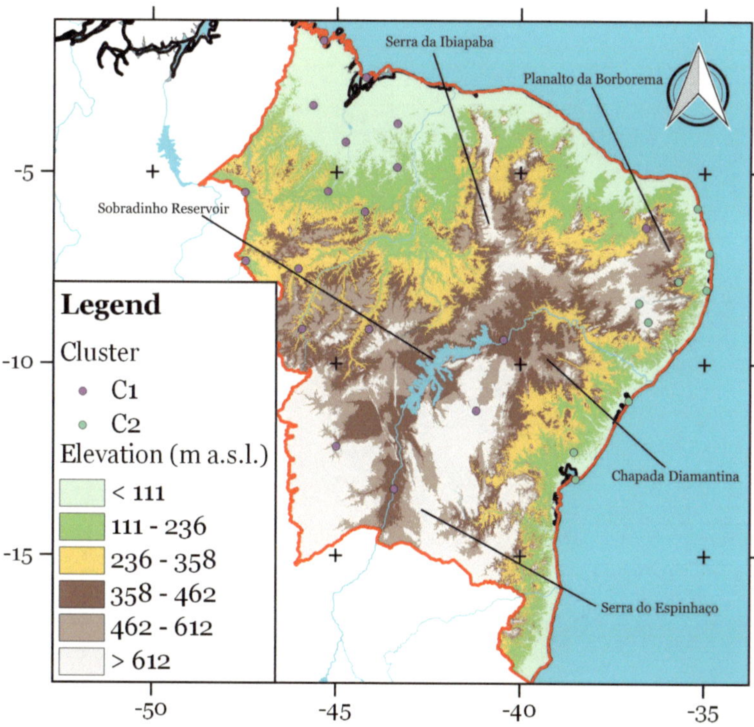

Figure 6.
Clustered stations according to their continuous and categorical scores at annual time scale. A 250-m digital elevation model derived from SRTM images is shown.

months. Similar to the continuous scores, these scores did not exhibit a long-term trend but a high temporal variation (i.e., 64.69%, 42.13%, and 63.97% for POD, FAR, and TR, respectively).

3.3 Clustering-based spatial performance

The previous statistical approaches provide a limited interpretation of the performance of CHIRPS, because they do not offer information about the degree of similarity among the selected stations in terms of their performance scores. Therefore, to identify the similar stations according to their scores, a medoid-based cluster analysis was applied. In order to adequately capture the spatiotemporal variability of the performance scores, an annual time scale was considered (i.e., **Figures 3a–c** and **4a–c**). The spatial distribution of the clustered stations is shown in **Figure 6** (N1, 18 stations; N2, 9 stations), while **Figure 7** displays the performance scores grouped by cluster.

Visual inspection of **Figure 7** reveals that the C1 stations showed the best performance in terms of R, ubRMSE, PBIAS, POD, and TS. The FAR values were similar in both clusters, indicating that CHIRPS tends to forecast false alarms in the

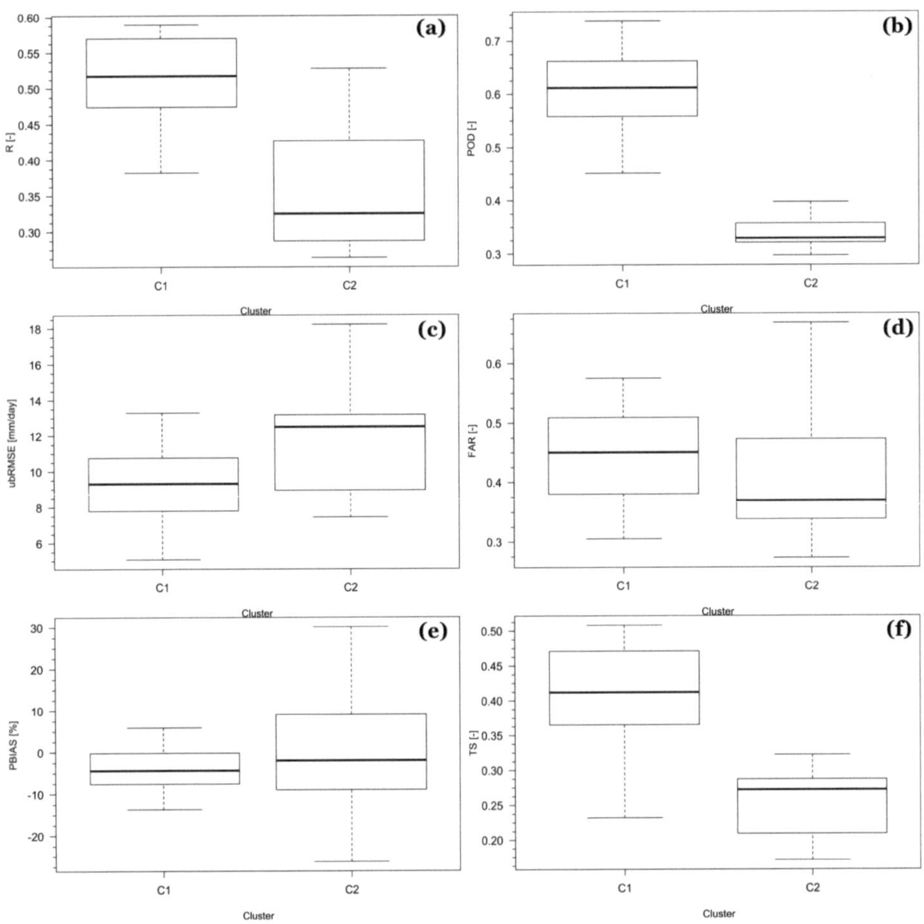

Figure 7.
Boxplots for (a) R (dimensionless); (b) POD (dimensionless); (c) ubRMSE (mm/day); (d) FAR (dimensionless); (e) PBIAS (%); and (f) TS (dimensionless) at annual time scale grouped by cluster, where the thick line depicts the median, while the other horizontal lines of the box depict the maximum, upper quartile, lower quartile, and minimum. For clarity the outliers were omitted.

entire NEB (i.e., CHIRPS estimates to occur a rainfall event, but did not occur), which is also evident in **Figure 4**. It is interesting to note that the C2 stations were mostly concentrated near the coast.

A more detailed comparison, considering the auxiliary data sets (see Section 2.3), showed that there were no significant differences between both clusters in terms of average annual precipitation and terrain elevation (test based on Wilcoxon's t-statistic at the 5% level was used). This means that these local factors did not affect the performance scores. However, regardless of the land cover, most of the C1 stations are located in open flatlands (i.e., terrain slope $< 7\%$) with tropical savanna climate (i.e., Aw), which seem to be favorable surface conditions for better performance of CHIRPS.

4. Discussion

Several performance scores were used to evaluate the CHIRPS rainfall product against gauge observations in Northeast Brazil during the period from January 1981 to June 2019. This region is characterized by large interannual rainfall variations and severe droughts [6, 15]. In line with previous studies [22–24], the CHIRPS data set captured relatively well the spatiotemporal pattern of rainfall across NEB, showing acceptable accuracies (see **Figures 3** and **4**), thanks to the blending process to merge the CHIRP data set derived from IR brightness temperature and TRMM, with ground-based observations [16].

CHIRPS exhibited poorer performance at daily time scale in terms of R (R median: 0.49) than that obtained with monthly time scale (R median: 0.94, reported by Paredes et al. [22]), indicating that increasing temporal aggregation leads to better agreement between CHIRPS and ground-based observations in NEB. This was expected because errors at daily scale time showed closely symmetric characteristics (see **Figure 5**); therefore, they tend to cancel each other during the temporal aggregation [32]. By contrast, this procedure did not provide a significant improvement on the performance in terms of PBIAS (PBIAS median: -4.10% and -3.58% [22] for daily and monthly time scales, respectively), likely due to its high variability at daily time scale (about 700%).

These first results are consistent with the previous findings in other regions with similar climatic features such as South Sudan [33], where CHIRPS became more accurate in terms of R and RMSE as the duration of the integration time increased from months to years. It is important to note, however, that this characteristic is not unique to CHIRPS. Most of the satellite-based rainfall products tend to improve their general performance as the aggregation period increases owing to the effect of cancelation of errors [34, 35].

Overall, CHIRPS showed the best (worst) performance with the (lowest) highest of R and POD and the (highest) lowest bias and FAR during the (driest) wettest months of the year (see **Figures 3** and **4**). This result is consistent with the findings of Paredes-Trejo et al. [24] and Nogueira et al. [23], who found that CHIRPS tends to overestimate low and underestimate high rainfall values in NEB. Likewise, it should be mentioned that the PBIAS and R values were highly sensitive to drought conditions, such as those observed from 2012 to 2015, where CHIRPS showed lower R values (about 0.20) and higher overestimation of the rainfall amount (see **Figure 5a** and **e**). The degradation of the performance under extreme droughts may be attributed to the evaporation processes of raindrops in the dry atmosphere before reaching the surface [20]. In this context, CHIRPS forecasts a rainfall event, but does not occur. According to the equations listed in **Table 2**, this phenomenon leads to higher PBIAS values and near-zero values for R, POD, and TS.

The sub-cloud evaporation plays an important role in the overestimation of rainfall occurrence over different semiarid and arid regions in the world [19, 32, 36]. Therefore, it can help to explain the poor performance of CHIRPS over the driest region of NEB (i.e., the Sertão region), especially in autumn and winter (see **Figures 3** and **4**) and during drought years induced by climate anomalies from the tropical Pacific Ocean (i.e., El Niño-Southern Oscillation) [37]. When this occurs, the air in the lower atmosphere is drier and hotter than usual conditions over the Sertão region [4]. Then, an intensification of the sub-cloud evaporation processes might be expected.

On a seasonal time scale, the reliability of the CHIRPS product was evident in reproducing the seasonal rainfall pattern with results comparable with the ones previously published by Melo et al. [30] for the TRMM 3B42V7 rainfall product, which is its parent rainfall product [16] (see Section 2.2). Similar to TRMM, it was found that CHIRPS exhibits poorer performance over those stations near the coast than the ones located in inland regions of NEB (see **Figures 6** and 7), particularly in winter (see **Figures 3** and **4**). The reason behind this can be attributed to the prevalence of warm-top stratiform cloud systems along the coastal region [38, 39]. Under these conditions, CHIRPS may not detect rainfall because the cloud tops tend to have a value warmer than the IRP CCD threshold value (i.e., 235 K) [19], leading to a large underestimation in the daily precipitation and poor detection of rainfall events.

As can be seen from **Figure 6**, the landscape at most of the stations is characterized by high topographic complexity, where warm-rain processes induced by orographic lifting are dominant [40, 41]. Similar to the warm-top stratiform cloud systems in the coastal areas mentioned above, CHIRPS has limitations in reproducing the orographic rainfall due to the adoption of a fixed IRP CCD threshold value (i.e., 235 K), leading to classify warm orographic clouds as nonprecipitating [19]. Even though orographic clouds are relatively warm, they can produce substantial amounts of rain [15].

Interestingly, although the number of stations used in the CHIRPS blending process as anchor stations showed a gradual temporal decrease in NEB during the period January 1981 until June 2019 (see https://bit.ly/2ZZFAvA), there was no statistically significant trend in their performance scores (see **Figure 5**). For this study, at least 12, 19, and 21 rain gauges not included as anchor stations for the calculation of CHIRPS rainfall estimations during 1981–1998, 1999–2013, and 2014–2019, respectively, were used. One implication of this situation is that it can be considered a relatively independent validation.

5. Conclusions

The synergetic use of ground-based rainfall observations and satellite-based rainfall estimates is of paramount importance in semiarid regions such as Northeast Brazil. CHIRPS is a state-of-the-art satellite rainfall data set characterized by its blending procedure using thermal infrared satellite observations, TRMM 3B42-based rainfall estimates, monthly precipitation climatology, and atmospheric model rainfall fields from NOAA CFS, with ground-based rainfall measurements [16]. This study set out with the aim of evaluating the performance of CHIRPS against ground-based observations in NEB. The analysis was performed on a pixel-to-station basis at daily time scale and during the period 1981–2019. The major novelty of this study with respect to previous studies [22, 23, 42] is the use of the newest in situ data from the INMET meteorological stations. The main conclusions reached are the following:

1. The CHIRPS rainfall data set exhibits better performance in inland regions with open flatlands than near the coast (see **Figures 6** and 7).

2. The accuracy of CHIRPS is better in the wettest months (i.e., summer) than in the driest months (i.e., winter) (see **Figures 3** and **4**). In general, CHIRPS underestimates (overestimates) high (low) rainfall amounts.

3. CHIRPS appears to be sensitive to the precipitation from the warm-top stratiform cloud systems (e.g., near to the coast), the warm-rain processes induced by orographic lifting (e.g., the mountain areas of NEB), and the sub-cloud evaporation processes (e.g., the Sertão region). The first and second are mainly attributed to a fixed IRP CCD threshold (i.e., 235 K) used by CHIRPS (see Section 2.2), which may be too cold for regions where the warm-rain processes are dominant [34], while the third is a usual phenomenon in semiarid regions [19].

Based on the abovementioned conclusions, CHIRPS can serve as an alternative source of data for operational applications that require rainfall data, especially over the inland regions of NEB (see the C1 stations in **Figure 6**), during the wettest months of the year (see **Figures 3** and **4**), and at monthly or annual time scales taking advantage of the cancelation of errors of CHIRPS rainfall estimates as the duration of the integration increases [34]. However, future investigations are needed to adequately choose the operational applications of CHIRPS for each sub-region of the NEB.

Acknowledgements

This work was funded by the Coordination for the Improvement of Higher Education Personnel (CAPES) and the National Council for Scientific and Technological Development (CNPq) (Grant no. 88887.091737/2014-01: Edital Pró-Alertas no 24/2014 under project Análise e Previsão dos Fenômenos Hidrometeorológicos Intensos do Leste do Nordeste Brasileiro). We acknowledge to the National Institute of Meteorology (INMET) and the University of California Santa Barbara's Climate Hazards Group (CHG) for providing data that made this study possible.

Conflict of interest

The authors declare no conflict of interest.

Author details

Franklin Paredes-Trejo[1], Humberto Alves Barbosa[2*], Tumuluru Venkata Lakshmi Kumar[3], Manoj Kumar Thakur[4] and Catarina de Oliveira Buriti[5]

1 University of the Western Plains Ezequiel Zamora, San Carlos, Venezuela

2 Laboratory for Analyzing and Processing Satellite Images, Federal University of Alagoas, Brazil

3 Atmospheric Science Research Laboratory, Department of Physics, SRM Institute of Science and Technology, Tamilnadu, India

4 Tribhuvan University, Kathmandu, Nepal

5 National Semi-Arid Institute (INSA), Ministry of Science, Technology, Innovations and Communications (MCTIC), Brazil

*Address all correspondence to: barbosa33@gmail.com

References

[1] Beck HE, Vergopolan N, Pan M, Levizzani V, Van Dijk AIJM, Weedon GP, et al. Global-scale evaluation of 22 precipitation datasets using gauge observations and hydrological modeling. Hydrology and Earth System Sciences. 2017;**21**(12): 6201-6217. DOI: 10.5194/hess-21-6201-2017

[2] Zambrano F, Wardlow B, Tadesse T, Lillo-Saavedra M, Lagos O. Evaluating satellite-derived long-term historical precipitation datasets for drought monitoring in Chile. Atmospheric Research. 2017;**186**:26-42. DOI: 10.1016/j.atmosres.2016.11.006

[3] Funk CC, Peterson PJ, Landsfeld MF, Pedreros DH, Verdin JP, Rowland JD, et al. A quasi-global precipitation time series for drought monitoring. US Geological Survey Data Series. 2014; **832**(4):1-12. DOI: 10.3133/ds832

[4] Marengo JA, Bernasconi M. Regional differences in aridity/drought conditions over Northeast Brazil: Present state and future projections. Climatic Change. 2015;**129**(1–2): 103-115. DOI: 10.1007/s10584-014-1310-1

[5] Brito SSB, Cunha APMA, Cunningham CC, Alvalá RC, Marengo JA, Carvalho MA. Frequency, duration and severity of drought in the Semiarid Northeast Brazil region. International Journal of Climatology. 2018;**38**(2):517-529. DOI: 10.1002/joc.5225

[6] Barbosa HA, Lakshmi Kumar TV. Influence of rainfall variability on the vegetation dynamics over Northeastern Brazil. Journal of Arid Environments. 2016;**124**:377-387. DOI: 10.1016/j.jaridenv.2015.08.015

[7] Michaelides S, Levizzani V, Anagnostou E, Bauer P, Kasparis T, Lane JE. Precipitation: Measurement, remote sensing, climatology and modeling. Atmospheric Research. 2009; **94**(4):512-533. DOI: 10.1016/j.atmosres.2009.08.017

[8] Bai L, Shi C, Li L, Yang Y, Wu J. Accuracy of CHIRPS satellite-rainfall products over mainland China. Remote Sensing. 2018;**10**(3):362. DOI: 10.3390/rs10030362

[9] Kidd C, Becker A, Huffman GJ, Muller CL, Joe P, Skofronick-Jackson G, et al. So, how much of the Earth's surface is covered by rain gauges? Bulletin of the American Meteorological Society. 2017;**98**(1):69-78. DOI: 10.1175/BAMS-D-14-00283.1

[10] Brocca L, Filippucci P, Hahn S, Ciabatta L, Massari C, Camici S, et al. SM2RAIN–ASCAT (2007–2018): Global daily satellite rainfall data from ASCAT soil moisture observations. Earth System Science Data. 2019;**11**(4):1583-1601. DOI: 10.5194/essd-11-1583-2019

[11] Raghavan S. Radar Meteorology. Dordrecht: Springer; 2013. p. 549. DOI: 10.1007/978-94-017-0201-0

[12] Maggioni V, Sapiano MRP, Adler RF. Estimating uncertainties in high-resolution satellite precipitation products: Systematic or random error? Journal of Hydrometeorology. 2016; **17**(4):1119-1129. DOI: 10.1175/JHM-D-15-0094.1

[13] Loew A, Bell W, Brocca L, Bulgin CE, Burdanowitz J, Calbet X, et al. Validation practices for satellite-based Earth observation data across communities. Reviews of Geophysics. 2017;**55**(3):779-817. DOI: 10.1002/2017RG000562

[14] Kumar TVL, Barbosa HA, Thakur MK, Paredes-Trejo F. Validation

of satellite (TMPA and IMERG) rainfall products with the IMD gridded data sets over monsoon core region of India. In: Rustamov RB, editor. Satellite Information Classification and Interpretation. Rijeka: IntechOpen; 2019. p. 13. DOI: 10.5772/intechopen.77202

[15] Correia Filho WLF, De Oliveira-Júnior JF, De Barros SD, De Bodas Terassi PM, Teodoro PE, De Gois G, et al. Rainfall variability in the Brazilian northeast biomes and their interactions with meteorological systems and ENSO via CHELSA product. Big Earth Data. 2019;**3**(4):315-337. DOI: 10.1080/20964471.2019.1692298

[16] Funk C, Peterson P, Landsfeld M, Pedreros D, Verdin J, Shukla S, et al. The climate hazards infrared precipitation with stations—A new environmental record for monitoring extremes. Scientific Data. 2015;**2**: 150066. DOI: 10.1038/sdata.2015.66

[17] Paredes Trejo FJ, Barbosa HA, Peñaloza-Murillo MA, Alejandra Moreno M, Farías A. Intercomparison of improved satellite rainfall estimation with CHIRPS gridded product and rain gauge data over Venezuela. Atmosfera. 2016;**29**(4):323-342. DOI: 10.20937/ATM.2016.29.04.04

[18] Prakash S. Performance assessment of CHIRPS, MSWEP, SM2RAIN-CCI, and TMPA precipitation products across India. Journal of Hydrology. 2019;**571**: 50-59. DOI: 10.1016/j.jhydrol.2019.01.036

[19] Dinku T, Funk C, Peterson P, Maidment R, Tadesse T, Gadain H, et al. Validation of the CHIRPS satellite rainfall estimates over Eastern Africa. Quarterly Journal of the Royal Meteorological Society. 2018;**144**(S1): 292-312. DOI: 10.1002/qj.3244

[20] Rivera JA, Marianetti G, Hinrichs S. Validation of CHIRPS precipitation dataset along the Central Andes of Argentina. Atmospheric Research. 2018; **213**:437-449. DOI: 10.1016/j.atmosres.2018.06.023

[21] Zambrano-Bigiarini M, Nauditt A, Birkel C, Verbist K, Ribbe L. Temporal and spatial evaluation of satellite-based rainfall estimates across the complex topographical and climatic gradients of Chile. Hydrology and Earth System Sciences. 2017;**21**:1295-1320. DOI: 10.5194/hess-21-1295-2017

[22] Paredes F, Barbosa HA, Lakshmi-Kumar T. Validating CHIRPS-based satellite precipitation estimates in Northeast Brazil. Journal of Arid Environments. 2016;**139**:26-40. DOI: 10.1016/j.jaridenv.2016.12.009

[23] Nogueira SMC, Moreira MA, Volpato MML. Evaluating precipitation estimates from Eta, TRMM and CHRIPS data in the south-southeast region of Minas Gerais state-Brazil. Remote Sensing. 2018;**10**(2):313. DOI: 10.3390/rs10020313

[24] Paredes-Trejo F, Barbosa H, Rossato L. Assessment of SM2RAIN-derived and state-of-the-art satellite rainfall products over Northeastern Brazil. Remote Sensing. 2018;**10**(7): 1093. DOI: 10.3390/rs10071093

[25] Instituto Brasileiro de Geografia e Estatística. 2010 Census (Censo 2010). [Online] IBGE. Available from: https://bit.ly/2Nufhsu [Accessed: 07 March 2019]

[26] Junquas C, Li L, Vera CS, Le Treut H, Takahashi K. Influence of South America orography on summertime precipitation in Southeastern South America. Climate Dynamics. 2016;**46**(11–12):3941-3963. DOI: 10.1007/s00382-015-2814-8

[27] Molion LCB, Bernardo S. de O. Uma revisão da dinâmica das chuvas no nordeste brasileiro. Revista Brasileira de Meteorologia. 2002;**17**(1):1-10

[28] Houghton RA, Bontemps S, Peng S, Lamarche C, Li W, MacBean N, et al. Gross and net land cover changes based on plant functional types derived from the annual ESA CCI land cover maps. Earth System Science Data Discussions. 2017;**10**(1):1-23. DOI: 10.5194/essd-2017-74

[29] Beck HE, Zimmermann NE, McVicar TR, Vergopolan N, Berg A, Wood EF. Present and future Köppen-Geiger climate classification maps at 1-km resolution. Scientific Data. 2018;**5**: 180214. DOI: 10.1038/sdata.2018.214

[30] Melo D d CD, Xavier AC, Bianchi T, Oliveira PTS, Scanlon BR, Lucas MC, et al. Performance evaluation of rainfall estimates by TRMM multi-satellite precipitation analysis 3B42V6 and V7 over Brazil. Journal of Geophysical Research. 2015;**120**(18):9426-9436. DOI: 10.1002/2015JD023797

[31] Arora P, Varshney S, et al. Analysis of k-means and k-medoids algorithm for big data. Procedia Computer Science. 2016;**78**:507-512. DOI: 10.1016/j.procs.2016.02.095

[32] Ayehu GT, Tadesse T, Gessesse B, Dinku T. Validation of new satellite rainfall products over the Upper Blue Nile Basin, Ethiopia. Atmospheric Measurement Techniques Discussions. 2017;**11**(November):1-24. DOI: 10.5194/amt-2017-294

[33] Basheer M, Elagib NA. Performance of satellite-based and GPCC 7.0 rainfall products in an extremely data-scarce country in the Nile Basin. Atmospheric Research. 2019;**215**:128-140. DOI: 10.1016/j.atmosres.2018.08.028

[34] Belay AS, Fenta AA, Yenehun A, Nigate F, Tilahun SA, Moges MM, et al. Evaluation and application of multi-source satellite rainfall product CHIRPS to assess spatio-temporal rainfall variability on data-sparse western margins of Ethiopian highlands. Remote Sensing. 2019;**11**(22):1-22. DOI: 10.3390/rs11222688

[35] Dembélé M, Zwart SJ. Evaluation and comparison of satellite-based rainfall products in Burkina Faso, West Africa. International Journal of Remote Sensing. 2016;**37**(17):3995-4014. DOI: 10.1080/01431161.2016.1207258

[36] Saeidizand R, Sabetghadam S, Tarnavsky E, Pierleoni A. Evaluation of CHIRPS rainfall estimates over Iran. Quarterly Journal of the Royal Meteorological Society. 2018;**144**: 282-291. DOI: 10.1002/qj.3342

[37] Marengo JA, Torres RR, Alves LM. Drought in Northeast Brazil—Past, present, and future. Theoretical and Applied Climatology. 2017;**129**(3–4): 1189-1200. DOI: 10.1007/s00704-016-1840-8

[38] Rozante J, Vila D, Barboza Chiquetto J, Fernandes A, Souza AD. Evaluation of TRMM/GPM blended daily products over Brazil. Remote Sensing. 2018;**10**(6):882. DOI: 10.3390/rs10060882

[39] Fedorova N, Levit V, Fedorov D. Fog and stratus formation on the coast of Brazil. Atmospheric Research. 2008; **87**(3–4):268-278. DOI: 10.1016/j.atmosres.2007.11.008

[40] Anders AM, Nesbitt SW. Altitudinal precipitation gradients in the tropics from Tropical Rainfall Measuring Mission (TRMM) precipitation radar. Journal of Hydrometeorology. 2015;**16**(1): 441-448. DOI: 10.1175/JHM-D-14-0178.1

[41] Giovannettone JP, Barros AP. Probing regional orographic controls of precipitation and cloudiness in the central Andes using satellite data. Journal of Hydrometeorology. 2009; **10**(1):167-182. DOI: 10.1175/2008JHM973.1

[42] Paredes-Trejo F, Barbosa H, dos Santos CAC, Paredes-Trejo F, Barbosa H, dos Santos CAC. Evaluation of the performance of SM2RAIN-derived rainfall products over Brazil. Remote Sensing. 2019;**11**(9):1113. DOI: 10.3390/RS11091113

Chapter 9

Improved Narrow Water Extraction Using a Morphological Linear Enhancement Technique

Wu Bo, Zhang Jinmu and Zhao Yindi

Abstract

An improved water extraction method using a morphological linear enhancement technique is proposed to improve the delineation of narrow water features for the modified normalized difference water index (MNDWI) derived from remote sensing images. This method introduces a morphological white top-hat (WTH) transforming operation on the MNDWI to extract multi-scale and multidirectional differential morphological profiles and constructs a morphological narrow water index (MNWI). The MNWI can effectively enhance the local contrast of linear objects, allowing narrow water bodies to be easily separated from mountain shadows and other features. Furthermore, to accurately delineate surface water bodies, a dual-threshold segmentation method was also developed by combining an empirical threshold segmentation with the MNDWI for wide water bodies and an automatic threshold segmentation with the MNWI for narrow water bodies. This method was validated using three experimental datasets, which were taken from two different Landsat images. Our results demonstrate that narrow water bodies can be sufficiently identified, with an overall accuracy of over 90%. Most narrow streams or rivers keep a continuous shape in space, and the boundaries of the water bodies are accurately delineated as compared with the MNDWI method. Finally, the proposed method was used to extract the entire inland surface water of Fujian province, China.

Keywords: narrow water extraction, white top-hat transform, MNWI, dual-threshold segmentation

1. Introduction

Surface water is one the most vital earth resources undergoing changes in time and space as a consequence of land use/cover (LULC) changes, climate change, and other forms of environmental changes in many parts of the world. Timely and accurate monitoring and delivery of data of the dynamics of surface water are, therefore, critically important in various scientific disciplines, such as the assessment of present and future water resources, climate models, agricultural suitability, river dynamics, wetland inventory, watershed analysis, surface water surveys, and environmental monitoring [1].

Remote sensing at different spatial, spectral, and temporal resolutions provides an enormous amount of data for mapping water resources and its dynamics at local to global scales. At a result, it has become a routine approach for the monitoring of land surface water bodies, since the acquired data can provide macroscopic, real-time, dynamic, and cost-effective information, which is substantially different from conventional in situ measurements. Various approaches for water body extraction from multispectral images have been developed in the past decades [2–4], which can be broadly grouped into three categories: spectral band segmentation, image supervised classification, and water indices. Among all these methods, of particular interest is the spectral water index-based method, as it is a reliable and cost-effective method. This type of method takes advantage of reflectivity differences of each involved band for water body extraction based on the analysis of signature differences between water and other surfaces.

One of the most widely used indexes is the normalized difference water index (NDWI) [4], which utilizes the green (band 2) and near-infrared (band 4) of Landsat TM to delineate open water features. However, Xu found that the NDWI cannot efficiently suppress the signal from built-up surfaces and therefore proposed an improved one, called modified normalized difference water index (MNDWI) [5], where the NDWI was modified by replacing band 4 with band 5 of Landsat TM/ETM. The MNDWI has been validated as one of the most widely used water indices for various applications, though it is still difficult to obtain a high accuracy of water extraction in complex circumstances. Carleer and Wolff [6], among others, have found that the land cover classifications of water and shadow can often be confused. This issue often arises in environments where a large amount of shadow and water regions exist, such as urban and mountainous landscapes. The identification of narrow water bodies (such as narrow streams, canals, ponds, small reservoirs, etc.) can be a difficult task when using NDWI or MNDWI images, because the shallow and narrow water pixels may generate unstable spectral profiles or characteristics, due to the mixed reflectance caused by sediment and/or adjacent land covers. Narrow water is typically defined as a water body with an apparent width less than or equal to three pixels in an image. Therefore, narrow water features often contain mixed pixels, and the extraction of them from NDWI or MNDWI images generally exhibits a discontinuous shape in space.

To remedy this problem, past studies have attempted to identify narrow water features by combining different procedures. Yang et al. proposed a method of extracting initial water information via a user-defined specified water index, and then they performed a series of operations. These operations include morphological dilation, image filtering, and thinning techniques applied to the water index image to recover the continuity of narrow rivers or streams [7]. This method can be effective in the extraction of narrow water bodies if the water disruption is short; however, it may increase false water identification when water disruption is large, since it is dependent on the morphological dilation operation to reconnect the narrow rivers. Such simple threshold techniques are not often a sufficient solution to identify narrow water bodies; therefore, Li et al. suggested an object-oriented method of small water body extraction [8]. They first extracted textural and shape-related features from images as supplementary information to spectral bands and then performed a segmentation operation on the images using an optimal scale to identify the potential water bodies. Yet, their method is not an automatic process, since it involves multiple user-defined parameters in image segmentation, which prohibits its use in large areas. An alternative approach was performed by Jiang et al. who extracted narrow water features via the enhancement of linear features in NDWI images [9]. However, their procedure involves multiple empirical thresholds, so it is not a cost-effective method for water feature extraction on large scale.

Attempting to improve on these previous approaches, here we propose an automatic water extraction method that constructs a novel narrow water index, denoted as morphological narrow water index (MNWI). The MNWI is constructed using multi-scale and multidirectional differential morphological profiles (DMPs) on a MNDWI image, and then water bodies are automatically extracted using a dual-threshold segmentation. The successful use of DMPs to extract various thematic information from images has been sufficiently validated, such as buildings in urban areas [10, 11], rare earth mining areas [12], and mapping of mangrove forests [13]. In this paper, we introduce a DMP technique to highlight the contrast of bright features as a way of narrow water recognition in the MNDWI images. This can be accomplished because water pixels have higher value than surrounding pixels in MNDWI images. Our approach is expected to improve the ability of narrow water feature identification by enhancing its spatially implicit characteristics using multi-scale morphological features, e.g., other land cover uses that have similar values in a MNDWI image, such as bare patches and shadows, can be easily identified. Our approach also involves a dual-threshold strategy which is adopted for wide and narrow water body extraction, since a simple threshold is not often an adequate solution [14]. An empirical threshold is used first to obtain possible water areas from a MNDWI image, followed by an automatic threshold which is determined by the maximum interclass variance criterion [15] used for extracting narrow water features from a MNWI image. Finally, a logical operation is performed by combining the two potential water features to identify the true water body boundary.

The remainder of this chapter is organized as follows. In Section 2, we describe the MNWI method of narrow water extraction. Experimental results are shown in Section 3 using multiple experiments on TM images, and we use the MNWI method in a practical application for extracting inland water features in Fujian province, China, in Section 4. Finally, Section 5 presents conclusions.

2. The proposed MNWI method

It is well accepted that open and wide water features can be easily separated from other land cover features by using the NDWI or the MNDWI methods, but extracting narrow water boundaries is generally a more difficult task due to its being confused with built-up areas, roads, hill shadows, etc. Therefore, the goal of our proposed MNWI method is the separation of narrow water and other land cover features by depicting the implicit spectral and structural characteristics of MNDWI. Narrow water usually exhibits strong linear shapes and continuous spatial curves; therefore, we propose a linear enhancement operation on a MNDWI image to form a MNWI image, according to the following steps:

Step 1: Generation of a MNWDI image. Level 1 T Landsat images in the study region are first collected, which are then corrected geometrically. Atmospheric and radiometric corrections were then applied using the 6S approach to transform the images into reflectance datasets. Because MNWDI performs better in the extraction of water features than NDWI [5], it is selected for initial water extraction and calculated as:

$$MNDWI = \frac{(Green - MIR)}{(Green + MIR)} \tag{1}$$

where Green and MIR are the image reflectance of the green band and medium-wave infrared band (which correspond to the TM/ETM+ band 5), respectively.

Step 2: Formulation of the MNWI. MNDWI can improve the local contrast between water and other land cover features, since most water bodies can easily be extracted using a threshold segmentation method. However, narrow water bodies are still difficult to extract, as they are easily confused with urban areas, roads, and mountain shadows because of mixed pixels. To alleviate this problem, we adobe a linear object enhancement technique using a white top-hat transform operation on a MNDWI image via the extraction of multi-scale and multidirectional differential morphological profiles to form a MNWI image, according to the following three sub-procedures:

1. **Define linear structures.** A narrow river or stream has clear linear features with two main directions, but the shapes of buildings and mountain shadows have polygon-like features. Hence, we introduce a DMP method to separate them. The use of DMPs involves the designing of a filtering operator (e.g., size and shape), known as a structural element (SE). This acts as a probe to extract or suppress specific structures by checking that each part of the SE fits within the objects in the image. A single-SE size approach is typically not suitable for complex structures; therefore, a series of linear structural elements are implemented to form DMPs, so that the size and directional bias of the narrow water features are identified clearly. Thus, the linear structure element was defined as se = strelem(d, s), where d represents the orientation of the linear structure (e.g., 0°, 45°, 90°, 135°) and s denotes the scale of the small water body.

2. **White-hat morphological reconstruction (white top-hat).** In general, an opening/closing operator can isolate bright or dark structures in an image when the objects are brighter or darker than the surrounding features. An opening operator can help separate water objects from other land cover features since water appears brighter in a MNDWI image. To isolate features that have a thinner support than a given SE, a common practice to use is a top-hat morphological transform in taking the residual of the opening, closing, and original images to ensure a better shape preservation [11–13]. The white top-hat reconstruction operation is formulated by subtracting the opening operation from the initial image using the same image. This enhances linear features with a structure smaller than the SE, and a morphological reconstruction of the white-hat MNDWI image is according to:

$$WTH(d,s) = MNDWI - \gamma_{MNDWI}(se) \tag{2}$$

where $\gamma_{MNDWI}(se)$ is the output image yielded by performing a closing operator to a MNDWI image. A closing operator can suppress smaller, darker objects and join adjacent objects together; thus we expect that small water bodies smaller than the structural elements will be highlighted after this reconstruction, and open water bodies and background objects larger than the structural elements should be suppressed. We previously defined a narrow water body as no larger than three pixels; thus, the linear structural elements in our algorithm are extended in four directions (0°, 45°, 90°, and 135°), and each direction was applied with three scales ($s_{min} = 1$, $s_{max} = 3$, $\Delta_s = 1$) to generate the WTH image.

3. **Construction of the morphological narrow water index.** The WTH process essentially suppresses the nonrelevant background, though there still may be small features (e.g., buildings or shadows) that must be removed. Narrow water features have linear features with two main orientations, and the difference between maximum and minimum values of the WTH value in different directions is relatively large. Conversely, the shapes of buildings and mountain shadows usually have polygon-like features, indicating that the

WTH difference is small. With this consideration in mind, we determine the contrast of the maximum and minimum values of the WTH in all directions to enhance the linear structure further. Thus, the MNWI is formulated as:

$$MNWI = WTH_{\mathrm{Max}} - WTH_{\mathrm{Min}} \tag{3}$$

where $\mathrm{WTH}_{\mathrm{Max}}$ and $\mathrm{WTH}_{\mathrm{Min}}$ are the maximum and the minimum values of the reconstruction of the white top-hat procedure in all directions, respectively.

Step 3: **Dual-threshold segmentation**. A simple threshold is not usually adequate to separate water features in large and complex regions from a MNDWI or MNWI image. Therefore, we employ a dual-threshold strategy to delineate water feature characteristics. A relatively large threshold was first determined empirically to separate the MNDWI image into possible water regions, denoted as W_1. Experimental results suggest that this threshold should be set at 0.2, such that all wide water bodies were extracted, and most other objects were excluded. However, many narrow water bodies may be missed in this procedure, and small rivers may be of a discontinuous shape in space. Thus, another threshold is determined by the Otsu method [15] performed on the MNWI image to extract possible narrow water bodies, denoted as W_2. The final water feature information is then delineated according to the following logical rule: If the possible water in W_2 is connected to any wide water extracted from the MNDWI image (i.e., W_1), it is then determined to be narrow water; otherwise it is categorized as not water.

Step 4: **Image post-processing**. Small trails might also display a high value in a MNWI image and may be misclassified as a narrow water body. To reduce this error, the difference built-up index (NDBI) [16] can also be used to refine the final result. In the present study, a threshold of NDBI >0.05 is used, so that most of the roads and trails are excluded from the final water determinations.

3. Method validations

To evaluate our method, three study regions are taken from two Landsat ETM+/OLI images with different water body types and different terrains, as shown in **Figure 1**. Study area 1 is a sub-scene image with 1000 by 1000 pixel size chosen from a Landsat 7 ETM+ image acquired on September 1, 2001, centered on the Panjiakou Reservoir near Tangshan and Chengde cities in Hebei province in northern China, which covers multiple branches of Luanhe River.

Study area 2 is also a sub-image of 1000 by 1000 pixels, which is located in Luoyuan County, Fujian province, which is acquired from the Landsat 8 operational Landsat image (OLI) on December 13, 2014. This region contains very narrow streams (**Figure 2b**), and it is used to test the extraction ability from mixed pixels. Study area 3 (**Figure 3c**) is a 1000 by 1000 pixel scene selected from the same OLI image as #2, which is located in southern Youxi County, Fujian province, and covers one of main branches of Minjiang River as well as other narrow streams (Qingyin, Qing, Wenjiang, etc.). Youxi County and several villages are in this study area, which contain multiple sources of possible background noise.

Actual water feature information for these study areas are not available, so the water bodies in the three images are manually digitized using high resolution spatial images to provide a basis map for comparison. High-resolution Google Earth™ images were also used as a complementary reference to assist in distinguishing water pixels that might be confused with background noise, such as mountain shadows, trails and built-up areas.

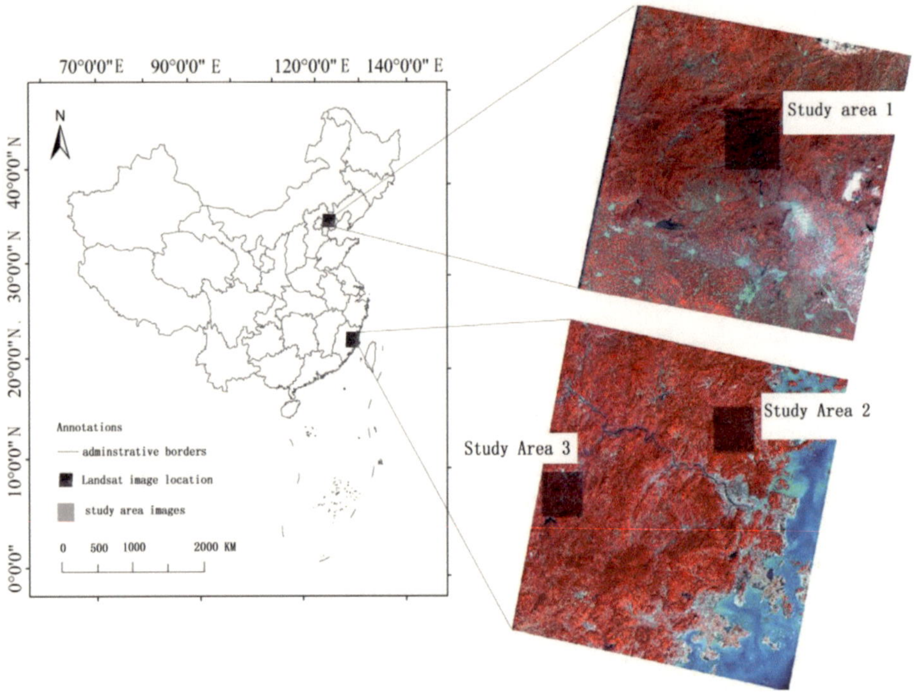

Figure 1.
The collected images and locations for the study areas.

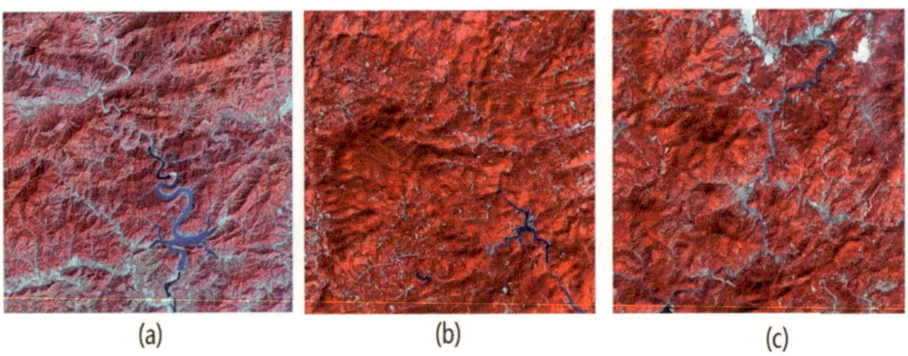

Figure 2.
The false-colored images for the three study areas used in our experiments. (a) study area 1, (b) study area 2, (c) study area 3.

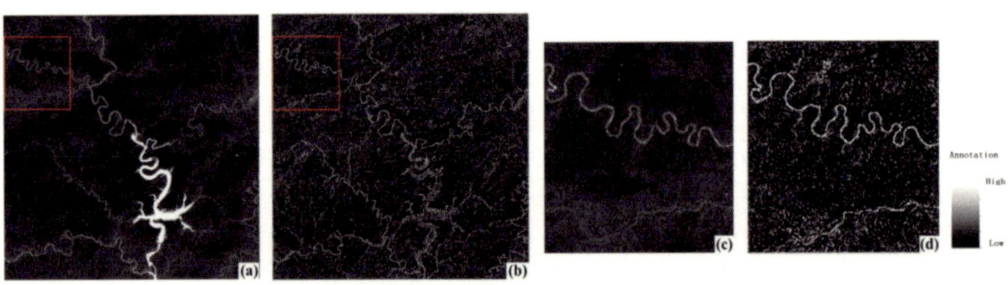

Figure 3.
Comparison of small bodies of water in the MNDWI and the MNWI for study area 1 where (a) and (b) are the extracted water bodies with the MNDWI and the MNWI indexes, respectively, and (c) and (d) denote the focused areas highlighted in red boxes.

3.1 Validation of MNWI features

A comparison between MNDWI and MNWI methods was first described. The two types of water index images for study area 1 are shown in **Figure 3a** and **b**, with focused areas highlighted in red boxes shown in **Figure 3c** and **d**. The pixels' values of water typically have higher values (white areas) than those of the other land cover features in these images, since both the MNDWI and the MNWI water indexes strongly enhance the water body signals. It is also observed in **Figure 3a** and **c** that the brightness difference between wide water and narrow water in the MNDWI image is larger, suggesting that a threshold segmentation on the MNDWI image cannot resolve entire water bodies. In contrast, the difference in the MNWI image is significantly reduced, though they maintain relatively higher values than other land cover features, as shown in **Figure 3b** and **d**. By looking at **Figure 3d**, it is seen that the local contrast between narrow water and other land cover features is significantly enhanced. Additionally, narrow water such as small rivers and branches maintains a continuous spatial shape, suggesting that narrow water features can be accurately extracted with a threshold implementation.

Furthermore, 1200 samples of typical land cover types from study area 3 were randomly selected from each category and were analyzed by calculating the maximal value, the minimal value, the mean value, and the deviation. The criteria for the sample selection are the following: (1) Each land cover has ~200 samples to keep a sample balance; and (2) each land cover contains several small patches from different locations to maintain a spectral variety. **Figure 4** reports the spatial distribution of the samples for study area 3 and their statistical information for six typical land cover types, i.e., wide open water, narrow water, vegetation, built-up area, roads, and shadow. As can be seen in **Figure 4**, the values of the wide water are very high for MNDWI. However, it is difficult to discriminate narrow water from other land cover types, especially for shadows, roads, and built-up areas. In contrast, the values of the narrow water in the MNWI method are relatively high compared with

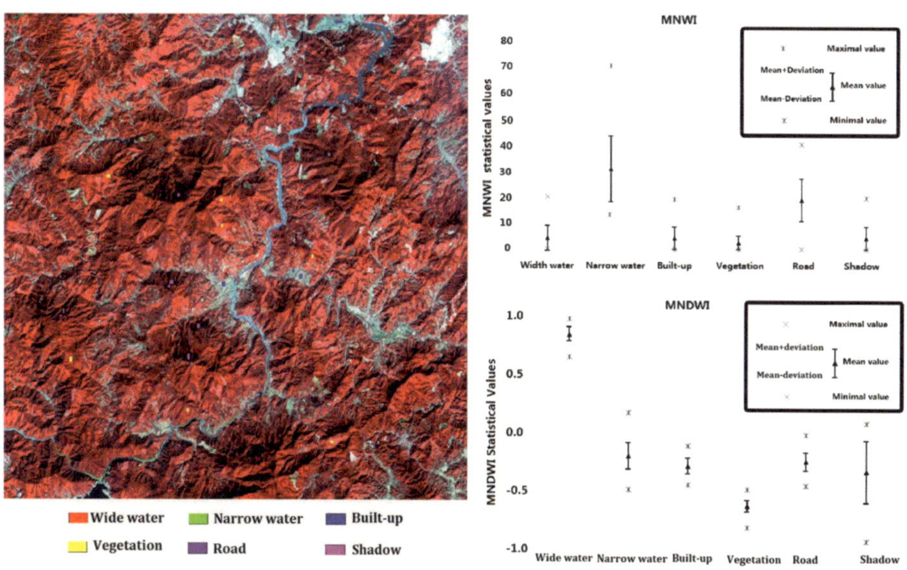

Figure 4.
The spatial distribution of 1200 selected samples for study area 3 (left) and the statistical information of six typical land cover types for MNDWI and MNWI images, respectively (right).

other land cover types, indicating that it is relatively easy to separate the narrow water bodies from other land cover types.

It can also be seen in **Figure 5** that the values of objects with polygon-like shapes, such as wide water, built-up areas, and shadows, are heavily suppressed in the MNWI image since they do not exhibit a linear structure. However, roads also show linear structural characteristics; thus they have high values in a MNWI image. This demonstrates that neither MNDWI nor MNWI can effectively identify entire water bodies using a threshold segmentation. To remedy this, we adopted a dual-threshold strategy. The first threshold is used for wide water extraction from MNDWI, and the second is employed to extract narrow water features from MNWI.

3.2 Validation of dual-threshold segmentation

An empirical threshold (0.2) was first used to perform a rough extraction of potential water features from a MNDWI image; then a second threshold determined by Otsu was determined from the MNWI image for possible narrow water features. Next, a combing procedure was carried out to extract entire water bodies using an "if-then" logic calculation according to the following rule: If two potential water regions are spatially connected, then they are determined to be water bodies; otherwise they are determined to be other land cover types.

As a demonstration of the dual-threshold segmentation method, comparisons between single threshold segmentation of a MNDWI image and dual-threshold segmentation of a MNWI image are shown in **Figure 6(a)–(c)**. It is seen that when a smaller threshold (T = 20) is adopted, a narrow stream keeps a relatively complete spatial shape, yet it also contains a trail (road) at the bottom of the image (**Figure 6a**). However, if we increase the threshold T to 40, this trail is no longer extracted, but the stream exhibits discontinuities along the stream. Thus, the use of a dual-threshold segmentation can better identify this stream as continuous and avoid the identification of the trail.

3.3 Visual assessment

Figure 7 presents the extracted water features using our proposed method for each of the three study areas. As a comparison, water information derived from the MNDWI image using an optimal threshold segmentation is also listed. For clarity, the corrected, misclassified, and omitted water information is labeled with different color schemes. Visual inspection shows that our method significantly outperforms the MNDWI method when using an optimal threshold segmentation. It can be seen that the majority of narrow rivers in each of the study areas are successfully extracted by our method. Six branches of the Luanhe River are

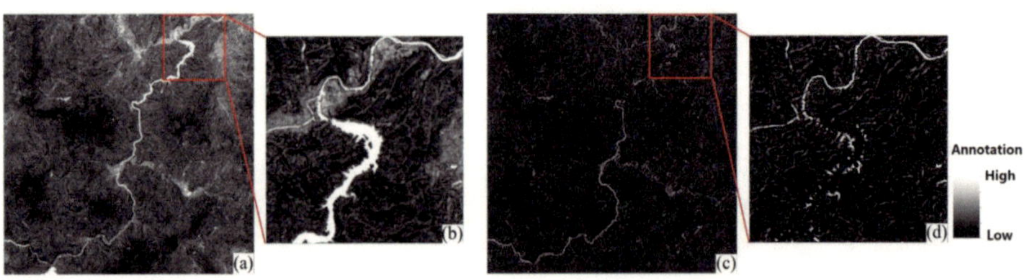

Figure 5.
Illustration of the linear enhancement and polygon compression of different land cover types in study area 3, where (a) and (c) are the MNDWI and the MNWI features, respectively, and (b) and (d) are the corresponding focused areas.

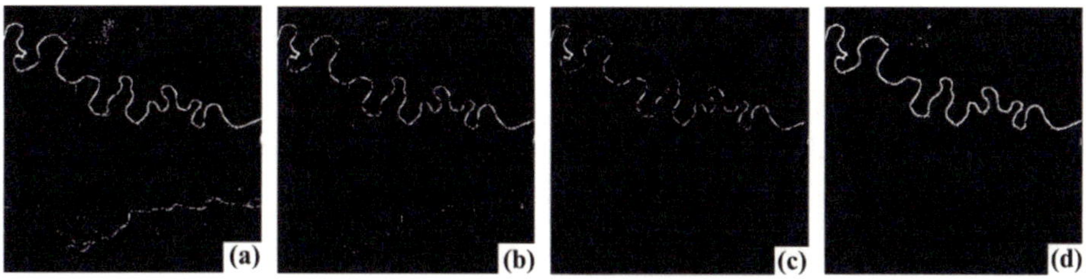

Figure 6.
Segmentation performed on the focused region of study area 1 using different thresholds, where (a), (b), and (c) are obtained by a threshold T equal to 20, 30, and 40, respectively, and (d) is the same image using a dual-threshold segmentation.

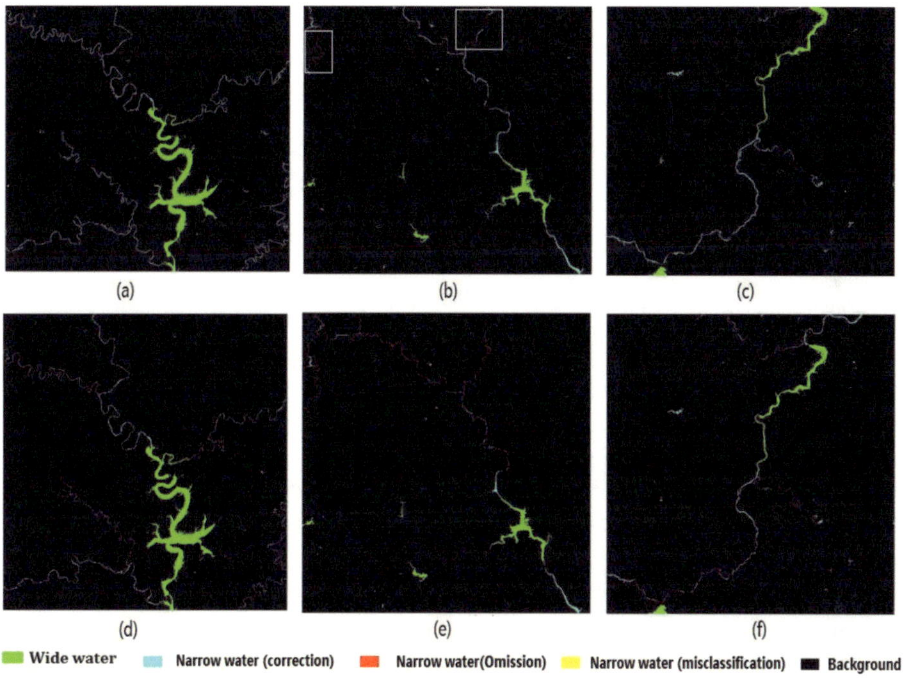

Figure 7.
Water extraction results from the three study areas, where the first and second lines are the results extracted by our method and the optimal threshold segmentation method, respectively.

clearly delineated in study area 1, and most narrow streams in study areas 2 and 3 are clearly extracted. However, a few narrow tributaries were misclassified by the MNDWI method, highlighted in red in **Figure 7**. A closer inspection of study area 2 shows that there are still two omissions which are highlighted with white rectangles, due to the width of two streams being too narrow (less than 10 m) to occupy a footprint, and the reflectance of these pixels are strongly mixed with other land cover types. Conversely, the results derived from the MNDWI image are less effective, as only small portions of the narrow rivers were extracted correctly. This is especially true for the narrow streams in the top region of the image, as most of them are ignored.

Another misclassification issue is the delineation of the sides of rivers, due to mixed pixel effects. Another experiment in the study area 3 was conducted to demonstrate this. These results are reported in **Figure 8**, where the water information that was corrected, omitted, and misclassified is shown in cyan, magenta, and red, respectively. It can be found that the boundary of the Youxi River can be accurately

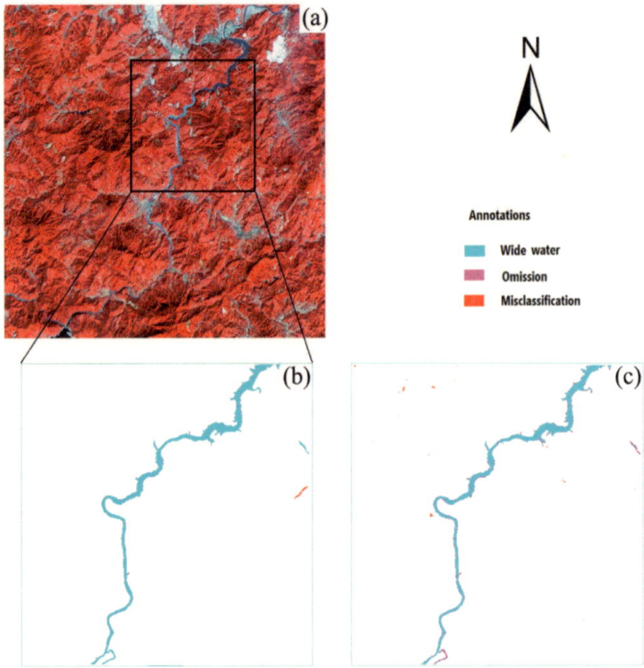

Figure 8.
The results of the river side misclassification experiment performed on study area 3. (a) the original image, (b) and (c) are the extracted water features in the focused area using out-proposed and the OT method, respectively.

extracted with the proposed method as shown in **Figure 8b**. However, using the MNDWI method with an optimal threshold, the omitted water pixels were along both sides of the river boundary (**Figure 8c**).

3.4 Quantitative evaluation

We now quantitatively evaluate our extracted results. Four measurements are used for comparison; the user and producer accuracy, the kappa coefficient, and the overall accuracy. Additionally, two recently developed methods for narrow water extraction, i.e., the method developed by Yang et al. [7] and the linear feature enhancement (LFE) developed by Jiang et al. [9], are also included for comparison. **Table 1** gives a pixel-by-pixel analysis of the classification accuracies for all datasets, with the best results highlighted in bold.

It can be seen in **Table 1** that the optimal threshold segmentation method was the least accurate, as it failed nearly completely in study area 2 with a 35% product accuracy and a kappa coefficient of 0.477. The method of Yang has some similar effects in narrow water extraction to our datasets, especially for mixed water features. The method of Jiang significantly improves the accuracy of narrow water extraction as compared to the optimal threshold segmentation method, as most of the narrow streams are well extracted in each of the study areas and had a comparable accuracy to our method. However, it should also be noted that this method is not an automatic method, since many parameters need to be tuned, which prevents it from being effectively used in larger areas. Overall, our method outperforms all others in terms of measurements except for producer accuracy in study area 3, where Yang's method achieves a relatively higher accuracy.

Methods		Optimal threshold	Yang	Jiang	Ours
Study area 1	Producer accuracy	53.8%	75.1%	92.1%	**92.3%**
	User accuracy	79.8%	85.3%	92.8%	**95.6%**
	Overall accuracy	67.3%	79.5%	91.3%	**93.6%**
	Kappa	0.631	0.748	0.907	**0.924**
Study area 2	Producer accuracy	35.2%	43.6%	82.8%	**83.9%**
	User accuracy	83.9%	57.6%	99.1%	**99.6%**
	Overall accuracy	55.4%	48.5%	89.2%	**90.7%**
	Kappa	0.477	0.453	0.872	**0.885**
	Producer accuracy	56.3%	55.5%	**87.1%**	86.9%
Study area 3	User accuracy	87.9%	73.8%	96.6%	**98.9%**
	Overall accuracy	71.2%	65.8%	90.3%	**91.6%**
	Kappa	0.696	0.627	0.881	**0.905**

Table 1.
Comparison of accuracies for different water extraction methods, with the best results given in bold text.

4. Extraction of inland water of Fujian province

The aforementioned experiments demonstrate that the MNWI is the most efficient algorithm for narrow water extraction, but it is still interest to address whether it is applicable to large-volume data in an actual scenario. Therefore, we apply our method to extract inland water features in Fujian province, China, which is a relatively large region that covers an area of about 121,000 km^2. Fujian is a mountainous province, located on the southeast coast of China and facing Taiwan across the Taiwan Strait. It has significant vegetation cover because of high mean precipitation and warm annual temperatures. Topographically, Fujian is a very mountainous region, having abundant water resources, rivers, lakes, and reservoirs. Many rivers run through these mountains, of which the most important is the Min River, as its drainage area covers over 50% of the province. The upstream Jin River, Futun River, and Shaowu River all converge into the Min River. The Jiulong River flows south of the Min River, reaching the sea at Xiamen city, and the Ting River runs across Fujian's southwestern border. It is thus an appropriate region to test our methods in a large area. To cover the entirety of Fujian province, 13 Landsat 8 OLI images were collected. Fujian is usually cloudy and rainy in spring and summer, so we collected all the images in winter to avoid cloud interference. The acquisition information is summarized in **Table 2**. Note that the quality of all acquired data is relatively good, with cloud cover less than 10%.

Path/row	118/041	118/042	119/041	119/042	119/043	120/040	120/041
Acquisition time	November 17	December 3	October 23	October 23	October 23	December 1	December 1
Path/row	120/042	120/043	120/044	121/041	121/042	121/043	
Acquisition time	December 1	December 1	December 1	October 5	October 5	October 5	

Table 2.
The information of collected Landsat 8 OLI images covering Fujian province in 2013.

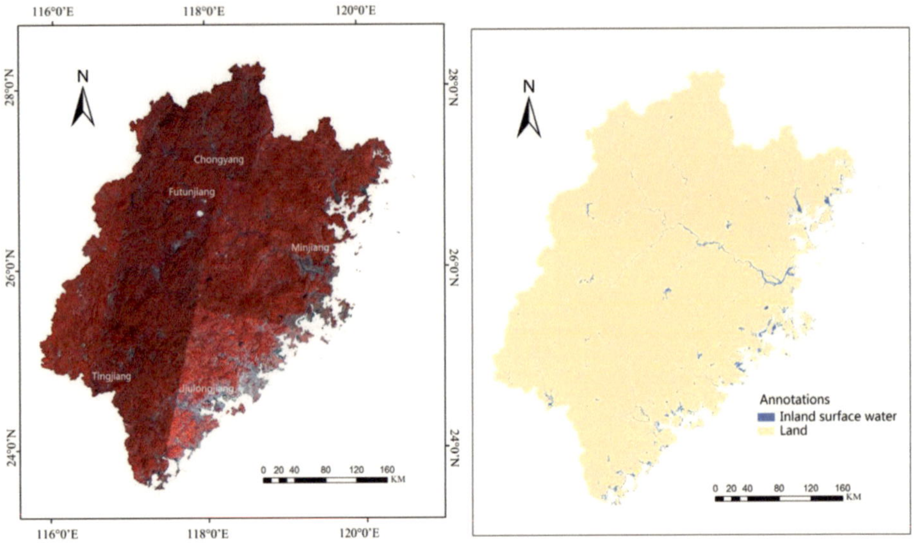

Figure 9.
The mosaic image and the final result of inland water bodies in Fujian province, China.

All the images were matched and stitched without altering their spectral color (**Figure 9**, left), and the final inland water information for the Fujian province (**Figure 9**, right) shows that 2,494,988 pixels were classified into inland surface water. It can be calculated that the total inland water area of Fujian province is about 2245.49 km^2 in the winter of 2013. Visually, our method can extract the most of perceptible water bodies with a high accuracy, where the main rivers, such as Min River, Jiurong River, Ting River, etc., are all correctly delineated with clear river boundaries. Moreover, most of the spatial shapes of small tributaries were continuous and so were the lakes and reservoirs. To evaluate the quantitative classification accuracy, 6800 samples of water and non-water features were randomly selected. The water samples included 3340 pixels, of which wide water to narrow water ratio was about 2:1, as the non-water samples were 3460 pixels, chosen from possibly confused land cover types such as forest, built-up areas, and rare soil. The producer accuracy, user accuracy, overall accuracy, and kappa coefficient calculated from these samples are 94.33%, 98.7%, 96.61%, and 0.932, respectively, indicating that the proposed method can achieve a high accuracy for water extraction in a large and complex area and it is an effective optional tool for practical water extraction.

5. Conclusions

Most water indices can perform well for the extraction of wide water features from remote sensing images, but they are normally ineffective in the extraction of narrow water features. This chapter has described a new method using a morphological top-hat transform to form a novel narrow water index, denoted as MNWI, which improves the extraction accuracy of narrow water from Landsat images. Experimental results demonstrated impressive performances of our method of the narrow water extraction. A case study in Fujian province suggests that it is an effective and practical tool for large area inland water.

Acknowledgements

Funding was provided by the Natural Science Foundation of China (Grant No. 41571330, 41830108).

Author details

Wu Bo[1]*, Zhang Jinmu[2] and Zhao Yindi[3]

1 School of Geography and Environment, Jiangxi Normal University, Nanchang, China

2 Key Laboratory of Poyang Lake Wetland and Watershed Research, Ministry of Education, Jiangxi Normal University, Nanchang, China

3 School of Environment and Spatial Informatics, China University of Mining and Technology, Xuzhou, China

*Address all correspondence to: mywubo@fzu.edu.cn

References

[1] Sun F, Sun W, Chen J, Gong P. Comparison and improvement of methods for identifying water bodies in remotely sensed imagery. International Journal of Remote Sensing. 2012;**33**:6854-6875

[2] Frazier PS, Page KJ. Water body detection and delineation with Landsat TM data. Photogrammetric Engineering and Remote Sensing. 2000;**66**:1461-1468

[3] Rokni K, Ahmad A, Solaimani K, Hazini S. A new approach for surface water change detection: Integration of pixel level image fusion and image classification techniques. International Journal of Applied Earth Observation and Geoinformation. 2015;**34**:226-234

[4] McFeeters S. The use of the normalized difference water index (NDWI) in the delineation of open water features. International Journal of Remote Sensing. 1996;**17**:1425-1432

[5] Xu H. Modification of normalised difference water index (NDWI) to enhance open water features in remotely sensed imagery. International Journal of Remote Sensing. 2006;**27**:3025-3033

[6] Carleer AP, Wolff E. Urban land cover multi-level region-based classification of VHR data by selecting relevant features. International Journal of Remote Sensing. 2006;**27**(6):1035-1051

[7] Yang S, Xue S, Liu T. Method for automatically extracting fine water body by using TM image. Journal of Surveying and Mapping (in Chinese with English Abstract). 2010;**39**(6):611-617

[8] Li Y, Ding J, Yan R. Study on extraction method of small water bodies in mountainous area based on GF-1 remote sensing images. Resource Science (in Chinese with English Abstract). 2015;**37**(2):408-416

[9] Jiang H, Feng M, Zhu Y. An automated method for extracting rivers and lakes from Landsat imagery. Remote Sensing. 2014;**6**(6):5067-5089

[10] Beneditsson JA, Pesaresi M, Arnason K. Classification and feature extraction for remote sensing images from urban area based on morphological transformations. IEEE Transaction on Geosciences and Remote Sensing. 2003;**41**(9):1940-1949

[11] Huang X, Zhang L. A multidirectional and multi-scale morphological index for automatic building extraction from multispectral GeoEye-1 imagery. Photogrammetric Engineering & Remote Sensing. 2011;**77**(7):721-732

[12] Wu B, Fang C, Yu L, Huang X, Zhang Q. A fully automatic method to extract rare earth mining area from Landsat image. Photogrammetric Engineering and Remote Sensing. 2016;**82**(9):55-64

[13] Huang X, Zhang L, Wang L. Evaluation of morphological texture features for mangrove forest mapping and species discrimination using multispectral IKONOS imagery. IEEE Geoscience and Remote Sensing Letters. 2009;**6**(3):393-397

[14] Ji L, Zhang L, Wylie B. Analysis of dynamic thresholds for the normalized difference water index. Photogrammetric Engineering and Remote Sensing. 2009;**75**(11):1307-1317

[15] Otsu N. A threshold selection method from gray-level histograms. IEEE Transactions on Systems, Man and Cybernetics. 1979;**9**(1):62-66

[16] Zha Y, Gao Y, Ni S. Use of normalized difference built-up index in automatically mapping urban areas from TM imagery. International Journal of Remote Sensing. 2003;**24**(3):583-594

Section 4

Biodiversity of Inland Waters

Chapter 10

Stream Invertebrate Zoology

Kenneth W. Cummins

Abstract

For over a century, there has been strong interest in freshwater streams and rivers. Since the inception of studies on running waters, invertebrates have been a central theme. Early descriptive work in Scandinavia and New Zealand was followed by work in Europe, England, and then North America and Australia. Presently, there is a very significant interest worldwide including Asia, Central and South America, and Africa in freshwater invertebrates. Throughout, insects have dominated the focus on invertebrates. Although the major marine invertebrate groups are present in freshwaters, there are essentially no marine insects. A clear picture of the habitat and food requirements of running water invertebrates shows that they serve as important indicators of water quality and fisheries. Major paradigms, such as the River Continuum and functional feeding groups, have provided frameworks for studies of running water (lotic) invertebrates. Once stream and river research achieved an international status by separation from lake domination of the limnology discipline, there has been an avalanche of running water invertebrate research.

Keywords: stream and river ecology, lotic invertebrates, functional feeding groups (FFG), River Continuum, FFG ecosystem surrogate ratios

1. Introduction

For over a century, there has been significant interest in stream and river (lotic) ecology. A major foundation fueling this interest has been the aquatic invertebrates. From the beginning, focus has been on certain marine-derived groups and on insects. Lotic macroinvertebrate communities are usually dominated by insects [1], but some marine taxa, such as annelids, mollusks, and crustaceans, are often abundant as well [2, 3]. There are essentially no marine insects, the argument being that by the time insects evolved, all the marine ecological niches were filled. Macroinvertebrates conventionally have been defined as those individuals greater than 1 mm in size. However, many present-day studies include all invertebrates retained on a 0.25 -mm mesh screen as macroinvertebrates (1). The far less studied and much smaller microinvertebrates include taxa also found in the marine ecosystem such as protozoans, rotifers, and annelids. But, some very small insect taxa (Diptera, *Chironomidae*) and the first instar of most aquatic insects are also in this arbitrarily small size category.

Examples of earlier investigations of running water invertebrates can be found in Shelford [4] and Shelford and Edy [5] (North America), Moon [6, 7] (Great Britain), Wessenberg-Lund [8] (Denmark), and Allen [9] (New Zealand). The North American references [4, 5] contain early descriptive work on the components of lotic invertebrate populations and their habitats. The British publication [6] established the fundamental classification of flowing water habitats as either erosional (riffles) or depositional (pools). This basic view endures to the present time and is similar to the lake (lentic) designations of littoral or profundal [10]. The basic morphological and behavioral adaptations of lotic invertebrates, to either erosional or depositional habitats, are discussed below. The Scandinavian volume by Wessenberg-Lund contains a treasure trove of biological and ecological information on freshwater insects including habitat descriptions [8]. Someone could contribute significantly to the study of lotic invertebrates by translating this book from Danish German into English. The famous New Zealand publication by Allen [9] used unique illustrations of the fauna to represent relative densities of the stream macroinvertebrates. A wide-ranging geographical scope of lotic invertebrate study was, and is, an important component of the broad development of the field of stream and river ecology. The discussion on running water invertebrates that follows is greatly informed by recent advances in taxonomy, biology, and ecology [2, 3], especially of insects. Examples are the new (5th) edition of the aquatic insects of North America (Berg et al. [1]), DNA barcoding to validate taxonomic affinities [11, 12], stable isotope (carbon 13, nitrogen 15) analysis of food webs [13, 14], and functional feeding group (FFG) characterization of trophic relationships [15]. It is now possible to identify most genera of North American aquatic insects using morphological characters [1]. It should be noted that many of the North American families and genera occur on other continents.

2. Taxonomic invertebrate groups of lotic invertebrates

2.1 Macroinvertebrates

The typical macroinvertebrate groups and their characteristics in headwater streams are summarized in **Table 1**. The conterminous continent-wide US study of selected river basins, the River Continuum Project [16], developed a paradigm linking position along a river basin channel network that was described as stream order [17]. Energy sources for the component communities of macroinvertebrates were predicted along the continuum. Headwater streams (orders 1–3) receive their energy supply from streamside (riparian) terrestrial vegetation (plant litter) along the stream channel. This is termed an allochthonous energy source.

Wider mid-sized stream/rivers are less shaded by riparian trees, allowing much more light to reach the channel. This additional light input drives in-stream primary production by algae and aquatic vascular plants. This is termed an autochthonous energy source. The macroinvertebrate communities of the mid-sized stream/river ecosystems are populated with taxa that utilize the greatly increased beds of rooted aquatic vascular plants, especially herbivore shredders. Also, these order 4–6 running waters have fauna which contains headwater taxa derived from terrestrial ancestors (insects) together with taxa of marine ancestral origin (e.g. mollusks) (**Table 2**) [16].

The dominant energy source for larger rivers (orders 7–10 or greater) is input from the upstream channel network (orders 1–7) plus periodic return flow from the floodplain [16, 18–21]. These larger rivers are usually turbid, and, although adequate light reaches the surface of the water, penetration to the bottom is poor and primary production is restricted. The dominant food resource for the invertebrates is fine particulate organic matter (FPOM). FPOM consists of particles >1 mm in size, organic and mineral particles surface-colonized by bacteria, and algae and microinvertebrates in suspension. The primary mode of feeding for the

Taxa	Primary energy source; and habitat	Food resource category	Habit (mode of attachment, concealment, movement)	FGG
Oligochaeta (segmented worms)	Autochthonous FPOM produced by invertebrate feeding, mechanical breakage of CPOM, and from the stream bank, pools, backwaters, and margins	FPOM (organic particles, coated and colonized by bacteria, including invertebrate feces); deposited on or in the bottom sediments	Burrowers, mostly in fine sediments	**GC**
Gastropoda (snails)	Autochthonous attached nonfilamentous algae; in riffles	Periphyton: attached nonfilamentous algae and associated FPOM and microinvertebrates	Clingers, on coarse sediments in riffles	SC
Crustacea Amphipoda, *Gammarus* (scuds) [22], Isopoda, *Assellus* (*sow bugs*)	Allochthonous CPOM riparian plant litter; accumulations against obstructions, in backwaters, pools, and stream margins	Conditioned (colonized by microbes, especially aquatic hyphomycete fungi) CPOM plant litter	Burrowers, in plant litter accumulations	DSH
Ephemeroptera (mayflies) Heptageniidae, Ephemerellidae, *Drunella*	Autochthonous attached nonfilamentous algae; in riffles	Periphyton: attached nonfilamentous algae and associated FPOM and microinvertebrates	Clingers, on coarse sediments in riffles	SC
Baetidae, *Baetis* Leptophlebiidae, *Paraleptophlebia*	Autochthonous FPOM produced by invertebrate feeding, mechanical breakage of CPOM, and from the stream bank, pools, backwaters, and margins	FPOM (organic particles, coated and colonized by bacteria, including invertebrate feces); deposited on or in the bottom sediments	Swimmers, in pools, backwaters, and margins, occasionally moving through riffles	GC
Ephemeridae *Ephemera*	Autochthonous produced by invertebrate feeding, mechanical breakage of CPOM, and from the stream bank; gravel riffles	FPOM (organic particles), including invertebrate feces, coated and colonized by bacteria in the size range that can be pumped through burrows	Burrowers, in gravel riffles where they pump water through burrows	FC
Trichoptera (caddisflies) Limnephilidae[1], *Hydatophylax*, *Pycnopsyche*	Allochthonous riparian plant litter; accumulations against obstructions in the current and in backwaters or pools	Conditioned (colonized by microbes, especially aquatic hyphomycete fungi) CPOM plant litter	Burrowers, in plant litter accumulations	DSH[1] (SC)
Hydropsychidae, Philopotamidae	CPOM, FPOM, and small invertebrates in the appropriate size that can be retained in filtering nets; retreats fastened to coarse substrate in riffles	FPOM (organic particles, including invertebrate feces, coated and colonized by bacteria and small invertebrates in the appropriate size that is caught by capture nets)	Clingers, on coarse sediments in riffles	FC
Hydroptilidae	Autochthonous filamentous algae	Individual filamentous algal cell contents	Climbers, in filamentous algal colonies	PC
Rhyacophilidae	Autochthonous in-stream invertebrate prey; riffles or plant litter accumulations	Invertebrate prey of appropriate size	Clingers on coarse sediments or sprawlers in plant litter accumulations	P

Taxa	Primary energy source; and habitat	Food resource category	Habit (mode of attachment, concealment, movement)	FGG
Coleoptera (beetles) Psephenidae (water pennies) Elmidae (adults)	Autochthonous attached nonfilamentous algae; in riffles	Periphyton: attached nonfilamentous algae and associated FPOM and microinvertebrates	Clingers, on coarse sediments in riffles	SC
Dytiscidae (larvae and adults) Gyrinidae (larvae and adults) Hydrophilidae (larvae)	Autochthonous macroinvertebrate prey in pools and backwaters	Invertebrate prey of appropriate size	Sprawlers (larvae), swimmers (adults)	P
Hydrophilidae (adults)	Autochthonous FPOM detritus settled in pools and backwaters	FPOM organic particles plus microbes	Swimmers, in pools and backwaters	GC
Diptera (true flies) Tipulidae, *Tipula* Orthocladiinae, *Brillia*	Allochthonous plant litter accumulations against obstructions, in backwaters, pools, and at stream margins	Conditioned (colonized by microbes, especially aquatic hyphomycete fungi) CPOM plant litter	Burrowers, in plant litter accumulations	DSH
Chironomidae (midges) Chironomini Orthocladiinae genera	Autochthonous FPOM deposited in sediments in pools, backwaters, and slow riffles	FPOM (organic particles, including invertebrate feces, coated and colonized by bacteria); deposited on or in the bottom sediments	Burrowers, in fine sediments	GC
Tanytarsini	Autochthonous FPOM in transport in habitats with moderate flow	FPOM (organic particles), including invertebrate feces, coated and colonized by bacteria of appropriate size to be captured	Clingers, on substrates in moderate current	FC
Simuliidae (blackflies)	Autochthonous clingers; coarse sediments, or wood in riffles	FPOM (organic particles), including invertebrate feces, coated and colonized by bacteria in the size range that can be captured by the filtering head fans; suspended in the passing water column	Clingers, on coarse sediments in riffles	FC
Chironomidae Tanypodinae Ceratopogonidae (biting midges, no-see-ums)	Autochthonous small invertebrate prey	Small prey (e.g. midges, blackflies)	Clingers, on coarse sediments in riffles or in plant litter accumulations	P

Allochtonous energy has source from outside the stream channel (riparian zone); autochthonous energy source within the stream; CPOM is coarse particulate organic matter >1 mm size [23]; FPOM is fine particulate organic matter <1 mm size [24]. Conditioned CPOM is riparian plant litter (e.g. leaves and needles); conditioning involves colonization by microbes, especially aquatic hyphomycete fungi [25]; FFG is functional feeding group: SC = scrapers, DSH = detrital shredders; HSH herbivore Shredders, GC = gathering collectors; FC = filtering collectors, PC = algal cell piercers, P = predators [1, 16, 19, 22, 26–28].
[1]*Many genera have organic cases in the first four instars and are DSH but have mineral cases in the 5th instar and are scrapers.*

Table 1.
Typical North American macroinvertebrates of headwater streams (orders 1–3) [22, 27, 29].

Taxa	Primary energy source; habitat	Food resource category	Habit (mode of attachment, concealment, or movement)	FFG
Oligochaeta (segmented worms)	Autochthonous FPOM produced by invertebrate feeding, mechanical breakage of CPOM, and from the stream bank; gravel riffles	FPOM particle in the bottom sediments	Burrowers in the sediments	GC
Bivalvia (bivalve clams) *Sphaerium*, *Pisidium*	Autochthonous FPOM (transport from upstream and from river banks, invertebrate feces, and mechanical breakage of CPOM); bottom sediments	FPOM organic particles of appropriate size to be filtered through incurrent siphon of material in transport at water sediment interface	Burrowers (with incurrent siphon above sediment surface to allow for filtering of FPOM)	FC
Crustacea Decapoda (crayfish)	CPOM detritus; rooted aquatic plant beds, pools, backwaters, side channels, and river margins where CPOM detritus accumulates	Fragmenting and decomposing rooted vascular plant tissue (and some live vascular plant tissue)	Sprawlers (in accumulations of CPOM surface and rooted plant beds)	DSH HSH
Crustacea Amphipoda, *Hyalella*	Autochthonous rooted aquatic plant beds; stems of rooted plants	Periphyton (algae and associated detritus and microarthropods on rooted plant stems)	Climbers (on rooted aquatic vascular plant stems)	SC
Ephemeroptera Ephemeridae	Autochthonous FPOM produced by invertebrate feeding, mechanical breakage of CPOM from bank and riffles	FPOM in the size range that can be pumped through burrows	Burrowers in river bed gravel, sand, mud sediments with sufficient flow to provide FPOM to be pumped through burrow tube	FC
Ephemeroptera Baetidae	Autochthonous FPOM, settled in depositional areas, especially rooted vascular plant beds	FPOM (organic particles, coated and colonized by bacteria, including invertebrate feces)	Swimmers (among rooted aquatic plant beds and backwaters)	GC
Hemiptera Corixidae	Autochthonous rooted aquatic plant beds; stems of rooted plants	Periphyton algae on rooted plant stems	Climbers (on rooted aquatic vascular plant stems)	SC
Ephemeroptera Baetidae	Autochthonous FPOM, settled in depositional areas, especially rooted vascular plant beds	FPOM (organic particles, coated and colonized by bacteria, including invertebrate feces)	Swimmers (among rooted aquatic plant beds and backwaters)	GC
Trichoptera Hydropsychidae	Autochthonous cell contents of filamentous algae	Individual cell contents of filamentous algae		FC

Allochtonous is energy source from outside the stream channel (riparian zone); autochthonous is energy derived within the stream; CPOM is coarse particulate organic matter >1 mm size; FPOM is fine particulate organic matter <1 mm size; conditioned CPOM is riparian plant litter (e.g. leaves and needles); conditioning involves colonization by microbes, especially aquatic hyphomycete fungi; FFG is functional feeding group: SC = scrapers; GC = gathering collectors; FC = filtering collectors; DSH = detrital shredders, HSH = herbivore shredders; PC = algal cell piercers; P = predators [1, 3, 19, 22, 26, 28, 29].

Table 2.
Typical North American macroinvertebrates of mid-sized rivers (orders 4–6).

Taxa	Primary energy source; habitat	Food resource category	Habit (mode of attachment, concealment, or movement)	FFG
Oligochaeta (segmented worms)	Autochthonous FPOM transported from upriver tributaries and settled on or trapped in the bottom sediments	FPOM benthic organic particles colonized by microbes	Burrowers, in the sediments	GC
Bivalvia (large bivalve clams)	Autochthonous FPOM transport from upriver tributaries in the water column past the bottom sediments where the clams reside	FPOM consisting of organic particles colonized by microbes and phytoplankton and zooplankton in suspension	Burrowers, in sediments with siphon above the surface allowing capture and filtration of FPOM in transport	FC
Crustacea (zooplankton) Cladocera, Copepoda	Autochthonous phytoplankton, bacteria, rotifers, and protozoans produced in situ and micro- FPOM in the water column	FPOM consisting of phytoplankton, bacteria, rotifers, and protozoans and microorganic particles	Swimmers, in the water column (limited directed movement, easily carried by any current)	FC
Megaloptera (Dobsonflies) Corydalidae	Autochthonous invertebrate prey on large woody debris along river bank or against point bars	Prey consisting of micro- and macroinvertebrates cohabiting large woody debris (e.g. Diptera Chironomidae)	Clingers, on large woody debris	P
Trichoptera Hydropsychidae	Autochthonous FPOM in transport in the water column from upriver tributaries in the water column past the capture nets on large woody debris where the larval retreats are attached	FPOM consisting of organic particles colonized by microbes and phytoplankton and zooplankton in suspension	Clingers, on large woody debris	GC
Coleoptera (beetles) Dytiscidae, Gyrinidae (larvae and adults), Hydrophilidae (larvae)	Autochthonous macroinvertebrate prey in pools and backwaters	Invertebrate prey of appropriate size	Sprawlers (larvae), swimmers (adults), in backwaters	P
Hydrophilidae (adults)	Autochthonous FPOM detritus plus microbes in backwaters	FPOM detritus consisting of dead organic matter plus microbes	Swimmers, in backwaters	GC
Diptera Simuliidae (blackflies)	Autochthonous FPOM in transport in the water column from upriver tributaries past boulders and large cobbles in rapids and large woody debris surfaces	FPOM consisting of organic particles colonized by microbes in suspension	Clingers, in rapids and on large woody debris	FC
Chironomidae (midges) Chironomini	Autochthonous FPOM in from upriver tributaries and deposited in the	FPOM consisting of deposited organic particles colonized by microbes	Burrowers, in sediments and crevices in large woody debris	GC

Taxa	Primary energy source; habitat	Food resource category	Habit (mode of attachment, concealment, or movement)	FFG
	sediments and crevices on large woody debris			
Tanytarsini	Autochthonous FPOM in transport in the water column from upriver tributaries past large woody debris surfaces along river bank or against point bars	FPOM consisting of organic particles colonized by microbes in suspension	Clingers, on the surface of large woody debris	FC

Allochtonous is energy source7–10i derived from upstream channel network (orders 1–6); FFG i = functional feeding group: SC = scrapers; GC = gathering collectors; FC = filtering collectors; DSH = detrital shredders, HSH = herbivore shredders; PC = algal cell piercers; P = predators [1, 16, 19, 20, 30, 31].

Table 3.
Typical North American macroinvertebrates of large rivers (orders 7–10).

microinvertebrates (zooplankton) and macroinvertebrates (e.g. clams) of the rivers is filtering (filtering collectors) (**Table 3**) [16].

Thus, the River Continuum model predicts that the small headwater streams will be dominated by invertebrate taxa that are dependent on an allochthonous energy [16]. The most common macroinvertebrates are Detrital Shredders utilizing coarse particulate organic matter (CPOM) plant litter: (DSH) scuds (Amphipoda), several stonefly (Plecoptera), and caddisfly (Trichoptera) families and crane flies *Tipula* (Diptera, *Tipulidae*) are the macroinvertebrates that feed on the plant litter inputs. Certain mayflies (Ephemeroptera) and midges (Diptera, *Chironomidae*) supported by FPOM generated by FFG DSH taxa. Macroinvertebrates of mid-sized rivers utilize autochthonous food resources, especially aquatic plants. Large river invertebrate food chains depend on autochthonous FPOM in suspension delivered from the upstream channel network and the floodplain.

2.2 Microinvertebrates (zooplankton)

The very small lotic microinvertebrate taxa are shared with marine environments, such as Protozoa, Rotifera, Nematoda, and micro-Crustacea (Cladocera and Copepoda) [32]. The unofficial definition of microinvertebrates is individuals smaller than 1 mm. Defined this way, early stages of macroinvertebrates need to be included. The difference is that "true" microinvertebrates do not grow beyond the 1 mm size-defined category. In the vast majority of studies in running waters (almost exclusively in rivers), the term microinvertebrates can be replaced by zooplankton. Zooplankton are small invertebrates that live suspended in the water column and have limited ability to control their location [2]. Studies of running water macroinvertebrates vastly exceed those of zooplankton. The River Continuum project [16] recognized zooplankton as a dominant group in rivers of orders 7–10 together with benthic macroinvertebrate such as oligochaetes, bivalve gastropods, and some micro-crustaceans [1]. Studies of river zooplankton have focused on their role in food chains of fish and organic matter cycling. Although most of the zooplankton inhabit the water column of the river where they filter feed on suspended FPOM [32] the benthic forms filter FPOM from the water column and on depositional FPOM [1, 2, 24].

3. Habitats of lotic invertebrates

3.1 Erosional habitats

Erosional habitats encompass coarse sediments (boulders, cobbles, and gravel) and large wood debris in fast-flowing water (riffles and runs). Because these habitats are well oxygenated, they normally support invertebrate populations that are the most sensitive to degradation of water quality. Microbes in the organic waste can reduce dissolved oxygen levels sufficient to stress-sensitive invertebrates [33, 34]. Erosional habitats are normal features of headwater streams and mid-sized rivers but occur less frequently in large rivers. In larger rivers, erosional habitat is found primarily in sections having a significant change in grade.

Macroinvertebrates adapted to erosional habitats (clingers) include Gastropoda (snails, e.g. Sulcospiridae, *Juga*), Ephemeroptera (mayflies, e.g. Heptageniidae), Trichoptera (stone case-bearing and net-spinning caddis), Plecoptera (stoneflies, predaceous Perlidae), and Coleoptera (Psephenidae, water pennies). The ecological tables in Berg et al. [1] identify erosional habitat adaptations by taxonomic group. Structures, such as suckers, hook, or claws of various sorts, silk that fastens down their retreats or provides anchors, and body shape and behavior that avoids the major force of the current are the main adaptations to erosional conditions [35].

3.2 Depositional habitats

Depositional habitats are drop zones where fine sediments settle out. Substrates of sand, silt, and clay are found in pools, backwaters, and along channel margins. FPOM, and in some cases. CPOM plant litter lotter [23], also accumulates in depositional habitats. These depositional habitats are dominated by sprawlers and burrowers that move across the soft substrate or are concealed beneath it [1]. Some of the Ephemeroptera sprawlers have modified first abdominal gills that cover the remaining gills to protect them from smothering by depositional silt (e.g. Caenidae, Tricorythidae) [1]. Burrowing depositional taxa include Oligochaeta, Ephemeroptera (Ephemeridae), Diptera (Chironomidae midges), and predator Odonata dragonflies (Gomphidae) [1].

3.3 Rooted aquatic vascular plants

Aquatic vascular plants occur in both erosional or depositional habitat but are more common in the depositional areas, especially in larger rivers. The macroinvertebrates associated with vascular plants feed on floating leaves such as Lepidoptera (moth larvae, e.g. Noctuidae) and Coleoptera (beetles, e.g. Chrysomelidae, *Galeracella*) and some Diptera, Chironomidae (midges). These are all herbivore shredders (HSH) that mine leaves or burrow into stems, or feed on roots, especially of *Nuphar*, where they penetrated foot tissue and extract oxygen (Coleoptera, Chrysomelidae, *Donacia*)) [1].

In the Lepidoptera ecology table [1], all genera are described as herbivore shredders (HSH) occurring in lentic (standing water) habitats. Two families (Cosmopterigidae, 180 sp., and Noctuidae, 12 sp.) are in a category "generally lentic" because they also occur in lotic systems. The emergent and floating-leaf plant beds are in backwaters, along margins and in areas of slow current of mid-sized and larger rivers. Some streams of orders 2 and 3 support floating-leaf plants in erosional habitats (e.g. *Valsineria*).

4. Functional feeding groups (FFG)

Investigation of lotic invertebrates has been taxonomy based since its inception [1]. In the 1980s, Robert Penna, a major authority on freshwater invertebrates of the day [2], argued that ecology of freshwater invertebrates must be based on *species*-level taxonomy (personal communication). This goal was, and still is a valid one, but then, and even now, rarely possible to achieve. There are very few groups of lotic invertebrates for which complete species inventories have been accomplished. For example, the most recent compilation of the taxa of aquatic insects of North America [1] allows identification of almost all known genera, but not species, collected in lotic samples. The emphasis is on immature nymphs and larvae. Coleoptera adults, which are aquatic, and adult females of some aquatic insects that enter the water in order to deposit eggs are also keyed [1].

In 1973 and 1974, Cummins [36, 37] argued that while efforts will continue toward expanding the taxonomy of freshwater invertebrates (especially insects), it should be possible to address ecological study of lotic macroinvertebrates by employing analyses of their morphological and behavioral adaptations. The proposal was to use five functional feeding group (FFG) adaptation categories and match them to the five basic food resource categories available in varying amounts in streams and rivers [15, 16, 26, 36, 38]. In this FFG analysis, scrapers are matched with periphyton (attached nonfilamentous algae and associated material), gathering collectors with FPOM deposited on or in the sediments and filtering collectors with FPOM transported in the water column, shredders with CPOM (conditioned plant litter or live vascular plants), and predators with their prey [15, 25, 26, 39, 40].

More recently, a refinement of two more FFG matching categories has been utilized [41]. The shredder category is divided into detrital shredders (DSH) matched with CPOM conditioned plant litter or wood [42] and herbivore shredders with live vascular aquatic plant tissue [41]. Piercers are matched with filamentous algae (**Tables 1–3**) [26, 38]. Taxa that share morphological (e.g. moth parts, body structure, color pattern) and/or behavioral (e.g. movement patterns, silk net-spinning, case construction) adaptations can be grouped in the same FFG. Through parallel or convergent evolution, they share features that result in the same modes of food acquisition. An example is the striking similarity between North American Heptageniidae and Brazilian Leptophlebiidae mayfly nymphs which are both scrapers [26, 38]. Also, as shown in [26], caddisfly genera in three different families (Glossosomatidae, *Glossosoma*; Helicopsychidae, *Helicopsyche*; and Uenoidae, Neophylax) and a beetle genus *Psephenus* scrape substrate surfaces in riffles.

A simple picture key can be used to sort macroinvertebrates collected in lotic field samples into FFG categories with an 80% or greater accuracy. This can be accomplished using structural and behavioral characters that can be readily observed in the field on live specimens with the unaided eye or a simple hand lens. For example, case-bearing Trichoptera can be separated based on the materials used in case construction: larvae with organic cases constructed of leaf or wood pieces are detrital shredders (DSH), and those with mineral cases made of sand or fine gravel are scrapers (SC). An example of a key used to separate Ephemeroptera nymphs (lateral abdominal gills) from Plecoptera nymphs (no lateral abdominal gills), separate mayfly scraper nymphs (clinger flat body shape) from mayfly Gathering Collector nymphs (swimmer cylindrical body shape). Dragonfly and damselfly nymphs can be separated from mayfly and stonefly nymphs by an extendible grasping labium (**Figure 1**) [15, 26, 40].

When separating live macroinvertebrates collected in a stream field sample into functional feeding groups, individuals in different taxa that share similar

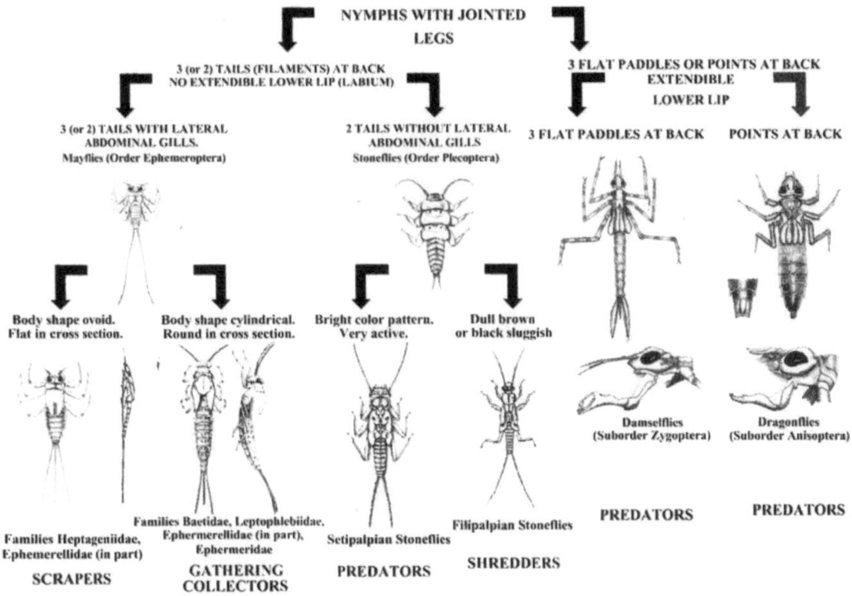

Figure 1.

A simple picture key for separating nymphal stream insects into functional feeding groups (FFG). Mayflies and stoneflies are separated by the presence or absence of lateral abdominal gills. Mayflies are separated into scrapers or gathering collectors by body shape. Stoneflies are into detrital shredders or predators by color pattern and activity level. Dragonflies and damselflies are all predators and are separated from all the other nymphs by having extensible, grasping labia (lower lips). Modified from [38].

adaptations are enumerated together. All muffin tin with the wells is marked by FFGs (e.g. SC, GC, etc.). For example, all snails, dorsal-ventrally flattened mayfly nymphs, mineral case-bearing caddisfly larvae, and water-penny beetle larvae are sorted into the scraper (SC) category [40]. Because scrapers usually feed on surfaces in riffles, they are adapted to maintain their location; they are clingers on the substrate surface. Common North American scraper taxa found in running waters are Gastropoda snails that have a rasping radula; Ephemeroptera mayfly Heptageniidae and Ephemerellidae *Drunella* nymphs that are dorsal-ventrally flattened; Trichoptera caddisfly Limnephilidae, *Hydatophylax*, *Pycnopsyche*, Uenoidae, Glossosomatidae, and Helicopsychidae larvae; and flattened Coleoptera water-penny beetle Psephenidae larvae (**Tables 1–4**) [1, 40]. Similar taxonomic groupings for gathering and filtering collectors, detrital shredders, herbivore shredders, or predarorsare given in **Table 4** and [40].

5. Macroinvertebrates used for evaluation of lotic ecosystem condition

The known North American aquatic insect species in genera for each order found in lotic habitats are listed in the ecological tables in Merritt et al. [43]. No lentic (standing water) genera are included in **Table 4**. The lotic genera are also assigned to FFGs in **Table 4**. Every FFG entry in **Table 4** is divided into % obligate or facultative number of species. The obligate category is defined as having a maximized % conversion of ingestion to growth. Obligate taxa are predicted to be most affected by environmental changes that alter their food resource. By contrast, the facultative forms are predicted to have flexible food requirements and to be better adapted to adjust to changes in food supplies, but the conversion of ingestion

Taxa	Functional feeding group (FFG) categories														
	Scrapers			Gathering collectors			Filtering collectors			Shredders			Predators		
	Fac.	Obl.	Total	Fac.	Obl.	Total	Fac.	Obl.	Total	Fac.	Obl.	Total	Fac.	Obl.	Total
Ephemeroptera															
Heptageniidae	0	49.5	49.5	49.5	0	49.5							0	1.0	1.0
Leptophlebiidae	15.2	3.3	18.5	51.0	44.0	95.0				30.3	0	30.3			
Baetidae	27.0	11.2	38.2	74.5	43.9	43.9	4.0	0	4.9						
Ephemeridae				50.0	0	50.0	0	50.0	50.0						
Ephemerellidae	32.0	280	60.0												
Isonychidae							0	100	100						
Plecoptera															
Pteronarcyidae	10.0	0	10.0							10.0	40.0	59.0	49.0	0	40.0
Peltoperlidae	11.2	2.0	13.2	22.0	0	22.0				49.3	11.3	60.6	4.4	0	4.4
Nemouridae	2.6	0	2.6	8.3	0	8.3				89.1	0	89.1			
Leuctridae				8.3	0	8.3				0	74.8	74.8			
Capniidae										0	100	100			
Taeniopterygidae	22.5	0	22.5	10.8	0	10.5				22.5	44.1	66.5			
Perlidae				13.3	0	13.3							0	86.7	86.7
Perlodidae	1.4	0	1.4	28.1	0	28.1							0	70.5	70.5
Chloroperlidae	2.1	0	2.1	31.9	0	31.9							0	66.0	66.0
Trichoptera															
Glossosomatidae	0	71.0	71.0	29.0	0	29.0									
Helicopsychidae	0	100	100												
Uenoidae	0	53.6	53.6	46.4	0	46.4									

Taxa	Functional feeding group (FFG) categories														
	Scrapers			Gathering collectors			Filtering collectors			Shredders			Predators		
	Fac.	Obl.	Total	Fac.	Obl.	Total	Fac.	Obl.	Total	Fac.	Obl.	Total	Fac.	Obl.	Total
Limnephilidae	6.4	4.1[1]	10.5	33.9	4.1	38.0				3.5	48.0[2]	51.5			
Lepidostomatidae										0	100	100			
Calamoceratidae				33.3	0	33.3				0	66.7	66.7			
Leptoceridae				0	22.9	22.9				39.6	0	39.6	8.3	29.2	37.5
Rhyacophilidae	0.8	0	0.8										0	99.2	99.2
Brachycentridae				38.8	0	38.8	0	33.8	33.8						
Hydropsychidae			0	21.8	21.8	0	0	78.2	78.2						
Philopotamidae							0	100	100						
Psychomyiidae	33.8	20.0	53.8	24.6	21.5	46.1									
Polycentropidae							42.6	5.8	48.4	3.6	0	3.6	7.2	40.6	47.8
Megaloptera															
Corydalidae													0	100	100
Sialidae													0	100	100
Odonata															
Anisoptera													0	100	100
Zygoptera													0	100	100
Hemiptera															
Corixidae	100	0	100												
Nepidae													0	100	100
Naucoridae													0	100	100

DOI: http://dx.doi.org/10.5772/intechopen.88283

Taxa	Functional feeding group (FFG) categories														
	Scrapers			Gathering collectors			Filtering collectors			Shredders			Predators		
	Fac.	Obl.	Total	Fac.	Obl.	Total	Fac.	Obl.	Total	Fac.	Obl.	Total	Fac.	Obl.	Total
Lepidoptera															
Crambidae, Noctuide										0	100	100			
Coleoptera															
Larvae and Adults Dytiscidae, Gyrinidae Hydrophilidae Larvae													0	100	100
Hydrophilidae Adults				0	100	100									
Diptera															
Tipulidae, *Tipula Holorusia*										100	0	100			
Dicranota, Pedicia, *Hexatoma*													100	0	100
Chironomidae Chironomini				0	84.2	84.2	12.2	0	12.2						
Tanytarsini				26.3	11.3	37.6		0	73.7	73.7					

For each taxon (column 1), the % species for each taxon is apportioned among the FFG categories representing that taxon. Fac. = facultative taxa, that is, these taxa are listed in the ecological tables in [43] as having several alternative FFG classifications. The Fac. values given in the table are the FFG that appears first (as the most likely) of the alternatives of the several presented. If the accompanying obligatory (Obl.) entry for that taxon is 0, this means there were no Fac. possibilities listed in [43]. Similarly, if there is an Obl. entry in the table and a 0 given for the accompanying Fac., this means that no Fac. alternatives were given in entries for that taxon in [43]. If there are % entries in both columns, there are values given for both categories in [43]. When selecting the % to be used to assign the proportion of each FFG to each taxon, the clearest approach would be to use only the Obl. designations throughout; the most conservative approach would be to use the total % values, that is, combining those taxa that are restricted to a given FFG with the most probable Fac. species per genus. The % values in this table will be subject to some changes when [1] is published.
[1]*Early instars of larvae with organic cases are obligate Detrital Shredders (DSH).*
[2]*Last (5th) instar of larvae with mineral cases are obligate Scrapers (SC).*
[3]*Because of uncertain Chironomidae taxonomy, the FFG percent values are based on approximate species per genus numbers [43].*

Table 4.
Percent species in genera of North American lotic macroinvertebrates [43].

to growth would be less efficient. Facultative taxa would be predicted to better survive environmental changes [15, 43]. When selecting the % to be used to assign the proportion of each FFG into each taxon, the least ambiguous approach would be to use only the obligate designations throughout; the most conservative approach would be to use the total % values, that is, combining those obligate taxa which are restricted to a given FFG with the most probable facultative species per genus. The % values in **Table 4** undoubtedly will be subject to some changes when [1] is published.

Using counts of numerical abundance of macroinvertebrates in field samples, ratios of the % numerical abundance of FFGs, like those in **Table 4**, can be used to calculate ratios of the FFGs. These ratios have been used as surrogates for stream and river ecosystem attributes [1, 26, 40, 41]. Because such ratios are dimensionless numbers, the resulting calculated ratios are essentially independent of sample size. For example, the FFG ratio from one riffle (coarse sediment) sample produces the same ratio as five samples, or one plant litter sample is the same as five.

A number of FFG ratios that serve as surrogates for running water ecosystem attributes are summarized in **Table 5**. Thresholds for evaluating the ratios are also proposed ([1], Table 6E). The ecosystem attributes can be measured directly, but this usually requires significant equipment, time, and direct tending by researchers. In addition, the actual measurements represent only a fraction of the temporal and special scales at which the processes occur. By contrast, the macroinvertebrates continuously monitor ecosystem conditions over their life stages in the water, at least weeks and usually annual or semiannual periods.

Arguably, the most all-encompassing and informative ratio is gross primary production (P) compared to community respiration. The P/R ratio also reflects the relative dominance of autotrophy (energy source within the stream or river relative to energy input from outside the aquatic ecosystem) (**Table 5**) [44–46]. The surrogate macroinvertebrate P/R ratio is all FFGs that depend on autochthonous primary production (algae and vascular plants) compared to all the FFGs that depend on FPOM and CPOM organic matter. That is, scrapers + herbivore shredders + algal cell piercers to detrital shredders + gathering collectors + filtering collectors (**Table 5**) [1, 10, 41, 47]. The P/R ratio that corresponds to a directly measured P/R, using closed, recirculating chambers that monitor dissolved oxygen, of P/R > 1.0 is a macroinvertebrate P/R > 0.75 (**Table 5**) [19, 44, 45]. The other surrogate ratios described compare detrital shredders available CPOM storage (Detrital Shredder index), relative abundance of filtering collectors to FPOM in transport (Filtering Collector Index), macroinvertebrates that require stable attachment or clinging sites compared to substrate stability, and predator abundance relative to prey available (Predator Index) (**Table 5**).

The ecological tables in [1, 43] also include US Environmental Protection Agency values for macroinvertebrate susceptibility/resistance that are indicators of pollution. As a general rule, the EPT Index will indicate the vulnerability of macroinvertebrates to stream and river water quality degradation. This index compares the abundance of Ephemeroptera (mayflies) + Plecoptera (stoneflies) + Trichoptera (caddisflies) to the rest of the macroinvertebrate fauna; the more dominant the EPT, the less polluted the stream or river is rated [33, 48].

Organic pollution reduces dissolved oxygen (DO) levels in freshwater due to the large oxygen demand by microbial respiration [33]. Significant reduction in DO is a major stressor for aerobic (DO requiring) invertebrates. This includes those with gills or some with cutaneous respiration: Mollusca, Crustacea, Ephemeroptera, Plecoptera, Odonata, Trichoptera, Megaloptera, some Lepidoptera, some Coleoptera larvae, and some Diptera. Oligochaeta and some Chironomidae have biochemical adaptations that allow them to tolerate low DO levels [1].

Lotic ecosystem attributes	FFG ratios	Symbols	Proposed thresholds	Descriptions
Autotrophic to heterotrophic Index (P/R, primary production/ community respiration)	Scrapers + herbivore shredders to detrital shredders + gathering collectors + filtering collectors	SC + HSH to DSH + GC + FC	P/R > 0.75	A P/R > 0.75 corresponds to P/R > 1 when primary production and community respiration are measured directly
Shredder Index (CPOM/FPOM) [27, 28]	Detrital shredders to gathering collectors + filtering collectors	DSH to GC + FC	CPOM/FPOM >0.5 (fall–winter); CPOM/FPOM >0.25 (spring–summer)	The CPOM/FPOM >0.5 during fall–winter when fast processed deciduous plant litter enters streams; > 0.25 in spring–summer when slower processed needles and wood remain in streams
Filtering collector Index (suspended FPOM to deposited FPOM)	Filtering collectors to gathering collectors	FC to GC	FC/GC > 0.5	FC > 0.5 favors filtering collectors capturing FPOM in transport; impaired lotic systems usually have values much higher than 9.5
Substrate stability Index (stable coarse substrates to unstable fine sediments)	Scrapers + filtering collectors + herbivore shredders to detrital shredders + gathering collectors	SC + FC + HSH to DSH + GC	Stable substrates/ unstable substrates >0.5	Stable substrates (bedrock, boulders and cobbles, large wood, and rooted plants) provide attachment and clinger sites greater than unstable fine sands and clay. Channel disturbance can reduce stable substrates or flush out fines
Predator Index (top-down to bottom-up control of macroinvertebrate communities)	Predators to scrapers + detrital shredders + herbivore shedders + filtering collectors + gathering collectors	P to SC + DSH + HSH FC + FC	Predators/all other FFGs present =0.10–0.15	The abundance of predators between 10% and 15% of the lotic macroinvertebrate community indicates sufficient prey (turnover to support predators)

Table 5.
Functional feeding group (FFG) ratios as surrogates for running water lotic ecosystem attributes and proposed thresholds.

However, there are stream and river macroinvertebrates that are adapted to breathe atmospheric oxygen (AO) by returning to the water surface to obtain air, such as some Diptera (the best known example being Culicidae mosquitoes). Others like some Coleoptera adults trap air in body surface hairs or under elytra that they carry under the surface and from which they extract DA. An index that would predict poor ability of lotic macroinvertebrate taxa to survive under declining DO conditions would be taxa with gills as a proportion of those taxa with adaptations, structural, behavioral, and/or biochemical, that allows them to breath AO.

The most widely used method to obtain qualitative samples in streams and shallow rivers is the D-frame dip net. The net usually has a 1.0 or 0.5 mm mesh size. However, nets with a 0.25 mesh are recommended to retain midges (Chironomidae and Ceratopogonidae) and early stages or instars of macroinvertebrate taxa. The samples collected with the D-frame net can be considered semiquantitative when fixed time (e.g. 30 s) sampling is employed. This method has been used to compare stream/river reaches, stream/river habitats, seasons, etc. D-frame net sampling has also been used to collect composite samples, that is, an effort to collect widely from all habitats in a stream or river reach [41, 47]. In such complete reach survey samples, it is useful to keep the major habitat samples (riffles, pools, CPOM plant litter accumulations, large wood) separate and compute a composite value by calculating a total value by combining the data after the samples have been processed. This method is even more useful if a percent of the stream/river bottom in the sample reach is covered by each habitat type. This is particularly helpful if the derived FFG ratios described above are to be used as surrogates for lotic ecosystem attributes. If the same person is used to collect the D-frame samples, the results are more comparable; if that same person is also used to collect quantitative Surber net samples, the variances are similar. Ratios and proposed threshold values are presented in **Table 5.**

6. Quantitative sampling

The two most commonly used quantitative devices used for sampling stream macroinvertebrates are the Surber and Hess samplers [3, 49]. Both of these confine an area of stream bottom and rely on the stream current to transport material disturbed by the hand into attached collection nets. The Surber net has a metal frame that delineates the bottom area to be collected with an erect frame at the back that holds the collection net. The net is washed into a sorting tray (usually a white enamel pan) for partial or complete processing in the field, or the contents are rinsed directly into a bottle and preserved in 70% ETOH. The Hess sampler is a cylinder that defines the bottom area sampled. It has mesh side panels and a collection net to retain material disturbed from the bottom similar to the Surber net. With both collection devices, the sample collected is better handled if cobbles are removed individually and scrubbed into the net and discarded. As with qualitative sampling, the net mesh size is a significant issue. Both Surber and Hess samplers are available with 1 or 0.5 mm nesh. However, as stated previously, a 0.25 -mm mesh is better because it will retain early instars and smaller species are lost with the coarser mesh sizes. If the samples are to be sorted for FFGs, it is highly recommended that this be done with live specimens following collection.

7. Concluding remarks

In the past and present, and predictably in the future, abundance and composition of stream and river invertebrate communities have been, and will be, the primary measuring biological tool used to evaluate ecosystem condition and to predict environmental change and vulnerability. These animals, because they continuously monitor the stream and river environment throughout their aquatic life, can provide better insight than the spatially and temporally and limited physical and chemical grab samples or even recording electrodes. Unlike algal cells and microbes, macroinvertebrates can be observed with the naked eye and a simple hand lens. They are far less migratory than fish that respond to an environmental stressor by leaving. Because running water macroinvertebrates are ubiquitous, easily collected, and observed and can be classified into meaningful categories thatchronical the condition of freshwater resources, they are the perfect vehicle for use by basically and easily trained local volunteers. This yields a potential army of environmental stewards who can enlist our very best freshwater monitors—the invertebrates. Both conventional morphological taxonomy and DNA barcoding will undoubtedly continue to lead to ever-better answers to what is it (classification), but this is only the initial step to answering the ultimate question—what does it do (function)? So, students, researchers, and armature naturalists, let us continue on this promising track.

Author details

Kenneth W. Cummins
Cooperative Fisheries Research Unit, Humboldt State University, Arcata, CA, USA

*Address all correspondence to: kc8161@gmail.com

References

[1] Berg MB, Merritt RW, Cummins KW, editors. An Introduction to the Aquatic Insects of North America. Dubuque, IA: Kendall/Hunt; 2019

[2] Smith DG. Pennak's Freshwater Invertebrates of the United States: Porifera to Crustacea. New York: Wiley; 2001

[3] Hauer FR, Resh VH. Macroinvertebrates. In: Hauer FR, Lamberti GA, editors. Methods in Stream Ecology. Volume 1: Ecosystem Structure. London: Academic Press/ Elsevier; 2017. pp. 297-319

[4] Shelford VE. An experimental study of the behavior agreement among animals of an animal community. The Biological Bulletin. 1914;**20**:294-315

[5] Shelford VE, Edy S. Methods in the study of stream communities. Ecology. 1929;**10**:381-392

[6] Moon HP. Aspects of the ecology of aquatic insects. Transactions of the British Entomological and Natural History Society. 1939;**6**:39-49

[7] Cummins KW. What is a river: Zoological description. In: Ogelsby RT, Calson CA, NcCann JA, editors. River Ecology and Man. New York: Academic Press; 1972. pp. 33-52. 465p

[8] Wessenberg-Lund C. Biologie der Susswasser Insecten. Copenhagen, Denmark: Gylden & Springer; 1943. 682p

[9] Allen RH. The Horokiwi stream. New Zealand Marine Department Fisheries Bulletin. 1951;**101**:1-231

[10] Cummins KW, Merritt RW, Berg MB. Ecology and distribution of aquatic insects. In: Berg MB, Merritt RW, Cummins KW, editors. An Introduction to the Aquatic Insects of North America. Dubuque, IA: Kendall/Hunt; 2019

[11] Ball SL, Armstrong KF. DNA barcodes for insect pest identification: A test case with Tussock moths (Lepidoptera; Lymantridae). Canadian Journal of Forest Research. 2006;**36**: 337-350

[12] Cordero RD, Sanchez-Damirez S, Currie DC. DNA barcoding of aquatic insects reveals unforeseen diversity and recurrent population divergence patens through broad scale sampling in northern Canada. Polar Biology. 2017; **40**:1687-1695

[13] Hershey AE, Northinton JC, Finley C, Peterson BJ. Stable isotopes in stream food webs. In: Lamberti GA, Hauer FR, editors. Methods in Stream Ecology. Volume 2: Ecosystem Function. London: Academic Press/Elsevier; 2017. pp. 3-20

[14] Peterson BJ, Fry B. Stable isotopes in ecosystem studies. Annual Review of Ecology and Systematics. 2004;**16**: 273-320

[15] Cummins KW, Klug MJ. Feeding ecology of stream invertebrates. Annual Review of Ecology and Systematics. 1979;**10**:147-172

[16] Vannote RL, Minshall GW, Cummins KW, Sedell JR, Cushing CE. The river continuum concept. Canadian Journal of Fisheries and Aquatic Sciences. 1980;**37**:130-137

[17] Strahler AN. Hyposometric (area-altitude) analysis of erosional topography. Americans Bulletin. 1952; **63**:1117-1142

[18] Cummins KW. The ecology of running waters: Theory and practice. In: Baker DB, Jackson WB, Prater BL, editors. The Ecology of Running Waters. In: 1975 Proceedings of the

Sandusky River Basin Symposium on International Joint Communications, International GP. Great Lakes Pollution from Land Use Activities. Washington, DC: U. S. Government Printing Office; 1976. pp. 277-293. 475 p

[19] Cummins KW, Klug MJ, Ward GM, Spengler GL, Speaker RW, Ovink RW, et al. Trends in particulate organic matter fluxes, community processes, and and macroinvertebrate functional groups, along a Great Lakes drainage basin river continuum. Verhandlungen des Internationalen Verein Limnologie. 1981;**21**:841-849

[20] Junk WJ, Bayley PB, Sparks RE. The flood pulse concept in river flood-plain systems. In: Proceedings of the International Large River Symposium. Canadian Special Publication of Fisheries and Aquatic Sciences; 1989. p. 106

[21] Ward JV, Stanford JA. Ecological connectivity in alluvial river ecosystems and its disruption by flow regulation. Regulated Rivers. 1995;**11**:105-119

[22] Marchant R. The ecology of *Gammarus* running water. In: Lock MA, Williams DD, editors. Perspectives in Running Water Ecology. New York: Plenum Press; 1981. pp. 225-249

[23] Lamberti GA, Entrekin SA, Tiegs SD. Coarse particulate organic matter: Storage, transport, and retention. In: Lamberti GA, Hauer FR, editors. Methods in Stream Ecology. Volume 2: Ecosystem Function. London: Academic Press/Elsevier; 2017. pp. 55-69

[24] Hutchens, Wallace JB, Grubough JW. Transport and storage of fine particulate organic matter. In: Lamberti GA, Hauer FR, editors. Methods in Stream Ecology. Volume 2: Ecosystem Function. London: Academic Press/Elsevier; 2017. pp. 47-53

[25] Benfield EF, Fritz KM. Leaf litter breakdown. In: Lamberti GA, Hauer FR, editors. Methods in Stream Ecology. Volume 2: Ecosystem Function. London: Academic Press/Elsevier; 2017. pp. 71-82

[26] Cummins KW. Functional analysis of stream macroinvertebrates. In: Limnology—Some New Aspects of Inland Water Ecology. Rijeka, Croatia: IntechOpen; 2018. DOI: 105772intechopen.79913

[27] Grubbs SA, Cummins KW. Linkages between Riparian forest composition and shredder voltinism. Archiv für Hydrobiologie. 1996;**137**:49-58

[28] Wessell KJ, Merritt RW, Wilhelm JGO, Allan JD, Cummins KW, Uzarski DG. Biological evaluation of Michigan's non-wadeable rivers using macroinvertebrates. Aquatic Ecosystem Health & Management. 2005;**11**:35-351

[29] Jackson JK, Batzed DP, Resh VH. Chapter 3: Sampling aquatic insects: Collection devices, statistical consideration, and rearing procedures. In: Berg MB, Merritt RW, Cummins KW, editors. An introduction to the aquatic insects of North America. 5th ed. Dubuque, IA, USA: Kendall/Hunt Publ. Co.; 2019

[30] Minshall GW, Peterson RC, Bott TL, Cushing CE, Cummins KW, Vannote RL, et al. Stream ecosystem dynamics of the Salmon River, Idaho: An 8th order system. JNABS. 1992;**11**: 111-137

[31] Rugenski AT, Minshall GW, Hauer FR. Riparian processes and interactions. In: Lamberti GA, Hauer FR, editors. Methods in Stream Ecology. Volume 2: Ecosystem Function. London: Academic Press/Elsevier; 2017. pp. 83-111

[32] Schailendra S, Siddique A, Singh K, Chouhan M, Vyas A, Solnki C, et al.

Population dynamics and seasonal abundance of zooplankton community in Narmada River (India). Research. 2010;**2**:1-9

[33] Karr JR, Chu EW. Biological monitoring: Essential foundation for ecological assessment. Human and Ecological Risk Assessment. 1997;**3L**: 993-1004

[34] Novak MA, Bode RW. Percent model affinity: A new measure of macroinvertebrate community composition. JNABS;**11**:80-85

[35] Newberry R, Wand Bates DJ. Dynamics of flowing water. In: Hauer FR, Lamberti GA, editors. Methods in Stream Ecology. Volume 1: Ecosystem Structure. London: Academic Press/ Elsevier; 2017. pp. 71-87

[36] Cummins KW. Structure and function of stream ecosystems. BioScience. 1974;**24**:631-641

[37] Cummins KW. Trophic relations in aquatic insects. Annual Review of Entomology. 1973;**18**:183-206

[38] Cummins KW. Combining taxonomy and function in the study of stream macroinvertebrates. Journal of Limnology. 2016;**75**:235-241

[39] Cummins KW, Spengler GL. Stream ecosystems. Americans Bulletin. 1978; **100**:1-9

[40] Merritt RW, Cummins KW, Berg MB. Trophic relationships of macroinvertebrates. In: Hauer FR, Lamberti GA, editors. Methods in Stream Ecology. Volume 1: Ecosystem Structure. London: Academic Press/ Elsevier; 2017. pp. 413-434

[41] Mattson RA, Cummins KW, Merritt RW, McIntosh M, Campbell E, Berg MB, et al. Hydrological monitoring of benthic invertebrate communities of the marsh habitats in the upper and middle St. Johns River. Florida Scientist. 2014; **77**:144-161

[42] Gregory SV, Gunell A, Piegay H, Boyer K. Dynamics of wood. In: Lamberti GA, Hauer FR, editors. Methods in Stream Ecology. Volume 2: Ecosystem Function. London: Academic Press/Elsevier; 2017. pp. 114-126

[43] Merritt RW, Cummins KW, Berg MB, editors. An Introduction to the Aquatic Insects of North America. Dubuque, IA: Kendall/Hunt; 2008. 1158p

[44] King DK, Cummins KW. Autotrophic-heterotrophic community metabolism relationships of a woodland stream. Journal of Freshwater Ecology. 1989;**5**:205-218

[45] King DK, Cummins KW. Factors affecting autotrophic-heterotrophic community metabolism relationships of a woodland stream. Journal of Freshwater Ecology. 1989;**5**:219-230

[46] King DK, Cummins KW. Estimates of detrital and epilithon community metabolism from particle-sized riffle sediments of a woodland stream. Journal of Freshwater Ecology. 1989;**5**: 231-246

[47] Merritt RW, Cummins KW, Berg MB, Novak JA, Higgins MJ, Wessell KJ, et al. Development and application of a macroinvertebrate functional groups approach in the bioassessment of remnant oxbows in the Caloosahatchee River, Southwest Florida. JNABS. 2002; **21**:290-310

[48] Carter JL, Resh VH, Hannaford MJ. Macroinvertebrates as biotic indicators of environmental quality. In: Lamberti GA, Hauer FR, editors. Methods in Stream Ecology. Volume 2: Ecosystem Function. London: Academic Press/ Elsevier; 2017. pp. 293-318

[49] Jackson JK, Batzer DP, Resh VH. Sampling aquatic insects: Collection devices, statistical consideration, and rearing procedure. In: Berg MB, Merritt RW, Cummins KW, editors. An Introduction to the Aquatic Insects of North America. Dubuque, IA: Kendall/Hunt; 2019

Chapter 11

Review of Aquatic Biodiversity Dynamics in the Okavango Delta: Resilience in a Highly Fluctuating Environment

Belda Quetina Mosepele and Ketlhatlogile Mosepele

Abstract

Wetlands are key ecosystems of high biological diversity that provide valuable ecosystem services. These are particularly important in water stressed semi-arid countries, which enhances their vulnerability to degradation. The Okavango Delta, a key wetland in Botswana, is characterised by dynamic inter and intra specific interactions. There are dynamic biotic and abiotic interactions in the system that enhances its resilience. The flood pulse is the main factor mediating bio-physical dynamics in this system. Despite the various perturbations that have been experienced in the system, the Delta has always been able to absorb them and retain its character at the general ecosystem level. These notwithstanding, there have been some changes at the local scale where the Delta has shifted regimes and entered into altered states as a consequence of either channel or lagoon failure. Management of these systems should ensure that their dynamic characteristics are maintained, and this is enshrined within the panarchy concept. Adopting the resilience framework in natural resources management allows for flexibility in devising management strategies to respond to future unexpected events.

Keywords: ecosystem perturbations, Okavango delta, resilience, panarchy, management

1. Introduction

Ecosystems face a major challenge of consistently providing ecological benefits to humanity balanced with biodiversity conservation [1]. This is even more daunting in freshwater wetlands ecosystems, which are not only biological hotspots [2, 3] but are also sources of key ecosystem services that sustain livelihoods globally [2, 4–7]. Services provided by wetlands are not synergistic [2]. According to Maltby and Acreman [6], wetlands are invariably degraded when management enhances the provision of some services like food, which often happens at the expense of other services like regulating services. This scenario is accentuated in tropical areas where, according to Junk [8], politicians prioritise development over environmental protection. Generally, these development priorities are undertaken without any meaningful assessment of the resultant environmental degradation of the wetland [4]. Subsequently, this approach results in wetlands being undervalued "in decisions

IntechOpen

relating to their use and conservation" [5]. Therefore, wetlands have been lost, degraded or significantly modified worldwide [9].

Wetlands are among the most important ecosystems in the world [10]. They have high economic importance in dryland Africa [11] and contribute to the livelihoods of many people in Sub-Saharan Africa [12]. Deriving benefits from wetlands puts ecological pressure on them which may affect their integrity and long term functioning [13]. Furthermore, water security is a major concern in most parts of the world [2], which makes wetlands critical sources of water in arid countries. Because of these intrinsic attributes, freshwater wetlands are vulnerable systems. According to Adger [14], the key parameters of concern to assess vulnerability in ecosystem are (i) the stress to which an ecosystem is exposed, (ii) its sensitivity and (iii) adaptive capacity. "Exposure is the nature and degree to which a system experiences environmental or socio-political stress". "Sensitivity is the degree to which a system is modified or affected by perturbations" [14]. According to Walker et al. [15], adaptive capacity refers to processes in ecosystems when they undergo structural reorganisation and reformation driven by internal processes and external influences. Furthermore, Adger [14] argues that ecosystem vulnerability occurs within the wider framework of the political economy of resource utilisation driven by either deliberate or inadvertent human action.

1.1 The human footprint in the Okavango Delta

The Okavango Delta (OD) is one of the main perennial water bodies in northern Botswana [16]. Its rich biodiversity makes it part of some of the world's most important wetlands [16], where the Okavango Delta is not only one of the world's largest Ramsar sites [17], but is also the 1000th World Heritage Site [18]. Historically, this system has sustained human livelihoods [19]. This wetland is vulnerable to degradation due to increased human impacts for livelihoods due to increasing population growth and socio-economic pressures. Increasing economic activity, especially tourism, in the Delta poses a significant threat to the system [20]. Moreover, the OD faces growing threats from increased agricultural activities.

According to Skelton et al. [21, 22], "insecticide spraying, encroachment of cattle onto the seasonal floodplains, pollution from boat engines, disruption of ecosystem function, and alteration of the flood regime," are some of the potential threats facing the OD. This chapter highlights aquatic ecosystem dynamics of the OD. It then discusses management of this dynamic system within the resilience theory within perturbations that occur in the system.

2. Aquatic ecosystem dynamics

Ecosystem processes and dynamics occur at various spatio-temporal scales [23], and the same has been observed for the Okavango Delta (OD). According to Junk et al. [24], the seasonal flood pulse is the key driver of change in freshwater floodplains where key ecosystem processes are mediated along a hydrological gradient. Therefore, this suggests that the deltas hydrological regime is the main factor mediating biodiversity dynamics in the system.

2.1 Okavango delta

The OD (**Figure 1**) is divided into "a confined entry channel called the Panhandle, the permanent swamp, the seasonal swamp and the occasional

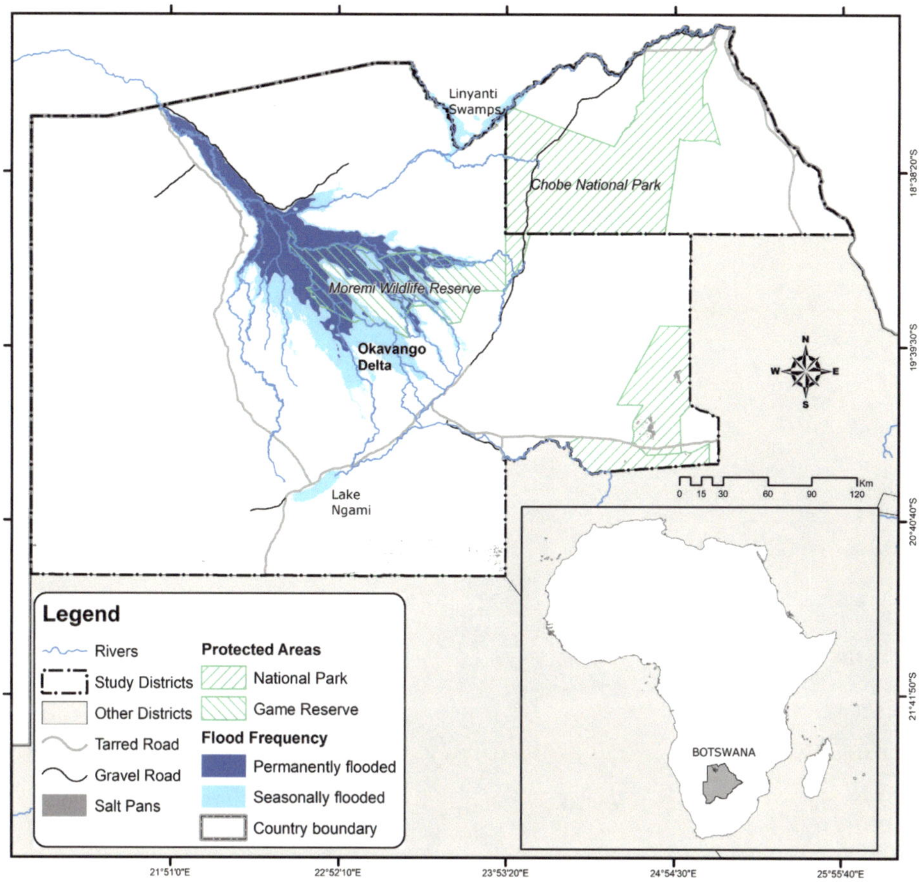

Figure 1.
Map of the Okavango Delta (Map produced by Mr. Masego Dhliwayo of the Okavango Research Institute GIS Laboratory).

swamp" [25, 26] and is one of the largest inland deltas in the world [27]. It is an alluvial fan subjected to annual flooding [28] composed of a mosaic of heterogeneous habitats (**Figure 2A, B, G, H**, [27, 29]) whose total flooded area expands and contracts at seasonal and annual scales. The total flooded area depends primarily on the magnitude of floods from Angola where the Okavango originates. According to Ashton et al. [30], the size of the delta ranges from 6000 to 8000 km^2 during the dry season to approximately 15,000 km^2 during the flood season. It is composed of permanent river channels, semi-permanent river channels, floodplains and lagoons which connect and disconnect due to seasonal flooding [30]. The Delta is also characterised by various isolated pools, ponds and puddles which receive inflow from the rest of the system at intermittent periods [16]. There are approximately 150,000 islands of variable size in the OD [25, 31], out of which approximately 60% of them were developed from termite mounds [25]. Termites are therefore a key ecosystem engineer in the delta [25]. These islands, in combination with the woody vegetation that characterises the islands' fringes, play a key role in removing salts from the OD's waters [31, 32]. Transpiration by woody vegetation creates an osmotic gradient between surface water and groundwater beneath islands and this causes salts to gradually concentrate underneath islands [32, 33] which results in a white salt encrusted centre (**Figure 2B**).

Figure 2.
Heterogeneous habitats of the OD with some of the key animals found in the system where (A) is seasonal grassland habitat with woody vegetation in the background, (B) islands in the OD showing white salt encrusted centres, (C) hippos in a floodplain lagoon, (D) hippos grazing on land, (E) an elephant, (F) herbivores grazing on a seasonal floodplain grassland, (G) river channel with Cyperus papyrus *(papyrus) on the water edge and (H)* Nymphaea nouchali *flowers in the foreground and* Phragmites australis *(reeds) in the background.*

2.2 Flooding dynamics

96% of the total inflow into the OD is lost through evapotranspiration, 2% flows out of the OD through terminal rivers, while the other 2% is lost through infiltration [32, 34]. Peak discharge in the OD occurs around February–March–April (**Figure 3A**, [30, 34, 35]), while maximum flooded extent in the system occurs

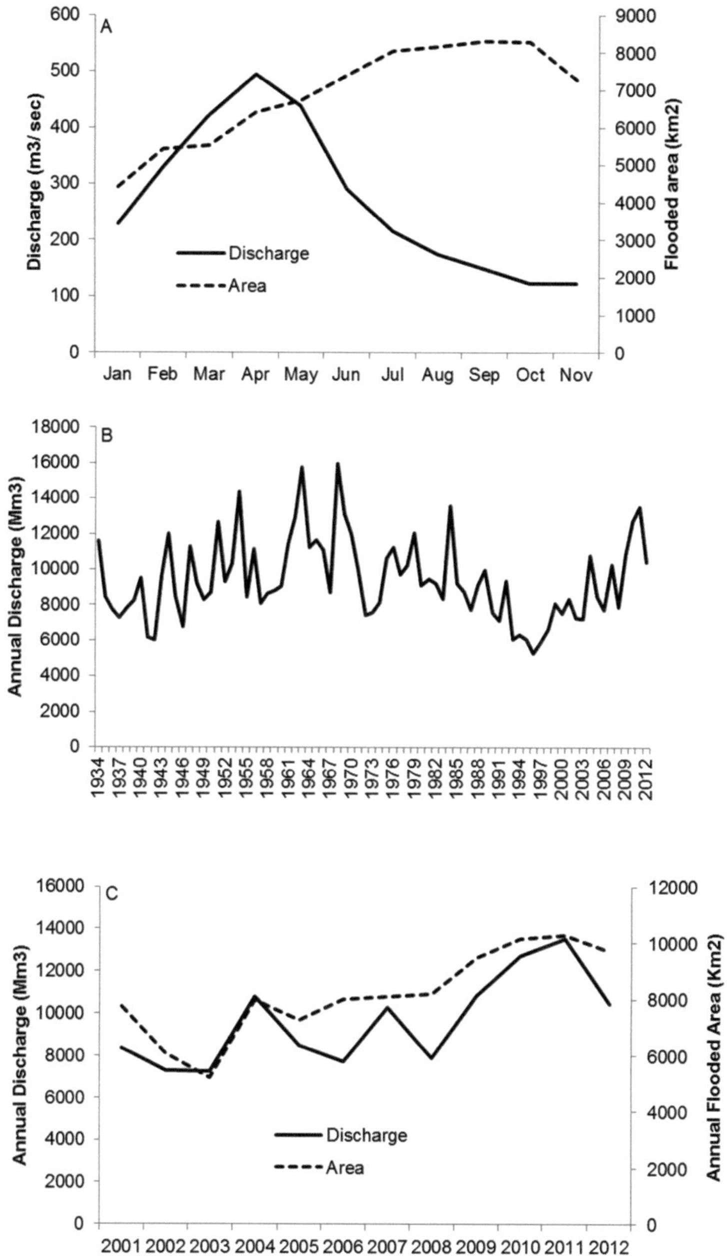

Figure 3.
Map of flooding dynamics in the Okavango Delta showing (A) seasonal discharge and flooded area, (B) annual discharge, (C) annual discharge and flooded area.

around July–August–September (**Figure 3A**, [28, 34, 35]), which is an approximately 5-month time lag between maximum discharge and maximum inundation [35]. The floods percolate slowly across the Delta and reach the mid Delta region around March before reaching the distal ends of the Delta in June–July, during the cold season [26, 30], 6 months after the rain that generated the flow in Angola [28]. On average, the annual food takes 4 months to travel from the inlet at Mohembo, to the outlet at Maun [26, 36, 37]. The amount of flooded area during any flooded season depends on antecedent conditions (**Figure 3C**, [34]), and local rainfall also plays a key role in flooding extent [38]. Inflows into the delta are subject to rainfall patterns in the Angolan catchment area [26]. Prolonged droughts in the 1980s and 1990s caused a decline in the long term average annual inflow into the Delta (**Figure 3B**, [39]). Because of the physiography of the OD, the permanent swamps experience seasonal fluctuations in water levels of approximately 0.2 m while the seasonal swamps undergo fluctuations in excess of 1.5 m [35, 37]. This suggests that the permanent swamps are more hydrologically stable than the seasonal swamps.

2.3 Water quality

The waters of the OD are generally clean and pristine [31, 40], despite the heavy sediment load that they transport annually [30, 36]. However, West [18] has revealed that surface water in some parts of the OD panhandle close to human settlements has high counts of Heterotrophic bacteria, total coliforms and faecal coliforms which makes it harmful to human health. Mogobe [41] found low concentrations of Fe, Mn, Ni, V, Zn, Pb, Cd, Cu, Cr and Co in that order in the OD, which occurred at spatio-temporal scales. Despite the low concentration of these trace metals in the OD, Mogobe et al. [41] highlighted that their concentrations were still a health risk to children in the age range of 6–12 months. Hart [40] also found spatio-temporal variability in several physico-chemical (e.g., conductivity and Si) attributes of surface waters in the delta's panhandle, where TSS values were high near some settlements. Spatio-temporal variations in Beryllium and Aluminium concentrations have also been observed in the panhandle [42]. While concentrations of these heavy metals are generally low in the panhandle, there are times when their limits exceed WHO standards which may cause acute health problems to the panhandle's riparian community [42]. Masamba and Muzila [43] also observed increasing concentrations of major cations (Na, K, Mg and Ca) along the OD which is attributed to evapo-concentration. This observation is consistent with Mosimane [44] who observed a similar trend for the same major cations and also dissolved silica (DSi) and dissolved Boron (B) in the OD. Conversely, concentrations of some trace metals (Co, Cr, Cu, Fe, Mn, Ni, Pb and Zn) generally decreased down the OD which suggests that the Delta acts as a filter for these metals [43]. Mosimane et al. [44] made similar observations for the same trace metals in the Delta. It is noteworthy that while Hart [40] fieldwork was done in 1986 and Mosimane et al. [44] in 2011, 25 years apart, the consistency of the results for conductivity between them suggests that there has been no significant change in the Delta water quality within that period. Generally, the water quality of the OD is good because the values of key parameters falls within international standards for potable water [45]. However, there is concern that some safari establishments might cause localised degradation of water quality as illustrated by depleted DO levels in the periphery of a safari camp in the OD [45].

Water quality in the terminal rivers of the OD is significantly degraded compared to upstream habitats. Tubatsi et al. [46] revealed that turbidity, *E. coli*, and *faecal streptococci* concentrations exceeded those set by the Botswana Bureau of Standards for potable water in the Boro-Thamalakane-Boteti river system, which are outlet rivers of the delta. This is consistent with Tubatsi et al. [47] who observed a spatial variability in water quality from upstream to downstream areas within the same area studied by Tubatsi et al. [46]. Masamba and Mazvimavi [48] also observed spatio-temporal variability in water quality of the Thamalakane-Boteti river system. Water quality was lowest at low flood levels and better at high flood levels [46–48] which reflects a concentration and dilution effect along a hydrological gradient [48]. Coincidentally, diarrheal cases among riparian communities increased significantly during the low flood period compared to the high flood period [46]. The poor water quality of the delta's outlet rivers is caused by various factors like, agricultural activities and pollution from tourism facilities along the river banks [48, 49].

Generally, there is a significant spatial variability in water chemistry in the Delta where the more permanent upstream habitats have lower conductivity and water temperature than downstream areas [20, 21]. Increasing conductivity down the Delta is indicative of evaporative water loss from the system [35]. Downstream habitats are shallower and the higher light penetration facilitates enhanced microbial degradation of organic matter which reduces DO levels in the water column [21]. Generally, there are no significant variations in water chemistry within the seasonal floodplains habitats [16]. However, there is variability in DO percent saturation in the seasonal floodplains which is attributed to higher primary production in some areas [21].

2.4 Nutrient dynamics and primary production

Nutrients into the OD are transported either through incoming water [36, 37, 40], aerosol deposition [36, 50], or herbivore dynamics in the seasonal food-plains [29, 36, 51]. The Delta's riverine habitats are oligotrophic, the swamps and floodplains vary between oligotrophic and mesotrophic while isolated water bodies are generally eutrophic [16]. This observation is consistent with McKay [20] that the OD is generally a nutrient poor system, where Ca is the most abundant cation in the system. Aerosol sediments/nutrients are estimated at 250000 tons year^{-1} in the OD [36] while approximately 450,000 tons year^{-1} are transported through incoming water [30]. Moreover, approximately 300,000 tons year^{-1} of suspended sediments (170,000 tons year^{-1} of aelion sand and 300,000 tons of kaolinite) are transported by incoming water [30]. Hart [40] estimated a total suspended solid load of 1×10^5 tons year^{-1}. Therefore, total nutrient loading in the system is an average of 725,000 tons year^{-1}. According to Garstang et al. [36], most of back-swamp habitats derive their nutrient input from aerosols and peat, and none from water. This makes aerosol nutrient deposition critical for the high productivity observed in the seasonal floodplains.

The shallow, seasonally flooded floodplains are the key engine of nutrient cycling in the OD [21, 44] where most of the primary production occurs [21, 31]. Initial flooding in the seasonal floodplains elicits increased concentrations of P and N which gradually decreases over time with decreasing flooded area [50, 52]. These initial nutrients were trapped in the soil and subsequently dissolved in the oncoming floodwaters. However, more nutrients, especially total N, nitrates and chlorides are added on the water through dust deposition as the flooding progresses [50]. Furthermore, Hoberg et al. [52] observed that both Chlorophyll *a* (μg l^{-1}) and primary production (μC l^{-1}) increase rapidly within the first week of new floods in the floodplains, followed by a gradual decrease towards the end of the first month of flooding.

The seasonal floodplains are also play a key role in DOM and DOC cycles in the Delta [16, 21, 26]. DOM and DOC fluxes are also mediated by the seasonal flood pulse in the Delta where DOM mobilisation is facilitated by annual flooding [26]. The new floods facilitate DOM microbial degradation which then results in an increase of vegetation derived DOC with increasing floods. The flood pulse in the floodplains also facilitates bacterial consumption of DOC [26]. There are higher DOC concentrations along the delta [26] and also between channel and floodplains habitats [53], which has been attributed primarily to evapo-concentration [26, 53]. Therefore, river-wetland interactions and evapo-concentration are the key drivers of carbon cycling in the Delta [26, 53]. The general trend is that the concentration of solutes increases from upstream habitats to the terminal rivers of the Delta [53]. According to Mladenov et al. [54], there are dynamic fluxes in DOC concentrations in the seasonal floodplain where DOM/DOC availability in the water column alternates between vegetation derived and microbial sources mediated by seasonal flooding. However, organic C derived from vegetation is a greater input to DOC in the floodplains than microbial sources within the floodplain. Overall, vascular plants are the main source of DOM in the system and the continuous input of freshly leached DOM from the floodplains is facilitated by inundation [55].

2.5 Micro-invertebrates

Forty-six micro-invertebrate taxa composed of 27 rotifers, 12 Cladocerans, 6 copepods and 6 ostracods have been identified in the temporary floodplains [56], distributed heterogeneously among the different micro-habitats. Conversely, 59 micro- invertebrate taxa were identified in the seasonal floodplains (i.e., seasonal, temporary and rarely flooded floodplains), composed of 35 rotifer species, 20 micro-crustacean species and four other taxa groups [57]. These dynamics are flood pulse driven [56, 57]. Micro-invertebrate diversity was highest within sedges while abundance was highest among floodplain grass [56]. Copepods were the most dominant taxa in the floodplains [56, 57]. Meanwhile Siziba et al. [57] revealed that micro-invertebrate diversity was high in seasonal floodplains while their density was significantly highest in the rarely flooded floodplains. Taxa of micro-invertebrates that emerged from frequently flooded floodplains sediments was higher than those that emerged from sediments of rarely flooded floodplains [58]. Therefore, while rarely flooded floodplains are key habitats for micro-invertebrate production in the OD, Siziba et al. [58] argue that high flooding frequency of seasonal floodplains is necessary to ensure that the integrity of propagules from this habitat is maintained.

According to Hoberg et al. [52], *Alona affinis*, *Ceriodaphnime quadrangula*, *Chydorus* sp., *Daphnia laevis*, *Macrothtrix* sp., *Moina micrura* and *Simocephalus vetulus* dominated the zooplankton community in the seasonal floodplains over a 3-month flooding season (June–August). Soil egg banks are the main inoculum of these zooplankton populations. These hatch from nesting eggs in soil egg banks [52, 58–60] where Cladocera, copepods and ostracods are the major groups [58]. The zooplankton biomass also fluctuated in relation to flooding in the seasonal floodplains where biomass peaked towards mid-June before almost becoming extinct in late August, at the end of the flooding season. Species succession characterised zooplankton population dynamics during the course of the flooding season [52, 59]. Initially, the zooplankton community in early June was dominated by *Moina micrura*, whose biomass was then surpassed by that of *Daphnia laevis* in early July [52, 59].

Overall, zooplankton species diversity is lowest in the permanent swamps and highest in the seasonal floodplain [60]. Species abundance and diversity also varies between habitats, where littoral habitats are less diverse and are dominated by *Caridina africana*. Some lagoons are dominated by *Tropodiaptomus kissi* while others by *Bosmina longirostris* [40]. Moreover, species diversity increases with increasing flood inundation in the seasonal floodplains [60]. Juvenile fish start appearing in the seasonal floodplains around July, which coincides with pronounced cyclomorphosis in *D. laevis.* A reduction in the *D. laevis* populations is then followed by an increase in *Chydorus* sp. towards the end of the flooding period in August [52]. This zooplankton production in the seasonal floodplains grazes down phytoplankton biomass whose abundance gradually decreases in concert with zooplankton populations [59]. Generally, the seasonal floodplains have higher abundance of desmids than the permanent swamps [27]. This agrees with Cronberg et al. [16] who also observed that desmid populations were more abundant and diverse in shallower parts of the OD. Ultimately, seasonal floodplains are a source habitat for the cladorecan meta-community where ephippia are dispersed from seasonal floodplains to other habitats in the delta, possibly by wind and mammals [60].

2.6 Seasonal vegetation dynamics and herbivore populations

Seasonal wetting and drying processes are key to enhancing primary production in the Delta. Drying processes release nutrients which are then trapped in the water column during the wetting phases [61]. At low water levels, herbivore herds (**Figure 2F**) enhance nutrient loading in the system through bioturbation and defecation [21, 29, 31, 51]. Flooding dynamics in the seasonal floodplains facilitate soil nutrients transport [62]. This is consistent with Bonyongo and Mubyana's [63] analysis that seasonal flooding is a key source of soil nutrients that sustains vegetation growth in the system. Flooding predictability and an extended feeding period have made the Delta one of the most productive systems globally, which maintains herbivores populations 4–8× more than similar wetlands [51]. Furthermore, hippos graze on nutrient rich soil vegetation like *Cynodon dactylon* and they end up enriching lagoons through defecation (**Figure 2C, D**, [29, 36, 64]). These nutrient enriched lagoons are therefore able to maintain high fish biomass production [64, 65]. Essentially, herbivores in the Delta create a nutrient loop where they feed on floodplain grass, deposit their dung which contributes nutrients to the system, which are trapped by new floods, which are released during the drying phase and are also used in grass production [29, 51].

2.7 Aquatic macro-invertebrates

Ninety-four Odonata species made up of 33 Zygoptera and 61 Anisoptera species have been described in the OD. Moreover, 37 micro-crustacea (16 Copepoda and 21 Cladocera) and 22 mollusca species (16 Gastropoda and 6 Bivalvae) have also been described for the OD [51]. The Delta's macroinvertebrate populations are relatively uniform across the Delta, but are structured along micro-habitats within the delta's macro-habitats [66, 67]. However, one exception to this is *A. caridinum* (freshwater shrimp) whose abundance decreases along a hydrological gradient from the upper to the lower Delta [66]. There are approximately 184 morpho-species taxa in the OD, which cover 63 families [67]. The dominant taxa in the OD are Hemiptera and Mollusca, while Oligochaeta, Lepidoptera and Acarina have only one family each [67]. Dallas and Mosepele [67] observed that marginal vegetation has the highest number of morpho-species per habitat while backwater detritus habitats in the OD have the lowest morpho-species richness. Apart from chironomids, sediments have a paucity of invertebrate taxa which Appleton [66] attribute to anoxic conditions caused by microbial degradation of organic matter.

2.8 Fish

There are 71 fish species in the OD [61] distributed heterogeneously among the different habitats [29]. Generally, fish species diversity is higher in the permanent swamps and lowest in the seasonal floodplains [68, 69]. Based on the Index of Relative Importance (%IRI), the fish community is dominated by *Clarias gariepinus* (**Figure 4J**), *Schilbe intermedius* (**Figure 4H**) and *Hydrocynus vittatus* (**Figure 4G**), respectively. *C. gariepinus* dominates the community during years of poor/low floods while *S. intermedius* dominates the fish community during years of good/high flood years [70]. *Clarias gariepinus* (sharp-tooth) is the biggest fish species in the OD while *Rhabdalestes maunensis* is the smallest species found in the delta [71]. Insectivores are the dominant feeding guild in the fish community [70] which attests

Figure 4.
Some key fish species in the OD showing (A) Oreochromis andersonii, *(B)* O. macrochir, *(C)* Marcusenius altisambesi, *(D)* Coptodon rendalli, *(E)* Hepsetus cuvieri, *(F)* Brycinus lateralis, *(G)* Schilbe intermedius, *(H)* Hydrocynus vittatus, *(I)* Tilapia sparrmanii, *and (J)* Clarias gariepinus.

to the importance of terrestrial food sources in the system. According to Mosepele [71], half of the fish species in the OD do not grow bigger than 17 cm (total length), while only 10% of the fish species growth bigger than 65 cm (Total Length). Therefore, the OD fish community is dominated by smaller sized fish species.

2.8.1. Community structure

The Delta fish community undergoes temporal variability along a hydrological gradient [70] and is also characterised by spatial variability [65, 72]. On a temporal scale, this variability is driven by fish longitudinal and lateral migrations [73].

Lagoons, which are a key fish habitat in the Delta, are dominated by the Cichlidae family [65]. According to Mosepele et al. [65], lagoons along a spatial gradient in the OD have different fish community structures, and morphometric attributes of some species are also significantly different among them. *Marcusenius altisambesi* (**Figure 4C**) and *Brycinus lateralis* (**Figure 4F**) contribute to the differences in fish species assemblages among the lagoons [72]. Mosepele et al. [65] also revealed that dry season fish biomass in seasonal floodplain lagoons was significantly higher than those from permanent swamp lagoons. This observation agrees with Mosepele et al. [72] who also showed that there are significant differences in fish abundance among lagoons in the Delta. Moreover, fish dynamics in the lagoons is driven by environmental variability [72]. Insectivores are the dominant fish feeding guild in the lagoons [65]. Notably, *Hydrocynus vittatus* (tiger-fish), which is one of the top piscivores in the system, is found mostly in the upper delta whilst *Hepsetus cuvieri* (African Pike, **Figure 4E**) also a top piscivore, is found mostly in the lower seasonal delta [69, 73, 74]. *H. vittatus* is a visual predator which occupies riverine habitats of the permanent swamps while *H. cuvieri* is an ambush predator which prefers sluggish backwater habitats found in the seasonal floodplains [64, 69].

2.8.2 Life history strategies

The Delta's seasonal flood pulse is a key driver of fish population dynamics in the system [75]. Most fish species undertake longitudinal migrations along the Delta's main channel as a reaction to the onset of the floods [68, 73]. However, some species then migrate into the seasonal floodplains once the new flood waters spill onto the floodplains [70]. Different species have developed plastic life history strategies as an adaptation to this dynamic system [68]. *S. intermedius* in the permanent swamps spawn in February just before peak floods, while the same species reaches peak spawning in October, almost 3 months after the floods' arrival in the seasonal floodplains [76]. *Oreochromis andersonii* (**Figure 4A**), *O. macrochir* (**Figure 4B**) and *Coptodon rendalli* (**Figure 4D**) between permanent and seasonal parts of the OD have different growth rates [74]. However, juvenile cichlids of *Tilapia sparrmanii* (**Figure 4I**) and *C. rendalli* grow faster in the permanent swamps lagoons than those from the seasonal floodplain lagoons [72]. Merron [68] showed that *O. andersonii* from the seasonal floodplains mature at a smaller size, grow faster and are less fecund than those from the permanent swamps. Therefore, fish species from the permanent swamps were *K* strategists while those from the lower delta were *r* strategists. This agrees with Bokhutlo [77] who established that there are two distinct populations of *C. gariepinus* between the permanent swamps and the seasonal floodplains. Their study showed that *C. gariepinus* from the seasonal floodplains grow faster and reach a smaller maximum size compared to those from the permanent swamps. Due to the lower DO levels in the seasonal floodplain habitats, *H. cuvieri* lays its eggs in an oxygen enriched foam nest to enhance the probability of survival for its young [68]. Moreover, several key cichlid species are mouth-brooders, which is also adaptation strategy to enhance the survivability of their young in a low DO environment [31].

2.8.3 Biology and ecology

Fish feeding [68, 73, 76, 78], spawning [70, 73, 76], growth [77, 78] and mortality [78] are all flood pulse driven [71] in the OD. The annual catfish run is perhaps the most visible impact of flooding on fish feeding ecology in the delta [68, 73]. In this phenomenon, *Clarias gariepinus* undertake seasonal feeding migrations at receding water levels when their prey species are back migrating from the drying out floodplains. Mosepele et al. [78] has also showed that the feeding ecology of some selected species is driven by the hydrological regime. They showed that *S. intermedius* diet is driven by discharge while that of *Marcusenius altisambesi* is driven by water depth. Mosepele et al. [78] showed that *S. intermedius* preyed on terrestrial insects. This agrees with Mosepele et al. [78] who revealed that diet for 1-year-old *S. intermedius* includes ants and bees at the onset of floods which were possibly drowned by advancing floods in the seasonal floodplains. Mice were found in the diet of 2-year-old *S. intermedius* at peak floods in the seasonal floodplains. This opportunistic feeding behaviour was also observed by Merron and Mann [76], who observed that terrestrial insects constitute 14% of *S. intermedius* diet. Mosepele et al. [78] showed that seeds were found in the diet of *Brycinus lateralis* and 1-year-old *Marcusenius altisambesi*. Mmusi et al. [79] then showed that *B. lateralis* ingests *Nymphaea nouchali* seeds which remain viable after passage through its gut. They concluded that *B. lateralis* might be one of the dispersal agents for *Nymphaea nouchali*, which makes *B. lateralis* a key ecosystem engineer. Generally, the Delta's fish community is characterised by trophic differentiation, diet flexibility and ontogenetic diets which minimises competition for food and maximises energy uptake [78].

2.8.4 Floodplain dynamics

Fish spawning behaviour is partitioned along a hydrological gradient in the OD [70]. Massive spawning among the delta fish species occurs just before maximum inundated area in the delta. The fingerlings then migrate out onto the seasonally flooded floodplains where they graze on zooplankton. However, seasonal floodplains are used not only by fish juveniles, but also by small sized fish species. Siziba et al. [80] observed that 38 small sized fish species belonging to 11 families used the seasonal floodplains. Species from the Poecilidae family dominate the primary and temporary flooded floodplains, while juveniles from the Cichlidae family dominate the rarely flooded floodplains [52, 80]. This is consistent with Siziba et al. [57] who observed that *temporary* floodplains are key nursery sites for juvenile fish. According to Siziba et al. [81], micro-crustacea dominated the diet of juvenile cichlids from rarely flooded floodplains than for those from frequently flooded floodplains. *M. spinosa*, *C. sphaericus*, and *M. micrura* dominated the diet of juvenile cichlids from rarely flooded floodplains while *Chydorus sphaericus* dominated the juvenile cichlid diet from frequently flooded floodplains.

2.9 Aquatic vegetation

There are approximately nine distinct plant communities in the Delta distributed along a hydrological gradient. Seven of these are wetland communities which range from permanently flooded swamps to the seasonally flooded floodplains. The remaining two communities are riparian woodlands that are not flooded but have species whose roots are in the water table of the both the permanent and seasonal floodplains [82]. *Cyperus papyrus* (papyrus) and *Phragmites australis* (and *P. mauritanius*) (reeds) dominate the permanent swamp habitats of the OD [31]. According to Ellery et al. [35], *C. papyrus* exists as unattached "semi-floating mats of rhizomes and roots". Papyrus enhances water velocity in the river channels through constriction of the channel by inward growth of channel margin vegetation [35]. Furthermore, the permanent swamps are characterised by aquatic grasses like *Vossia cuspidata* and *Echinocloa pyramidalis*, while the lower reaches are dominated by a patchy mosaic of aquatic, semi aquatic and terrestrial vegetation [82]. According to Mendelsohn et al. [31], dense growth of papyrus and reeds dominates the deeper waters of the permanent swamps while swamp grass dominates the shallower portions of the swamps. However, this vegetation does not encroach into the river channels as long as the water current is strong. The main channel river banks of the permanent swamps are "flanked by peat deposits" characterised by species like *Miscanthus junceus* and *Pennisetum glaucocladum* [35].

3. Theoretical framework: resilience in ecosystems

Resilience is a broad concept that incorporates the ecological character of ecosystems and the broader expanse of more complex socio-ecological systems [83, 84]. At the ecosystem level, resilience is the capacity of an ecosystem to maintain its identity amidst disturbance [15, 23, 85], or the rate at which species composition in an ecosystem returns to equilibrium following a major reduction in species [86]. Holling [87] defines resilience as the ability of systems to absorb change and disturbance and still maintain the same relationships between populations. Resilience has also been defined as the ability of an ecosystem to return to its reference state after a perturbation [84] or the capacity to deal with perturbations [88]. According to Angeler et al. [3], resilience is the capacity of ecosystems to absorb change without "moving to another stable state". This capacity lies in the regenerative ability of ecosystems to continue to provide services that are essential for human livelihoods [89]. Therefore, resilience is the capacity of ecosystems to absorb disturbance, or the buffer capacity of ecosystems that allows persistence [87, 90]. However, Peterson et al. [91] define resilience as the measure of the amount of change or disruption that is for a system to undergo regime shifts.

Based on these definitions, the three elements of resilience are (i) the amount of change that an ecosystem can undergo, (ii) the degree to which an ecosystem is capable of self-organisation, and (iii) the degree to which an ecosystem can build the capacity to learn and adapt [92]. Moreover, latitude, resistance, precariousness and panarchy are four crucial aspects of resilience identified by Walker et al. [85]. These various ecological definitions of resilience suggest that there are several ecological stable states of ecosystems which are occupied by resilient systems [3, 87]. Therefore, ecosystems can operate at different levels of what has been termed either "basins of attraction" [3, 85, 87] or "domains of attraction" [93]. However, the ability of an ecosystem to tolerate disturbance is finite [94], and when critical perturbation limits are exceeded, a different pattern of behaviour in the ecosystem may emerge [93, 95]. Therefore, when resilience is breached, system behaviour can in some cases change catastrophically from one basin of attraction to another [96].

Some dominant processes in ecosystems create "discontinuities" in the structural features of the system which are expected to persist despite the normal dynamics of the system [95, 97]. Changes in the structure of the system will only be observed if the system is pushed beyond the "limits of its resilience" [95, 98, 99]. It is therefore important to highlight the adaptive responses of socio-ecological systems to a transforming biosphere in order to understand ecosystem resilience and hence prevent collapse [97]. Generally, freshwater ecosystems are exposed to various pollutants from human impact which can alter the structure and functioning of these systems [94]. A concomitant exposure of aquatic wetlands to multiple stressors like physical hydrological alterations, climate change, species changes and pollution may enhance the vulnerability of these systems [3]. The combined effects of these stressors make ecosystems more vulnerable to changes that could previously be absorbed [90]. On the other hand, diversity in systems (i.e., genetic, species and landscape levels of biodiversity) enhances their ability to cope with shock and stress, which reduces their vulnerability [84, 98]. Vulnerable ecosystems have lost resilience, which implies loss of adaptability [90] which may result in an ecosystem collapse.

Resilience in ecosystems is enhanced by a suite of interactions between biotic and abiotic components of systems which create loosely structured hierarchical systems [98]. Resilience varies across scales, and this cross-scale structure, which is also described as a panarchy, is manifest as a nested set of adaptive cycles with clearly differentiated structures across scales [100, 101]. According to the panarchy approach, during reorganisation at a given scale within ecosystems, conservative structures at larger scales provide a system memory that allows for reorganisation around the same structures and processes instead of shifting to a different regime [84, 101]. Panarchy allows for adaptive management that enhances the resilience of ecosystems. According to Allen et al. [102], the three elements for managing systems for resilience are, taking action to prevent unwanted regime shifts from occurring, ensuring that the diversity of elements and feedback loops that keeps a system in the desired state is maintained, and reducing the likelihood of system crashes or flips to different states. These regime shifts result in shifts in ecosystems services, with consequent impacts on human societies [103]. That notwithstanding, Angeler et al. [104] has shown that these different stages are also as equally resilient as the original state. However, the new resilient state might not provide the ecosystem goods and services that were provided by the original ecosystem in its unaltered state. Therefore, systems may be ecological resilient but not socially acceptable [83], because of this loss in ecosystem services. Ultimately, the loss of resilience in ecosystems reduces their capacity to adapt to change [83].

The SES approach is central to the resilience framework theory [105]. According to Cummings [106], SES theory encapsulates ideas from resilience and vulnerability, among other disciplines. Subsequently, Carpenter [105], argue that important aspects of resilience in socio-ecological systems (SES) cannot be observed directly, but must be inferred. A key aspect of resilience theory in SES is its emphasis on adaptive capacity in analysing human/ecological relations [107]. Therefore adaptive dynamics are inherent to SES's. This argument agrees with Folke et al. [108] discussion that resilience within socio-ecological systems means the ability of an SES to continually change and adapt but remain within certain critical thresholds. Therefore continuous transformation and adaptation are key tenets of resilience theory in SES. Transformation and adaptation in ecosystems are encapsulated in the adaptive cycle theory which suggests that SES tend towards the four characteristic phases of (i) rapid growth and exploitation (*r*),

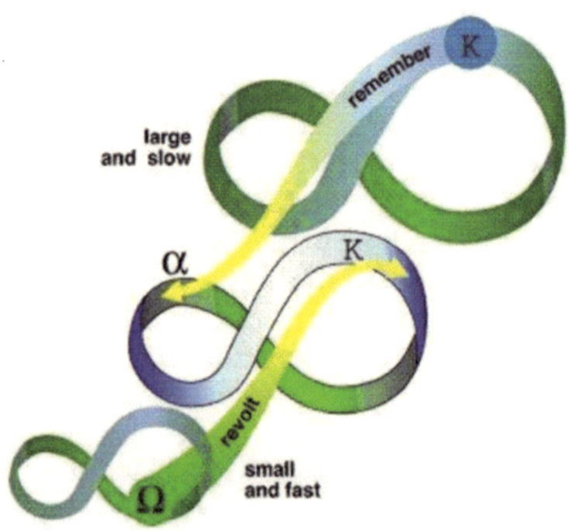

Figure 5.
A representation of a Panarchy (This figure was originally published in Panarchy: Understanding transformations in human and natural systems, edited by Lance H. Gunderson and C.S. Holling 2002. https://www.resalliance.org/panarchy) [Accessed: 15 May 2020].

(ii) conservation (*K*), (iii) collapse (Ω), and (iv) reorganisation (α) (**Figure 5**) [23, 85, 92, 100, 105, 109]. The period between the K and Ω phases is characterised by increased connectivity which results in decreased resilience in the ecosystem [93]. Resilience in the SES is high during the exploitative or the *r* phase [23]. This adaptive cycle shows two different loops where the first loop (*r* to *k*) is the slow gradual phase of growth and accumulation, while the second loop (Ω to α) is a rapid reorganisation phase [100, 109]. These define what Holling [98] defines as panarchy, "a concept that explains the evolving nature of complex systems" (**Figure 5**). It is the hierarchical structure in which SES undergo the adaptive cycle described above, which is represented in **Figure 5**.

4. Perturbations in the system: the past, present and future

Various bio-physical perturbations have been experienced in the OD, in the past, currently and also possibly in the future. These perturbations will invariably affect the resilience of the system, and may also result in regime shifts in the system from one basin of attraction to another. While these regime shifts may not result in ecosystem collapse, they may nonetheless result in some losses of ecosystem services. This might result in a system that is not socially acceptable. Nonetheless, most potential major impacts on the OD are external to the system [28] and can occur upstream in either Angola or Namibia. Most of these potential impacts are agricultural activities, water abstraction schemes or dams [30, 51]. However, any upstream developments that might affect the integrity of the OD will be confined within the framework of treaties and conventions at regional and global scales [39]. While OKACOM is the key regional entity mandated with management of shared water resources, it is currently "ill equipped" to deal with economic and socio-political pressures that underpin water use and allocation strategies in the region [110].

Climate change is another potentially major perturbation that can affect the overall ecosystem function and resilience of the OD. Some models predict increased temperatures over the region which may increase evapotranspiration rates from the delta [111], while other models indicate that there is still uncertainty on the overall impact of climate change on the delta [38]. Furthermore, tsetse spraying and channelling are some of the past major perturbations in the OD [51]. In fact, Moses [111] argues that climate change may cause ecosystem collapse which will result in loss of ecosystem services in the OD.

4.1 Tsetse fly spraying

Some key biological perturbations include what has been termed nuisance insects. "Nuisance insects" is one of the wetlands attributes that has created an "antagonistic relationship" between wetlands and humans [7]. These insects created a biological barrier that prevented cattle from accessing wetlands grass for cattle during the dry season [112]. According to Junk et al. [7], river-flooded grasslands associated with flood-pulsed wetlands are characterised by high productivity. Therefore, Tsetse fly could be classified as nuisance insects because they reduced/impeded socio-economic activity within the OD. This agrees with Ramberg et al. [61] who observed that the Tsetse fly (*Glossina morsitans*) protected the OD against farming, especially livestock production, until it was eradicated from the system.

Various control measures have been undertaken to eradicate the Tsetse fly from the delta. This included game destruction and bush clearing, fences to restrict wildlife movement [112, 113], large scale ground applications and aerial spraying of Dieldrin and Endosulfan, and ultimately spraying with Deltamethrin [112].

Insecticide applications have had some negative impacts on some of the OD's biodiversity like fish [114, 115], piscivorous birds [115] and aquatic invertebrates [116]. Direct fish mortalities were observed from Endosulfan spraying for Tsetse fly in the OD, even though no significant differences in diet were observed for *S. intermedius* and *M. lacerda* [114]. Kingfishers feeding rates decreased after fish populations within their vicinity died from aerial spraying with Endosulfan [115]. Moreover, The Deltamethrin spraying resulted in significant reductions in the abundance of some common aquatic invertebrate taxa, while other habitat specific taxa disappeared after the spraying [116].

However, aerial spraying with Deltamethrin had minimal impact on fish [117] while aquatic invertebrates at the community level recovered after 1 year [116]. According to Merron [117], some surface feeding fish species like *Aplocheilichthys johnstoni* and *Barbus haasianus* only showed some disorientation after Deltamethrin spraying. Moreover, the Atyidae and Pleidae families were negatively affected and their abundances were still low a year after Deltamethrin spraying. Furthermore, a morpho-species, Notonectidae, were also negatively affected by the Deltamethrin spraying 1 year later [116].

4.2 Channelling and other human impacts

One of the outlet rivers from the OD was dredged in the 1970s ostensibly to increase water flow to Maun [118, 119] and further downstream to the diamond mines in central Botswana [119, 120]. This involved "straightening, bunding and dredging the original river channel". However, this process resulted in a loss of the fringing wetland areas around the channel [119, 121]. Creating channels in the OD has been used extensively in the past to facilitate water flow in the system [122]. Most of these channels were created to bypass river blockages in the system that were created by river rafts made form papyrus that were used traditionally as water transport. These were discarded on and along river channels and they subsequently facilitated river channel failures because they provided initiation material for growth of emergent vegetation which ultimately blocked channels [122].

4.3 Plants

Biotic components of wetlands play a major role in regulating the hydrology of the OD [123]. Channel encroachment, especially by papyrus growth, contributes to channel failure in the OD [118, 123]. *Ficus verruculosa* is another key plant that has been associated with the failure of some prominent river channels in the OD [118] especially outlet channels [122]. Organic sudds form during summer in the OD's channels and lagoon surfaces. Some of these sudds then form a base for the growth of *Pycreus nitidus*. Other plant species that are associated with sudd formation are *T. capensis* and *Nympahea caerulea* [124]. Eventually, these facilitate dense growth of emergent vegetation [123] which results in channel width narrowing [124], that sometimes results in channel or lagoon failure.

4.4 Animals

Hippopotamus play a major role in river channel development in the OD [31, 123]. They open up channels leading into floodplain lagoons which enhances system connectivity. Hippos normally use the same track to move from lagoons for grazing which enhances water flow in the system [29]. Ultimately these hippo paths develop into river channels which can have a significant impact on the geomorphology of the OD. These new hippo channels can either facilitate flooding and linkages among water bodies on the delta's landscape or they can also contribute to lake failure [125]. Hippo channels also contribute to water loss through channel margins to peripheral floodplains. Therefore, hippos are key drivers of channel and lagoon dynamics in the OD [29, 125].

Elephants, which occur at high populations in the Delta [61], contribute to channel development thereby enhancing water flow [29]. Their heavy weight when they move over floodplains during the dry season creates path depressions which become flood pathways during flooding [29]. At the ecosystem level, these processes affect ecosystem resilience.

5. Discussion

In arid countries, there is a pronounced dry and wet season which acts as a key driver of change in wetlands [126, 127]. The Okavango Delta is flood pulse driven, which is a key driver of seasonal change [38, 61]. Other perturbations that have been identified in this study are exotic species [31], channelization [118], and pollution [30], whose impacts on the wetlands vary in space and time. The seasonal flood-pulse causes ecosystem wide perturbations in these systems, and causes constant ecosystem reorganisation. The flood-mediated change in the systems, where change occurs at various spatio-temporal scales, is a good illustration of the panarchy concept. In this case, at the primary production level, the seasonal flood pulse in the Delta creates boom and bust conditions at the onset of flooding in the seasonally flooded portions of the Delta. These production cycles scale up the ecosystem and ultimately provide goods and services that are characteristic of wetlands ecosystems.

According to Gunderson [23], regime shifts in ecosystems caused by disturbance can result in a resource crisis because the system could normally have a reduced ability to provide ecosystem services. Gunderson [23] discusses that the three management approaches within this scenario are, (i) to do nothing and wait for the system to return to its reference state, (ii) to actively manage the system in an attempt to return it to some desirable state, and (iii) to adapt to the new altered state.

5.1 The do-nothing approach

One example in the OD occurred in the 1980s when Lake Ngami dried up [128]. The lake supported a small scale commercial fishery when it had water [73]. However, a loss of water in the system meant that it stopped providing this ecosystem service. It was also a source of water for livestock in the region [118], and its desiccation meant a loss of this ecosystem service. The lower part of the OD also remained dry for several years [128], with a consequent loss of its ecosystem services. Several studies have shown that the OD is subject to an "80-year climatic oscillation" (**Figure 3B**, [34]) which affects water availability and distribution in the system. Tectonic shifts in the system also sometimes result in water flowing more to the eastern part of the delta, with none (or little) to the western portion [34, 129], which results in Lake Ngami not receiving water. Gumbricht et al. [34] further highlights that dense aquatic vegetation growth also had a compensatory effect of diverting more water to the eastern part of the delta, especially during years of low floods. Essentially, the system is constantly self-organising in concert with these flooding dynamics. In this respect, the desiccation of certain portions of the delta does not necessarily suggest that the system has irrevocably changed into a terrestrial environment. It has retained its flooding memory, and will certainly reflow. This happened to Lake Ngami which received floods in 2004 after being dry for over 20 years [128]. In this case, the "do-nothing" is the best management strategy because the system invariably springs back to its former character. This suggests that the system is still resilient because it was able to revert to its former shape after a perturbation.

5.2 Active management approach

Historically, many attempts have been made to manage the flow regime in the Okavango Delta through physical processes [118, 120, 122]. These attempts were aimed at either unblocking river channels to allow for unimpeded flows [122], or to simply direct water flow elsewhere for human use through dredging [119]. According to Bernard and Moetapele [130], the exclusion of local communities perhaps contributed to the desiccation of the Gomoti River channel, which is one of the outlet Rivers from the OD. They highlight that local people were historically involved in active burning in the Delta during low water levels to allow for free water movements when the floods arrived. Bernard and Moetapele [130] also highlighted that local people actively removed any vegetation build ups they encountered on the river channel during their forays in the Delta. Subsequently, the desiccation of this river channel resulted in a loss of ecosystem services that the peripheral riparian communities used to derive from this river. As highlighted earlier, channelization was common in the OD in the past [118, 122] and also active removal of blockages [122]. In some cases, Ellery and McCarthy [119] noticed significant encroachment of terrestrial vegetation onto the dredged channels. Overall, these active management strategies never served their intended purpose [131] but rather facilitated a shift in regimes of the ecosystem at a local scale.

Tsetse fly eradication is one major activity in the OD which illustrates active management of the system to facilitate access to its ecosystem services. The Tsetse fly is the vector for Trypanosoma which causes sleeping sickness in people and nagana in cattle. This disease is idiosyncratic to Sub-Saharan Africa which affects approximately 50 million people, 48 million cattle with estimated annual losses in cattle production ranging between USD 1–2 million [132]. The ecological impact of eradicating this species has not yet been evaluated in the OD at the ecosystem level. However, it is undoubtable that the presence of this fly restrained or regulated human encroachment to access most ecosystem services provided by the OD. The extensive use of pyrethroids as insecticides has increased their likelihood as aquatic pollutants [133], with unforeseen impacts on aquatic biodiversity. However, spaying for Tsetse fly generally resulted in localised ecological impacts of untargeted aquatic organisms. Eradicating the Tsetse fly using Deltamethrin had two major unforeseen consequences in the Delta, (i) short term impacts which resulted in aquatic macroinvertebrate mortalities in the Delta (ii) Long term impacts where the absence of the Tsetse fly opened up the Delta to a greater human footprint impact. This has resulted in some unsustainable tourism developments that have had negative environmental impacts. These will invariably affect the dynamicity of the system, will increase its vulnerability and imperil its resilience.

5.3 The new altered state approach

The western part of the OD has now transformed into a terrestrial environment from the aquatic ecosystem that used to exist. This was driven by changes in river flow either due to channel blockages by vegetation [118], or to plate tectonic shifts where most water in the system started to flow east of the OD [34]. In this scenario, the best management approach is to accept the new altered state of the system and manage it in the best possible way. Generally, the western portion of the OD has shifted from a floodplain system to a terrestrial habitat which occurred due to the failure for the main channel system in that part of the OD. This regime shift resulted in a grassland ecosystem which has been able to maintain large herbivore herds [134]. It has shifted from a stable aquatic state to a stable terrestrial state.

Ecosystems have been generally shaped and managed to derive services for human livelihoods. Therefore, natural ecosystems are changing rapidly, driven by socio-environmental conditions. Moreover, the speed of global changes through factors such as climate change, and the increasing human footprint are creating a "dynamic, uncertain" and unpredictable future which makes long term planning difficult [135]. Currently climate change is the main driver of ecosystem change that will might the resilience of the OD.

5.4 Ecosystem dynamics in the OD using the panarchy model

The OD is dynamic characterised by inter and intraspecific interactions as highlighted. (i) Rapid colonisation of seasonally flooded floodplains by zooplankton and aquatic macro-invertebrates characterises the *r* phase of ecosystem reorganisation (**Figure 5**). These *r* strategists trap terrestrial energy sources into the aquatic environment (ii) the species in the *K* phase are fish species, which have taken advantage of the *r* strategists to build up biomass in the system. (iii) The omega phase is characterised by floods recession in the delta. These floods recession are characterised by high fish mortalities when fish are stranded in drying out floodplain pools; it is also characterised by intense predation where top fish predators prey on smaller sized fish species back-migrating into the main channel from drying out floodplains; it is also characterised by increased competition for fish prey between piscivores in the system; it is characterised by intense grazing of newly grasses in the seasonal floodplain which are exposed by receding floods. (iv) The alpha phase of the reorganisation phase is characterised by plant regrowth in the seasonal floodplains. This is the time of innovation and creativity in the system where seeds dispersed take root, where the system can be restricted and new stable states created. Seeds dispersed by various agents in the aquatic system start to sprout in new areas in the Delta.

Destabilising forces in ecosystems are important in maintaining diversity [136], and this role is played by the seasonal flood-pulse in the OD. According to Baldwin and Mitchel [126], periodic wetting and drying processes in floodplains are essential for trapping nutrients in floodplain soils which enhances system productivity. This scenario creates a dynamic stable state in the OD, which could be a regime shift from a previous state. Essentially, the OD is constantly changing as a consequence of seasonal flooding. These changes are much more dramatic in seasonal flooding where changes in flood levels are more significant. The system even changes from completely dry ecosystems at temporary or rarely flooded floodplains, to aquatic systems during the flood season, or during years of exceptionally high floods. These extreme flooding events (terrestrial vs. aquatic ecotones) contribute to high ecosystem productivity and diversity. This resilience in the ecosystem also enhances high biodiversity because of the diverse micro-habitats in the system.

5.5 Adaptive management vs. "command and control"

"Command and control" top down management strategies in SES erode the resilience of these systems [23]. Top-down management approaches assume equilibrium conditions in ecosystem which is not a reflection of reality. Most fisheries management approaches in flood pulse systems is based on classical approaches which assume steady state conditions in the ecosystem [137]. Some of these management approaches are based on restrictions on fishing gear, fishing methods and mesh regulations [138]. However, these classical management approaches focus on short-term high yield scenarios [91], which is usually the case in most floodplain fisheries. Peterson et al. [91] argue that this management approach creates ecosystems that are less variable in time and space, which significantly reduces their resilience. Mosepele [71] has argued that classical fisheries management approaches in dynamic flood-pulsed fisheries makes them vulnerable to collapse. It can be argued therefore, that classical fisheries management approaches in floodplain fisheries erodes the resilience of floodplain fish populations.

However, one key approach based on adaptive management approaches is the balanced approach which advocates for a rational exploitation of the fish community across its various trophic levels using various fishing gears and methods [139]. This exploitation regime has been proposed for the OD [71] because it improves the resilience of ecosystems to perturbation [139]. Moreover, co-management regimes in fisheries management have been adopted in the OD [72], and their efficacy has been established in other systems [140]. The co-management approach adopted in the OD is based on an adaptive management framework which allows for management to be tailored to the dynamicity of the system [72]. These adaptive co-management strategies increase resilience in complex SES [141]. According to Olsson et al. [141], local ecological is an essential ingredient of co-management. This is relevant in the case of the OD because Mosepele [141] showed that local fishers have innate ecological knowledge that they use to exploit their preferred fish species in the system.

Wetlands management in most tropical countries have been placed under protected areas [8] which is also the case in Botswana, where the Okavango Delta is a Ramsar Site and parts of it is a game reserve. Despite these management interventions, Junk [8] highlights that protected wetlands in most countries are affected by multiple developmental pressures at "species, community and ecosystem levels". Junk [8] also argues that cattle in some wetlands outcompetes game animals and "significantly changes vegetation cover". Similarly, Verhoeven and Setter [142] highlight that wetlands are still in danger of degradation despite being Ramsar protected in 159 countries. It is possible that the eradication of the Tsetse fly from the Delta will result with increased cattle populations which may have a detrimental impact on the Delta's ecosystem.

Furthermore, development pressures in the developing world sees wetlands as an opportunity for primarily for agricultural development [6]. Conservation takes second place and it is this prioritisation that has resulted in the degradation of wetlands in most developing countries. This is the same philosophy that has resulted in the decimation of wetlands in the developed world. According to Maltby and Acreman [6], technological development and changing economic circumstances has resulted in the hydrological transformation of wetlands which has invariably reduced their resilience. The concepts of wise use and IWRM have now been incorporated into policy regarding wetlands utilisation globally. However, due to limited resources to implement and enforce these principles, these concepts remain aspirations only instead of realistic approaches to wetlands resources management [6].

5.6 Synthesis: resilience theory and panarchy in wetlands management

The biogeochemical, hydrological and ecological processes of wetland ecosystems will always be impacted by humanity at both local and scales [103]. What is important is for managers to work towards minimising these impacts [108]. Minimising the impacts should also be aimed at ensuring that the resilience of these wetlands is maintained. The key variable in wetlands management should be to maintain resilience and decrease the ability of the systems to shift their regimes from one state to another. Resilience theory acknowledges that ecosystems are constantly changing and reorganising, and does not necessarily focus on the stability of systems [84, 109]. Regime shifts are common in ecosystems due to human impacts which reduce resilience of ecosystems [103, 105]. These regime shifts are caused by various perturbations like pollution, climate change, hydrological changes, human exploitation and land use patterns [108]. These are difficult challenges because human population increase is inevitable, and population increases will bring these other factors into play in wetlands management. One of the best approaches towards management of these systems that has been identified in this study is to use the panarchy approach.

The panarchy model essentially acknowledges that systems are composed of sub-sets of self-organising systems at various scales which all contribute to the resilience of the system [100, 105]. Nested hierarchies underpin the theoretical framework of the panarchy and have a stabilising effect on ecosystems because they harbour system memories which the ecosystem uses to revert to its original state after a disturbance [136]. Panarchy argues that adaptive capacity of these systems should be maintained. At the social scale, this calls for adaptive management strategies, which should be progressive and proactive to respond to emergent threats. Therefore, governance of SES should mirror their dynamicity, instead of using steady state philosophical orientations in the utilisation of ecosystem services. Based on the panarchy, SES are constantly morphing in space and time, characterised by varying degrees of resilience. Management should account for these idiosyncrasies, instead of developing utilisation regimes focused on biomass build up only in systems. If utilisation of SES is asynchronous to the adaptive cycle that is modelled by the panarchy, then utilisation pressures may add untenable pressure to these systems, which would make them vulnerable to collapse. This will push them to different stables states, with a potential decrease or loss of services. One key attribute of the resilience theory within the anarchy model is that (i) change is inevitable and (ii) the adaptive cycles within the panarchy occur across scales [136]. This suggests that management should accept that change in wetland ecosystems is indeed inevitable, and this should be incorporated into management paradigms.

Disturbance in ecosystems is part of development where periods of rapid change and transition co-exist and complete each other [90]. Resource managers in SES have focused more on the *r* and *K* phases of the panarchy heuristic model of SES [90], which may invariably drive these systems towards new undesirable regimes. That notwithstanding, exploiting SES during the *r* phase makes sense because Gunderson [23] highlights that ecosystems can usually absorb a wide range of disturbances during this period. However Folke [90] advises that resource management should also focus on the release and reorganisation phases too because new opportunities (or services) may be opened up due to a system reorganisation. Gunderson [23] counters Folke [90] by arguing that while SES are stable at the release/reorganisation phase, this stability is very narrow and the system is generally vulnerable to small disturbances which may push it to a different trajectory. Resilience Alliance [105] also caution that management interventions aimed at reducing system variability and protecting it from disturbance may erode its resilience. Minimum disturbance in systems like the OD can shift them towards new or different stable states. Viewed holistically, Folke [90] highlights that resilience avails the opportunities opened up by disturbance in ecosystems during the system renewal and reorganisation phases. This agrees with Redman and Kinzig [136] who observed that disturbance helps to maintain "diversity, flexibility and opportunities" in ecosystems.

Therefore, wetlands resources management should be flexible, and adopt panarchy as its philosophical orientation. Panarchy provides a framework for natural resource managers on how to manage socio-ecological systems to ensure that they retain their resilience [99]. A mechanistic and deterministic management approach will only accelerate regime shift in wetlands [93, 109], to states that are socially undesirable. Panarchy allows for adaptive management of SES. According to Garmestani and Benson [99], adaptive management and governance of natural resources is the best vehicle for the operationalization of resilience theory.

6. Conclusion

It is commonly assumed that invasive species are transformative drivers that may reduce the resilience of systems and possibly shift them to an undesired regime [101]. According to Angeler et al. [143], ecosystems are hierarchically organised where lower level processes affect processes at higher levels. This process was observed in this study which is illustrated by "boom and bust" conditions at the primary production levels at the onset of the floods in the Okavango Delta. Furthermore, this agrees with Holling [144] who observed that spatio-temporal variability of ecological systems within ecosystems is mirrored in the structure of animal communities. Moreover, despite the disturbances in the system, the OD has largely retained its form and functioning, which suggests that it is still resilient.

Resilience science, which encapsulates adaptive management, adaptive governance and panarchy, should be integrated into environmental law [99] used in natural resources management. Adopting this approach will ensure that the resilient nature of the OD, which accounts for its dynamics in space and time, is maintained. Maintenance of the resilient nature of SES is critical because that will ensure that they continue to provide socially acceptable services. It will ensure that system vulnerability is kept to a minimum, which will also reduce the possibility of the system to shift regimes. Regime shifts, while ecologically stable, might not be desirable as already highlighted. According to Holling [87], adopting the resilience framework in resource management is an affirmation of our insufficient knowledge of natural ecosystems. The resilience framework places emphasis on regional processes and not local impacts, and emphasises heterogeneity in ecosystems.

Adopting the resilience framework in natural resources management allows for flexibility in devising management strategies to respond to future unexpected events. This is adaptive management, devised within the panarchy heuristic model. According to Resilience Alliance [105], adaptive management can enhance the resilience of ecosystems by "encouraging flexibility, inclusiveness, diversity and innovation". This agrees with Olsson et al. [141] who observed that adaptive management facilitates a philosophical communication between resilience and change which then has potential to create sustainable SES. Ultimately resilience is seen as a key paradigm for policy development and natural resource governance to preserve natural capital in this rapidly changing world [145]. Therefore, the Okavango Delta should be managed to maintain its dynamicity, which will ensure its resilience.

Conflict of interest

"The authors declare no conflict of interest."

Author details

Belda Quetina Mosepele* and Ketlhatlogile Mosepele
Department of Wildlife and Aquatic Resources, Faculty of Natural Resources, Botswana University of Agriculture and Natural Resources, Gaborone, Botswana

*Address all correspondence to: bmosepele@buan.ac.bw; bmoseps@gmail.com

References

[1] Folke C, Holling CS, Perrings C. Biological diversity, ecosystems, and the human scale. Ecological Applications. 1996;**6**(4):1018-1024

[2] Russi D, Brink P, Farmer A, Badura T, Coates D, Förster J, Kumar R, Davidson N. The Economics of Ecosystems and Biodiversity for Water and Wetlands. Final Consultation Draft. Gland, Switzerland: Ramsar Convention Secretariat; 2012

[3] Angeler DG, Allen CR, Birge HE, Drakare S, McKie BG, Johnson RK. Assessing and managing freshwater ecosystems vulnerable to environmental change. Ambio. 2014;**43**:113-125

[4] Barbier EB. Sustainable use of wetlands valuing tropical wetland benefits: Economic methodologies and applications. The Geographical Journal. 1993;**159**(1):22-32

[5] Brander LM, Florax RJGM, Vermaat JE. The empirics of wetland valuation: A comprehensive summary and a meta-analysis of the literature. Environmental and Resource Economics. 2006;**33**:223-250

[6] Maltby E, Acreman MC. Ecosystem services of wetlands: Pathfinder for a new paradigm. Hydrological Sciences Journal. 2011;**56**(8):1341-1359

[7] Junk WJ, An S, Finlayson CM, Gopal B, Květ J, Mitchell SA, et al. Current state of knowledge regarding the world's wetlands and their future under global climate change: A synthesis. Aquatic Sciences. 2013;**75**:151-167

[8] Junk WJ. Long-term environmental trends and the future of tropical wetlands. Environmental Conservation. 2002;**29**(4):414-435

[9] Reis V, Hermoso V, Hamilton SK, Ward D, Fluet-Chouinard E, Lehner B, et al. A global assessment of inland wetland conservation status. BioScience. 2017;**67**:523-533

[10] Mitsch WJ, Bernal B, Hernandez ME. Ecosystem services of wetlands. International Journal of Biodiversity Science Ecosystem Services and Management. 2015;**11**(1):1-4

[11] Adams WM. Indigenous use of wetlands and sustainable development in West Africa. The Geographical Journal. 1993;**159**(2):209-218

[12] Rebelo L-M, McCartney MP, Finlayson CM. Wetlands of sub-Saharan Africa: Distribution and contribution of agriculture to livelihoods. Wetlands Ecology and Management. 2010;**18**:557-572

[13] Pantshwa AO, Buschke FT. Ecosystem services and ecological degradation of communal wetlands in a South African biodiversity hotspot. Royal Society Open Science. 2019;**6**:181770. DOI: 10.1098/rsos.181770

[14] Adger WN. Vulnerability. Global Environmental Change. 2016;**16**:268-281

[15] Walker BH, Gunderson LH, Kinzig AP, Folke C, Carpenter SR, Schultz L. A handful of heuristics and some propositions for understanding resilience in social-ecological systems. Ecology and Society. 2006;**11**(1):13

[16] Cronberg G, Gieske A, Martins E, Nengu JP, Stenstrom I-M. Hydrobiological studies of the Okavango Delta and Kwando/Linyanti/Chobe River, Botswana. Botswana Notes and Records. 1995;**27**:151-226

[17] Secretariat of the Convention on Wetlands. The List of Wetlands of International Importance. 2020. Available from: www.ramsar.org/sites/

default/files/documents/library/sitelist.pdf [Accessed: 21 May 2020]

[18] West DT, van As JG, van As LL. Surface water quality in the Okavango Delta panhandle, Botswana. African Journal of Aquatic Science. 2015;**40**(4):359-372

[19] Kgathi D, Kniveton D, Ringrose S, Turton AR, Vanderpost C, Lundqvist J, et al. The Okavango; a river supporting its people, environment and economic development. Journal of Hydrology. 2006;**331**:3-17

[20] Mackay AW, Davidson T, Wolski P, Mazebedi R, Masamba WRL, Huntsman-Mapila P, et al. Spatial and seasonal variability in surface water chemistry in the Okavango Delta, Botswana: A multivariate approach. Wetlands. 2011;**31**:815-829

[21] Skelton PH, Bruton MN, Merron GS, van der Waal BCW. The fishes of the Okavango drainage system in Angola, South West Africa and Botswana: Taxonomy and distribution. Ichthyological Bulletin of the J. L. B. Smith Institute of Ichthyology. 1985;**50**:21

[22] Cawley KM, Wolski P, Mladenov N, Jaffé R. Dissolved organic matter biogeochemistry along a transect of the Okavango Delta, Botswana. Wetlands. 2012;**32**:475-486

[23] Gunderson LH. Ecological resilience—In theory and application. Annual Review of Ecology and Systematics. 2000;**31**:425-439

[24] Junk WJ, Bayley PB, Sparks RE. The flood pulse concept in river-floodplain systems. In: Dodge DP, editor. Proceedings of the International Large River Symposium. Canadian Special Publications in Fisheries and Aquatic Sciences, Vol. 106; 1989. pp. 110-127

[25] Gumbricht T, McCarthy J, McCarthy TS. Channels, wetlands and islands in the Okavango delta, Botswana, and their relation to hydrological and sedimentological processes. Earth Surface Processes and Landforms. 2004;**29**:15-29

[26] Mladenov N, McKnight DM, Wolski P, Ramberg L. Effects of annual flooding on dissolved organic carbon dynamics within a pristine wetland, the Okavango Delta, Botswana. Wetlands. 2005;**25**(3):622-638

[27] Coesel PEM, van Geest A. Taxonomic and biogeographical notes on Okavango desmids (Zygnematophyceae, Streptophyta). Systematics and Geography of Plants. 2008;**78**:27-46

[28] Andersson L, Gumbricht T, Hughes D, Kniveton D, Ringrose S, Savenije H, et al. Water flow dynamics in the Okavango River Basin and Delta—A prerequisite for the ecosystems of the Delta. Physics and Chemistry of the Earth. 2003;**28**:1165-1172

[29] Mosepele K, Moyle PB, Merron GS, Purkey D, Mosepele B. Fish, floods and ecosystem engineers; interactions and conservation in the Okavango Delta, Botswana. BioScience. 2009;**59**(1):53-64

[30] Ashton PJ, Nordin L-A, Alonso LE. Introduction to the Okavango Delta and the AquaRAP Expedition. In: Alonso LE, Nordin L, editors. A Rapid Biological Assessment of the Aquatic Ecosystems of the Okavango Delta, Botswana: High Water Survey. RAP Bulletin of Biological Assessment 27. Washington, DC: Conservation International; 2003. pp. 29-37

[31] Mendelsohn JM, van der Post C, Ramberg L, Murray-Hudson M, Wolski P, Mosepele K. Okavango Delta: Floods of Life. Windhoek: RAISON; 2010

[32] Gieske A. Vegetation driven groundwater recharge below the

Okavango Delta (Botswana) as a solute sink mechanism—An indicative model. Botswana Journal of Earth Sciences. 1996;**3**:33-37

[33] Krah M, McCarthy TS, Hannegarn H, Ramberg L. Airborne dust deposition in the Okavango delta, Botswana, and its impact on landforms. Earth Surface Processes and Landforms. 2004;**29**:565-577

[34] Gumbricht T, Wolski P, Frost P, McCarthy TS. Forecasting the spatial extent of the annual flood in the Okavango Delta, Botswana. Journal of Hydrology. 2004;**290**(3-4):178-191

[35] Ellery WN, McCarthy TS, Smith ND. Vegetation, hydrology, and sedimentation patterns on the major distributary system of the Okavango fan, Botswana. Wetlands. 2003;**23**(2):357-375

[36] Garstang M, Ellery WN, McCarthy TS, Scholes RJ, Swap RJ, Tyson PD. The contribution of aerosol and water-borne nutrients to the functioning of the Okavango Delta ecosystem, Botswana. South African Journal of Science. 1998;**94**:223-229

[37] McCarthy TS, Metcalfe J. Chemical sedimentation in the semiarid environment of the Okavango Delta, Botswana. Chemical Geology. 1990;**89**:157-178

[38] Wolki P, Murray-Hudson M. 'Alternative futures' of the Okavango Delta simulated by a suite of global climate and hydro-ecological models. Water SA. 2008;**34**(5):605-610

[39] Ashton P, Neal M. An overview of key strategic issues in the Okavango basin. In: Turton AR, Ashton PJ, Cloete TE, editors. Transboundary Rivers, Sovereignty and Development: Hydropolitical Drivers in the Okavango River Basin. Geneva: Green Cross International; 2003. pp. 31-63

[40] Hart RC. A limnological profile of the upper Okavango delta at low water level. Southern African Journal of Aquatic Sciences. 1997;**23**(2):21-33

[41] Mogobe O, Masamba WRL, Mosepele K. Trace metal concentrations in a pristine Ramsar site: The Okavango Delta. SN Applied Sciences. 2019;**1**:1604. DOI: 10.1007/s42452-019-1602-1

[42] Mogobe O, Masamba WRL, Mosimanyana E, Mosepele K. Occurrence of aluminum and beryllium in the Okavango Delta, Botswana: Human health risks. In: Gastescu P, Bretcan P, editors. Water Resources and Wetlands. Romania: Transversal Publishing House; 2016. pp. 270-277

[43] Masamba WRL, Muzila A. Spatial and seasonal variation of major cation and selected trace metal ion concentrations in the Okavango-Maunachira-Khwai channels of the Okavango Delta. Botswana Notes and Records. 2005;**37**:218-226

[44] Mosimane K, Struyf E, Gondwe MJ, Frings P, van Pelt D, Wolski P, et al. Variability in chemistry of surface and soil waters of an evapotranspiration-dominated flood-pulsed wetland: Solute processing in the Okavango Delta, Botswana. Water SA. 2017;**43**(1):104-115

[45] Mogobe O, Mosimanyana E, Masamba WRL, Mosepele K. Monitoring water quality of the Upper Okavango Delta. In: Revermann R, Krewenka KM, Schmiedel U, Olwoch JM, Helmschrot J, Jürgens N, editors. Climate Change and Adaptive Land Management in Southern Africa—Assessments, Changes, Challenges, and Solutions. Göttingen: Klaus Hess Publishers; 2018. pp. 106-111. DOI: 10.7809/b-e.00311

[46] Tubatsi G, Bonyongo MC, Gondwe M. Water use practices, water

quality, and households' diarrheal encounters in communities along the Boro-Thamalakane-Boteti river system, Northern Botswana. Journal of Health, Population and Nutrition. 2015;**33**:21. DOI: 10.1186/s41043-015-0031-z

[47] Tubatsi G, Bonyongo MC, Gondwe M. Water quality dynamics in the Boro-Thamalakane-Boteti river system, northern Botswana. African Journal of Aquatic Science. 2014;**39**(4):351-360

[48] Masamba WRL, Mazvimavi D. Impact on water quality of land uses along Thamalakane-Boteti River: An outlet of the Okavango Delta. Physics and Chemistry of the Earth. 2008;**33**(8):687-694

[49] Mazvimavi D, Mmopelwa G. Access to water in gazetted and ungazetted rural settlements in Ngamiland, Botswana. Physics and Chemistry of the Earth, Parts A/B/C. 2006;**31**(15):713-722

[50] Krah M, McCarthy TS, Huntsman-Mapila P, Wolski P, Annegarn H, Sethebe K. Nutrient budget in the seasonal wetland of the Okavango Delta, Botswana. Wetlands Ecology and Management. 2006;**14**:253-267

[51] Junk WJ, Brown M, Campbell IC, Finlayson M, Gopal B, Ramberg L, et al. The comparative biodiversity of seven globally important wetlands: A synthesis. Aquatic Sciences. 2006;**68**:400-414. DOI: 10.1007/s00027-006-0856-z

[52] Høberg P, Lindholm M, Ramberg L, Hessen DO. Aquatic food web dynamics on a floodplain in the Okavango delta, Botswana. Hydrobiologia. 2002;**470**:23-30

[53] Akoko E, Atekwana EA, Cruse AM, Molwalefhe L, Masamba WRL. River-wetland interaction and carbon cycling in a semi-arid riverine system: The Okavango Delta, Botswana. Biogeochemistry. 2013;**114**:359-380

[54] Mladenov N, McKnight DM, Wolski P, Murray-Hudson M. Simulation of DOM fluxes in a seasonal floodplain of the Okavango Delta, Botswana. Ecological Modelling. 2007;**205**:181-195

[55] Mladenov N, McKnight DM, Macko SA, Norris M, Cory RM, Ramberg L. Chemical characterization of DOM in channels of a seasonal wetland. Aquatic Sciences. 2007;**69**:456-471

[56] Siziba N, Chimbari MJ, Masundire H, Mosepele K. Spatial variations of micro-invertebrates across different microhabitats of temporary floodplains of lower Okavango Delta, Botswana. African Journal of Ecology. 2011;**50**(1):43-52

[57] Siziba N, Chimbari MJ, Masundire H, Mosepele K. Spatial and temporal variations of micro-invertebrates across temporary floodplains of the lower Okavango Delta, Botswana. Physics and Chemistry of the Earth. 2011;**36**(14-15):939-948

[58] Siziba N, Chimbari MJ, Mosepele K, Masundire H, Ramberg L. Inundation frequency and viability of micro-crustacean propagules in soils of temporary aquatic habitats of lower Okavango Delta, Botswana. Ecohydrology. 2012;**6**:722-730

[59] Lindholm M, Hessen D. Zooplankton succession on seasonal floodplains: Surfing on a wave of food. Hydrobiologia. 2007;**592**:95-104

[60] Lindholm M, Hessen DO, Ramberg L. Diversity, dispersal and disturbance: Cladoceran species composition in the Okavango Delta. African Zoology. 2009;**44**(1):24-35

[61] Ramberg L, Hancock P, Lindholm M, Meyer T, Ringrose S,

Sliva J, et al. Species diversity of the Okavango Delta, Botswana. Aquatic Sciences. 2006;**68**:310-337

[62] Omari K, Mubyana T, Matsheka MI, Bonyongo MC, Veenendaal E. Flooding and its influence on diazotroph populations and soil nitrogen levels in the Okavango Delta. South African Journal of Botany. 2004;**70**(5):734-740

[63] Bonyongo MC, Mubyana T. Soil nutrient status in vegetation communities of the Okavango Delta floodplains. South African Journal of Science. 2004;**100**:337-340

[64] Merron GS. The diversity, distribution and abundance of the fishes in the Moremi Wildlife Reserve, Okavango Delta, Botswana. South African Journal of Wildlife Research. 1993;**23**(4):115-122

[65] Mosepele K, Mosepele B, Bokhutlo T, Amutenya K. Spatial variability in fish species assemblage and community structure in four subtropical lagoons of the Okavango delta, Botswana. Physics and Chemistry of the Earth. 2011;**36**(14-15):910-917

[66] Appleton CC, Curtis BA, Alonso LE, Kipping J. Freshwater invertebrates of the Okavango Delta, Botswana. In: Alonso LE, Nordin L, editors. A Rapid Biological Assessment of the Aquatic Ecosystems of the Okavango Delta, Botswana: High Water Survey. RAP Bulletin of Biological Assessment 27. Washington, DC: Conservation International; 2003. pp. 58-68

[67] Dallas HF, Mosepele B. A preliminary survey and analysis of the spatial distribution of aquatic invertebrates in the Okavango Delta, Botswana. African Journal of Aquatic Science. 2007;**32**(1):1-11

[68] Merron GS. The ecology and management of the fishes of the Okavango delta, Botswana, with particular reference to the role of the seasonal floods [PhD thesis]. Rhodes: Rhodes University; 1991

[69] Merron GS, Bruton MN. Community ecology and conservation of the fishes of the Okavango Delta, Botswana. Environmental Biology of Fishes. 1995;**43**:109-119

[70] Mosepele K, Kolding J, Bokhutlo T. Fish community dynamics in an inland floodplain system of the Okavango Delta, Botswana. Ecohydrology and Hydrobiology. 2017;**17**(2):89-102

[71] Mosepele K. Dynamics of the seasonal floodplain fishery of the Okavango Delta, Botswana [PhD thesis]. Bergen: University of Bergen; 2019

[72] Mosepele K, Murray-Hudson M, Mosie I, Sethebe K. Lagoons' fish communities in flood-pulsed floodplains: Heterogeneity in a highly dynamic system? The case of the Okavango Delta. In: Mwinyihija M, editor. Lagoons: Habitat and Species, Human Impacts and Ecological Effects. New York: Nova Science Publishers; 2014. pp. 149-174

[73] Merron GS, Bruton MN. The Ecology and Management of the Fishes of the Okavango Delta, Botswana, with Special Reference to the Role of the Seasonal Floods. Investigational Report No. 29. Grahamstown: J.L.B. Smith Institute of Ichthyology; 1988

[74] Mosepele K, Mosepele B. Spatial and temporal variability in fishery and fish community structure in the Okavango Delta, Botswana: Implications towards fisheries management. Botswana Notes and Records. 2005;**37**:280-291

[75] Linhoss AC, Muñoz-Carpena R, Allen M, Kiker G, Mosepele K. A flood pulse driven fish population model for the Okavango Delta, Botswana. Ecological Modelling. 2012;**228**:27-38

[76] Merron GS, Mann BQ. The reproductive and feeding biology of *Schilbe intermedius* Riippell in the Okavango Delta, Botswana. Hydrobiologia. 1995;**308**:121-129

[77] Thethela B, Weyl O, Mosepele K, Wilson G. Age and growth of sharp-tooth catfish, *Clarias gariepinus* (Burchell 1822) (Clariidae), in the lower Okavango Delta, Botswana. Marine and Freshwater Research. 2015;**66**:1-9

[78] Mosepele K, Basimane O, Mosepele B, Thethela B. Using population parameters to separate fish stocks in the Okavango Delta fishery, Botswana: A preliminary assessment. Botswana Notes and Records. 2005;**37**:292-305

[79] Mmusi M, Mosepele K, Murray-Hudson M, Teketay D, Horn M. Role of *Brycinus lateralis* (Teleostei: Alestidae) in dispersal and germination of *Nymphaea nouchali* (Angiospermae: Nymphaeaceae) seeds on a seasonal floodplain of the Okavango Delta, Botswana. African Journal of Aquatic Science. 2016;**41**(4):489-494. DOI: 10.2989/16085914.2016.1244041

[80] Siziba N, Chimbari MJ, Mosepele K, Masundire H. Spatial and temporal variations in densities of small fishes across different temporary floodplain types of the lower Okavango Delta, Botswana. African Journal of Aquatic Science. 2011;**36**(3):309-320

[81] Siziba N, Chimbari MJ, Masundire H, Mosepele K, Ramberg L. Variation in assemblages of small fishes and micro-crustaceans after inundation of rarely flooded wetlands of the lower Okavango Delta, Botswana. Environmental Management. 2013;**52**(6):1386-1399

[82] Ellery WN, Tacheba B. Floristic diversity of the Okavango Delta, Botswana. In: Alonso LE, Nordin L, editors. A Rapid Biological Assessment of the Aquatic Ecosystems of the Okavango Delta, Botswana: High Water Survey. RAP Bulletin of Biological Assessment 27. Washington, DC: Conservation International; 2003. pp. 69-96

[83] Folke C, Colding, J, Berkes F. Synthesis: Building resilience and adaptive capacity in social–ecological systems. In: Folke C, Colding J, Berkes F. Navigating Social–Ecological Systems: Building Resilience for Complexity and Change. Cambridge University Press; Cambridge. 2003. p. 352 – 388.

[84] Berkes F. Understanding uncertainty and reducing vulnerability: Lessons from resilience thinking. Natural Hazards. 2007;**41**:283-295

[85] Walker B, Holling CS, Carpenter SR, Kinzig A. Resilience, adaptability and transformability in social–ecological systems. Ecology and Society. 2004;**9**(2):5. Available from: http://www.ecologyandsociety.org/vol9/iss2/art5

[86] Pimm SL. The complexity and stability of ecosystems. Nature. 1984;**307**:321-326

[87] Holling CS. Resilience and stability of ecological systems. Annual Review of Ecology and Systematics. 1973;**4**:1-23

[88] Scheffer M, Carpenter SR, Dakos V, van Nes EH. Generic indicators of ecological resilience: Inferring the chance of a critical transition. Annual Review of Ecology, Evolution, and Systematics. 2015;**46**:145-167

[89] Adger WN, Hughes TP, Folke C, Carpenter SR, Rockstrom J. Social-ecological resilience to coastal disasters. Science. 2005;**309**:1036-1039

[90] Folke C. Resilience: The emergence of a perspective for social–ecological systems analyses. Global Environmental Change. 2006;**16**:253-267

[91] Peterson G, Allen CR, Holling CS. Ecological resilience, biodiversity, and scale. Ecosystems. 1998;**1**:6-18

[92] Carpenter S, Walker B, Anderies JM, Abel N. From metaphor to measurement: Resilience of what to what? Ecosystems. 2001;**4**:765-781

[93] Garmestani AS, Allen CR, Cabezas H. Panarchy, adaptive management and governance: Policy options for building resilience. Nebraska Law Review. 2009;**87**:1036-1054

[94] Baho DL, Leu E, Pomati F, Hessen DO, Norberg J, Moe SJ, et al. Resilience of natural phytoplankton communities to pulse disturbances from micropollutant exposure and vertical mixing. Environmental Toxicology and Chemistry. 2019;**38**(10):2197-2208

[95] Allen CR, Gunderson L, Johnson AR. The use of discontinuities and functional groups to assess relative resilience in complex systems. Ecosystems. 2005;**8**:958-966

[96] Fatha BD, Cabezas H, Pawlowski CW. Regime changes in ecological systems: An information theory approach. Journal of Theoretical Biology. 2003;**222**:517-530

[97] Angeler DG, Drakar S, Johnson RK. Revealing the organization of complex adaptive systems through multivariate time series modelling. Ecology and Society. 2011;**16**(3):5. DOI: 10.5751/ES-04175-160305

[98] Holling CS. Understanding the complexity of economic, ecological, and social systems. Ecosystems. 2001;**4**:390-405

[99] Garmestani AS, Benson MH. A framework for resilience-based governance of social ecological systems. Ecology and Society. 2013;**18**(1):9. DOI: 10.5751/ES-05180-180109

[100] Garmestani AS, Allen CR, Gunderson L. Panarchy: Discontinuities reveal similarities in the dynamic system structure of ecological and social systems. Ecology and Society. 2009;**14**(1):15

[101] Allen CR, Angeler DG, Garmestani AS, Gunderson LH, Holling CS. Panarchy: Theory and application. Ecosystems. 2014;**17**:578-589

[102] Allen CR, Cumming GS, Garmestani AS, Taylor PD, Walker BH. Managing for resilience. Wildlife Biology. 2011;**17**:337-349

[103] Folke C, Carpenter S, Walker B, Scheffer M, Elmqvist T, Gunderson L, et al. Regime shifts, resilience, and biodiversity in ecosystem management. Annual Review of Ecology, Evolution, and Systematics. 2004;**35**:557-581

[104] Angeler DG, Allen CR, Rojo C, Alvarez-Cobelas M, Rodrigo MA, Salvador SC. Inferring the relative resilience of alternative states. PLoS One. 2013;**8**(10):e77338. DOI: 10.1371/journal.pone.0077338

[105] Resilience Alliance. Assessing Resilience in Social-Ecological Systems: Workbook for Practitioners. Version 2.0. 2010. Available from: http://www.resalliance.org/3871.php

[106] Cummings GS. Spatial Resilience in Social-Ecological Systems. London: Springer; 2011

[107] Cote M, Nightingale AJ. Resilience thinking meets social theory: Situating social change in socio-ecological systems (SES) research. Progress in Human Geography. 2012;**36**(4):475-489

[108] Folke C, Carpenter SR, Walker B, Scheffer M, Chapin T, Rockström J. Resilience thinking: integrating resilience, adaptability and

transformability. Ecology and Society. 2010;**15**(4):20. Available from: http://www.ecologyandsociety.org/vol15/iss4/art20/

[109] Holling CS, Gunderson LH. Resilience and adaptive cycles. In: Gunderson LH, Holling CS, editors. Panarchy: Understanding Transformations in Human and Natural Systems. Washington, DC: Island Press; 2002

[110] Mosepele K, Mogomotsi GEJ, Mogomotsi PK, Hambira W, Moses O, Dhliwayo M, et al. Water, ecosystem dynamics and human livelihoods in the Okavango River Basin (ORB); competing needs or balanced use? A review. In: Chandrasekaran PT, editor. Water and Sustainability. London: IntechOpen; 2018. pp. 41-64

[111] Moses O. Climate change in Lake Ngami, Botswana. In: Kgathi DL, Mosimanyana EN, Mbaiwa JE, Masamba WRL, editors. Ecosystem Services and Human Well-Being at Lake Ngami, Botswana. New York: Nova; 2018. pp. 61-69

[112] Davies JE, Bowles J. Ecology of the Tsetse fly in the Okavango Delta. In: Proceedings of the Symposium on the Okavango Delta and Its Future Utilization. Gaborone: National Museum; 1976. pp. 141-152

[113] Molefi KK. The impact of tsetse Fly in Ngamiland, 1916-1955. Botswana Notes and Records. 2008;**40**:35-45

[114] Gilmore C. The effects of ULV aerial applications of endosulfan on the feeding habits of "*Schilbe Mystus*" and "*Mormyrus Lacerda*" in the Okavango Delta. Botswana Notes and Records. 1977;**11**:107-114

[115] Douthwaite RJ. Changes in pied kingfisher (*Ceryle rudis*) feeding related to endosulfan pollution from tsetse Fly control operations in the Okavango Delta, Botswana. Journal of Applied Ecology. 1982;**19**(1):133-141

[116] Perkins JS, Ramberg L. Environmental Monitoring of Tsetse Fly Spraying Impacts in the Okavango Delta—2002, Final Report. Okavango Report Series No. 2. Maun: Harry Oppenheimer Okavango Research Centre; 2004. p. 150

[117] Merron GS. Tsetse Fly control and the environmental implications for fish in the Okavango Delta, Botswana. Botswana Notes and Records. 1992;**24**:49-56

[118] Brind WG. The Okavango Delta. Report on the 1951-1953 Field Surveys. Mafikeng, South Africa: Resident Commissioner; 1954

[119] Ellery WN, McCarthy TS. Environmental change over two decades since dredging and excavation of the lower Boro River, Okavango Delta, Botswana. Journal of Biogeography. 1998;**25**:361-378

[120] Kraatz DB. Restoration of flow in the Thaoge River. Technical Note No. 18. Rome, Italy: FAO; 1976

[121] Ashton PJ, Masundire H, Hart R, Prince-Nengu J, Botshelo O, Lekhuru M, et al. Water quality of the Okavango Delta, Botswana. In: Alonso LE, Nordin L, editors. A Rapid Biological Assessment of the Aquatic Ecosystems of the Okavango Delta, Botswana: High Water Survey. RAP Bulletin of Biological Assessment 27. Washington, DC: Conservation International; 2003. pp. 38-52

[122] Wilson BH. Some natural and man-made changes in the channels of the Okavango Delta. Botswana Notes and Records. 1973;**5**:132-153

[123] Ellery WN, Ellery K, Rogers KH, McCarthy TS, Walker BH. Vegetation,

hydrology and sedimentation processes as determinants of channel form and dynamics in North-Eastern Okavango Delta, Botswana. African Journal of Ecology. 1993;**31**:10-25

[124] Ellery K, Ellery WN, Rogers KH, Walker BH. Formation, colonization and fate of floating sudds in the Maunachira river system of the Okavango River system, Botswana. Aquatic Botany. 1990;**38**:315-329

[125] McCarthy TS, Ellery WN, Bloem A. Some observations on the geomorphological impact of hippopotamus (*Hippopotamus amphibius* L.) in the Okavango Delta, Botswana. African Journal of Ecology. 1998;**36**:44-56

[126] Baldwin DS, Mitchell AM. The effects of drying and re-flooding on the sediment and soil nutrient dynamics of lowland river–floodplain systems: A synthesis. Regulated Rivers: Research & Management. 2000;**16**:457-467

[127] Jenkins KM, Boulton AJ. Connectivity in a dryland river: Short-term aquatic micro-invertebrate recruitment following floodplain inundation. Ecology. 2003;**84**(10):2708-2723

[128] Wolski P, Murray-Hudson M. Recent changes in flooding in the Xudum distributary of the Okavango Delta and Lake Ngami, Botswana. South African Journal of Science. 2006;**102**:173-176

[129] McCarthy TS, Green W, Franey NJ. The influence of neo-tectonics on water dispersal in the northeastern regions of the Okavango swamps, Botswana. Journal of African Earth Sciences. 1993;**17**(1):23-32

[130] Bernard T, Moetapele N. Desiccation of the Gomoti River: Biophysical process and indigenous resource management in northern Botswana. Journal of Arid Environments. 2005;**63**:256-283

[131] McCarthy TS. Physical and biological processes controlling the Okavango Delta—A review of recent research. Botswana Notes and Records. 1992;**24**:57-86

[132] Ilemobade AA. Tsetse and trypanosomosis in Africa: The challenges, the opportunities. Onderstepoort Journal of Veterinary Research. 2009;**76**:35-40

[133] Galeb LAG, Ganeco LN, Fredianelli AC, Wagner R, Fam ALPD, Rocha DCC, et al. Acute intoxication by Deltamethrin in Jundia: Emphasis on clinical, biochemical and haematological effects. Global Advanced Research Journal of Environmental Science and Toxicology. 2013;**2**(3):060-067

[134] Ellery WN, Ellery K, McCarthy TS, Cairncross B, Oelofse R. A peat fire in the Okavango Delta, Botswana and its importance as an ecosystem process. African Journal of Ecology. 1989;**27**:7-21

[135] Messier C, Bauhus J, Doyon F, Maure F, Sousa-Silva R, Nolet P, et al. The functional complex network approach to foster forest resilience to global changes. Forest Ecosystems. 2019;**6**:21. DOI: 10.1186/s40663-019-0166-2

[136] Redman CL, Kinzig AP. Resilience of past landscapes: Ryesilience theory, society, and the *longue durée*. Conservation Ecology. 2003;7(1):14. Available from: http://www.consecol.org/vol7/iss1/art14

[137] Mosepele K. Classical fisheries theory and inland (floodplain) fisheries management; is there need for a paradigm shift? Lessons from the Okavango Delta, Botswana. Fisheries and Aquaculture Journal. 2014;**5**:101. DOI: 10.4172/ 2150-3508.1000101

[138] Mosepele K. Flood pulse in a subtropical floodplain fishery and the consequences for steady state management. In: Totolo O, editor. Water Resource Management; Science and Technology Innovation for Sustainable Development. Acta Press: Canada; 2008. pp. 56-62

[139] Law R, Plank MJ, Kolding J. On balanced exploitation of marine ecosystems: Results from dynamic size spectra. ICES Journal of Marine Science. 2012;**69**(4): 602-614. DOI: 10.1093/icesjms/fss031

[140] Sen S, Nielsen JR. Fisheries co-management: A comparative analysis. Marine Policy. 1996;**20**(5):405-418

[141] Olsson P, Folke C, Berkes F. Adaptive Co-management for building resilience in social–ecological systems. Environmental Management. 2004;**34**(1):75-90

[142] Verhoeven JTA, Setter TL. Agricultural use of wetlands: Opportunities and limitations. Annals of Botany. 2010;**105**:155-163

[143] Angeler DG, Allen CR, Barichievy C, Eason T, Garmestani AS, Graham NAJ, et al. Management applications of discontinuity theory. Journal of Applied Ecology. 2016;**53**:688-698

[144] Holling CS. Cross scale morphology, geometry and dynamics of ecosystems. Ecological Monographs. 1992;**62**(4):447-502

[145] Spears BM, Ives SC, Angeler DG, Allen CR, Birk S, Carvalho L, et al. Effective management of ecological resilience—Are we there yet? Journal of Applied Ecology. 2015;**52**:1311-1315

Chapter 12

Biodiversity and Environmental Integrity of Some Rivers in Derived Savannah Belt in Edo-North, Southern Nigeria

Abdul-Rahman Dirisu and John Ovie Olomukoro

Abstract

Investigative study on macroinvertebrates and pollution tolerance index of four rivers in the derived savannah belt of Edo-North in Southern Nigeria was carried out from January to December, 2010. The study involved monthly field sampling and laboratory analysis of macrofauna. The objectives of the study were to evaluate the biodiversity status and importantly, the environmental integrity of the four selected rivers. Two dominant macroinvertebrate families were recorded; Baetidae and Chironomidae while, 53 macroinvertebrate taxa were identified across the rivers. The highest number of species (12) was recorded for Diptera, while, Hirudinea, Nematoda and Arachnida each had 1 species in the population. General diversity indices ranged from 2.02 to 2.78 with the least and highest recorded in Edion and Etuno Rivers, respectively. The fauna evenness index indicated that the water bodies had values less than 1.0. Meanwhile, Pollution tolerance index revealed spatial and seasonal variations in water quality conditions but only Edion River exhibited poor water quality in all the months of study.

Keywords: macroinvertebrate taxa, water quality, species richness, pollution sensitive

1. Introduction

Over the years, degradation of water quality has remained a major issue in most parts of the world, including the Niger Delta region of Nigeria. This has continued to pose a threat to health and economic development in Nigeria and other countries alike [1], with about 60% of the rural and urban populace depending on ponds, streams and shallow wells for domestic water supply in some developing countries [2, 3]. The increasing concern for the level of pollutants in surface and groundwater makes water monitoring very essential [3–5].

Macroinvertebrates are highly useful bioindicators in understanding the ecological health of aquatic ecosystems, which could be more accurate than chemical and microbiological data, which exhibit short-term fluctuations [6–8]. Benthic macroinvertebrates include all bottom-dwelling plants and animals in water bodies and are found either crawling, burrowing or attached to various kinds of objects such as wood, stone, and organic matter [9]. These organisms play a vital role in the circulation and recirculation of nutrients in aquatic ecosystems.

Macroinvertebrates sampling is one technique used to determine the health of a stream, as they are subjected to day to day and longer term changes in pollution, oxygen and acidity levels. They can be used to monitor stream quality conditions over a broad area or they can be used to determine the effects of point source discharges, such as sewage treatment plants and factories, on a site-specific basis [3]. The advantages of this is that, unlike fish, benthos cannot move around as much, so they are less able to escape the effects of sediment and other pollutants that diminish water quality. Therefore, macroinvertebrates can give reliable information on stream, river and lake water quality. Their long life cycles allow studies conducted by aquatic ecologist to determine any decline in environment quality.

A Pollution Tolerance Index (PTI) is a method that is used to rate stream quality based on macroinvertebrate communities [10]. The three major tolerance categories of macroinvertebrate used in assessing the rivers water quality conditions are sensitive, facultative (or somewhat pollution tolerance) and pollution tolerant groups [11]. The aim of this study was to determine the quality of surface waters in Edo-North of Edo State in Southern Nigeria (using macrofauna as bioindicators), where anthropogenic factors and other human activities including uncontrolled waste disposal practices have placed immense stress on the quality of the water resources. The study determined the composition, abundance and the distribution of benthic macrofauna in the rivers; investigated the full biodiversity value of macrofauna as indices of pollution; and determined the Pollution Tolerance Index of the rivers, using macroinvertebrates.

2. Materials and methods

2.1 Study area

The sampled locations which are located within Edo State of Nigeria (**Figure 1**) were carefully chosen to monitor the water quality and pollution tolerance index of selected fresh water bodies in Edo-North. There are two distinct tropical seasons in this area, viz.; the rainy season, which occurs usually between April and October with its peak between July and August and dry the season, which occurs between November and March. The rainy season period is usually characterised by high relative humidity and low atmospheric temperatures. The four selected fresh water ecosystems locations are in the derived savannah zone. The spread was such that all ecological niches and areas of human activities were covered (**Table 1**). The GPS readings of the sampled sites were determined using GPS 12 Garmin model (serial number 36209488). Etuno River runs through Igarra community and takes its source from Kukuruku hill in Akoko-Edo area. There are cesspools in this catchment during the dry season, but the flood velocity increases in the wet season, ranging from 0.97 to 5.6 $m^2 s^{-1}$. On the other hand, Orli, Omodo and Edion Rivers flow through Agbede communities. The first two rivers are murky brown in colour particularly during the wet season. There are variations in their flow velocity throughout the year, which ranged from 0.1 to 14.71 $m^2 s^{-1}$. Two sampling points each (from upstream and downstream) were chosen along the stretch of the four rivers.

2.2 Sampling for macroinvertebrates

The substrates at the bottom of the selected stations were sampled for macrofauna which consists of invertebrates in the sediment and the roots of floating macrophytes. The modified grab [12] and kicking sampling techniques [13] were

Figure 1.
Map of Edo state indicating the sampled locations.

Station ID	Station/ river	Town	Coordinates		Land use pattern
1	Etuno	Igharra	N 7° 16′	E 6° 7′	Agricultural/ human settlements, stone crushing activities
2	Orlie	Agbede	N 6° 58′	E 6° 16′	Agricultural/human settlements
3	Edion	Agbede	N 6° 55′	E 6° 16′	Agricultural/ human settlements
4	Omodo	Agbede	N 6° 52′	N 6° 17′	Agricultural/ human settlements

Table 1.
GPS coordinates and land use pattern of sampled locations.

used to collect sediment cores. The kicking method caused vigorous disturbance and movement of substratum and emergent vegetation from upstream. The animals disturbed from the stream bed were washed by the current and collected by a hand-net held down-stream. Each "kicking activity" lasted for 10 minutes and collected samples were preserved using 10% formaldehyde. All samples were sorted in the laboratory using the American binocular dissecting microscope, and organisms were preserved in 4% formaldehyde in specimen bottles. Identifications of organisms were made using relevant key manuals and literature [14, 15].

2.3 Analysis of data

Using the Paleontological Statistics tool (PAST 1.99 version), basic diversity indices such as relative abundance, taxa richness and species diversity were computed to describe the macrofauna community structure [16]. These indices provide a convenient means of comparing differences within ecological communities and also for monitoring temporal changes. Pollution tolerance of macroinvertebrate organisms to pollutants was assessed using the Pollution Tolerance Index (PTI) model. Specimens were taken and examined for the presence and abundance of the different types of organisms. Their values were inputted in an equation, which gave an overall value to the stream. Tolerance values were assigned to the sensitive, facultative and pollution tolerant macroinvertebrate groups as adopted from Klemm et al. [17] and Olomukro and Dirisu [18].

3. Results

3.1 Macroinvertebrates and community structure

Fifty three macroinvertebrates taxa were identified during the study (**Table 2**). Ephemeroptera and Diptera were represented by four families each in the studied water bodies. The former was dominated by the family Baetidae and the latter group by the family Chironomidae. These two groups comprised over 70% of the macroinvertebrate communities in each water body.

The density of occurrence of Decapoda was highest in Omodo compared to the other rivers while, Atyidae was observed to be most dominant in the group. Other groups such as Odonata, Coleoptera, Trichoptera, Hemiptera and Mollusca were subdominant, common or rare in the various water bodies (**Table 2**).

3.2 Etuno river

Diptera was the dominant group in this river, while Ephemeroptera, Odonata (dragon fly), Coleoptera and Mollusca were the subdominant. Dipterans represented 61.8% of the total number of individuals (**Table 3**). Species richness index ranged from 0 to 2.84, with Dipterans having the highest richness among the groups.

3.3 Orlie river

Ephemeroptera (with 4 taxa) which is the dominant group in this station comprised Baetidae as the only family found, with 71% population density (**Table 3**). Species richness ranged between 0 and 2.23 indexes, with Diptera also having the highest species richness while Oligochaeta, Coleoptera, Trichoptera and Mollusca had the least diversity.

3.4 Edion river

A total of sixteen taxa were recorded for the macrofauna. The dominant and most frequently encountered was the Ephemeropterans which accounted for about 65% of the population followed by Decapoda (19.3%) and Diptera (13.7%). Species richness ranged between 0 and 1.44, with Trichoptera having the highest species richness and the least were Odonatans and Molluscans.

3.5 Omodo river

A total of 21 taxa were identified and categorised into five major groups as shown in **Table 3**. The dominant groups comprised Decapoda (42.6%) and Diptera

Nematoda (Aquatic Worms)	**Decapoda (Freshwater shrimp)**	**Ephemeroptera (Mayfly)**	**Odonata**	**Hemiptera (True Bug)**	**Coleoptera (Beetles)**	**Trichoptera (Caddis flies)**	**Diptera**	**Mollusca (Water snails)**
Rhabdolaimus spp.	*Caridina africana*	*Adenophlebriodes* spp.	*Anisoptera (Dragon Fly)*	*Lethocerus* spp.	*Acilius sulcatus*	*Leptocerid* spp.	*Chironomidae (Midge fly)*	*Ancycastrum* spp.
Other nematodes	*Caridina gaboneusis*	*Baetis* spp.	*Libellula* spp.	*Nepa* spp.	*Dytiscus* spp.	*Limnophilus* spp.	*Chironomus fractilobus*	*Hydrobia spp.*
OLIGOCHAETA *(Aquatic Worms)*	*Desmocaris trispinosa*	*Centroptilum* spp.	*Aphylla* spp.		*Dytiscus marginalis*	*Polycentropus* spp.	*Chironomus travaalensis*	*Hydrobia guyenoti*
Nais spp.	*Macrobrachium* spp.	*Cloeon* spp.	*Plathemis* spp.				*Chironomus sp.*	*Hydrobia lineate*
	Potamalpheops monodi	*Cloeon cyclindroculum*	Zygoptera (Damsel Fly)				*Pseudochironomus* spp.	*Potanorbis crista*
	Gammarus spp.	*Dicerconyzon* spp.	*Coenagrion spp.*				*Tanypus spp.*	*Pitar tumeus*
	Larva	*Ephemerella ignita*	*Enallagma* spp.				*Tanytarsus spp.*	*Tympanotonus radula*
	Cardiosoma spp. *(crab)*	*Habrophlebia* spp.	*Hesperangrion heterodoxum*				*Chinotanypus* spp.	
	Unidentified (crab)	*Psuedocloeon* spp.	*Lestes* spp.				*Cricotopus spp.*	
							Pentaneura spp.	
							Culex spp.	
							Insect larva	

Table 2.
The macrofauna found in four Rivers in Edo-north, Edo state, southern Nigeria.

		Etuno river			Orlie river	
Group	**Individuals/No**	**%Composition**	**Diversity index (D)**	**Individuals/No**	**%Composition**	**Diversity index (D)**
Nematoda	1(1)	0.9	0	—	—	—
Oligochaeta	2 (1)	1.8	0	1(1)	1	0
Hirudinea	—	—	—	—	—	—
Decapoda	—	—	—	—	—	—
Ephemeroptera	10 (2)	9.1	0.4343	71 (4)	71	0.7038
Odonata	9 (3)	8.2	0.91024	13 (6)	13	0.9494
Hemiptera	3 (2)	2.7	0.91024	—	—	—
Coleoptera	8 (2)	7.3	0.4809	2 (1)	2	0
Tricheotera	2 (1)	2.7	0	1(1)	1	0
Diptera	68 (13)	61.8	2.8439	6 (5)	6	2.2324
Mollusca	7 (1)	6.4	0	6 (1)	6	0
		Edion river			Omodo river	
Nematoda	—	—	—	—	—	—
Oligochaeta	—	—	—	6 (1)	4.4	0
Hirudinea	—	—	—	—	—	—
Decapoda	38 (3)	19.3	0.5498	58 (5)	42.6	0.9851
Ephemeroptera	128 (4)	65	0.6183	17 (4)	12.5	1.0587
Odonata	1(1)	0.5	0	7 (3)	5.1	1.0278
Hemiptera	—	—	—	—	—	—
Coleoptera	—	—	—	—	—	—
Tricheotera	2 (2)	1.0	1.4427	—	—	—
Diptera	27 (5)	13.7	1.2137	48 (8)	35.3	1.8082
Mollusca	1(1)	0.5	0	—	—	—

Table 3.
The number of individuals and diversity of the major taxonomic groups in the sampled locations (number of taxa in parenthesis).

Water body	No. of species	Abundance	Magalef (D)	Shannon_H	Evenness	Dominance
Etuno	24	104	4.95	2.78	0.67	0.08728
Orlie	20	88	4.24	2.46	0.59	0.118
Edion	16	219	2.78	2.02	0.47	0.189
Omodo	24	176	4.45	2.45	0.48	0.1532

Table 4.
Water bodies with their number of species, abundance and diversity indices.

(35.3%), and the subdominant were the Ephemeropterans with 12.5% density. Species richness ranged from 0 to 1.808 indexes, with Dipterans having the highest species richness and Oligochaetans the least.

3.6 Macroinvertebrate biodiversity

Biodiversity is the number and the distribution of species of living organisms. Change in biodiversity over time is the measure that can be used to assess the health of an ecosystem [10]. As biodiversity is lost, it is believed that the health and viability of an ecosystem declines. The highest biomass was reported in Edion River (219 individuals m^{-2}) while, Orlie and Omodo rivers had <105 individuals m^{-2} (**Table 4**). Shannon-Wiener diversity indices showed that values ranged from 2.02 to 2.78 with the highest in Etuno River at Igarra, while the least diversity index was in Edion River. The macrofauna evenness index indicated that most of the water bodies had values less than 1.00, the highest being for Etuno (0.67) while the least was recorded in Edion (0.472).

4. Pollution tolerance index (PTI)

Pollution Tolerance Index (PTI) values were determined at monthly interval for each rivers using the available macroinvertebrate populations in the aquatic ecosystems. There was spatial and seasonal variation in water quality conditions in the water bodies. The sampled rivers revealed moderate water quality in March, July and December 2010 most importantly, while the other months of the year indicated poor water quality (**Table 5**, **Figures 2–5**).

5. Discussion

A decrease in diversity and corresponding increase in abundance of a limited number of species is a common community response to environmental disturbance. The high diversity indices observed in the rivers indicated that many species had equal or near equal opportunity of co-existence. However, the macroinvertebrates were not evenly distributed across the sampled rivers, which is bound in most aquatic ecosystems [3, 19].

The abundance of Ephemeroptera group which is a clean water representative and is generally intolerant to contaminated aquatic ecosystems, clearly revealed that most of the rivers had moderate water quality with the exception of Etuno River. The water quality status was confirmed by the PTI estimation for the rivers. Monthly variations of PTI were observed in all the rivers. Unfortunately, there were no PTI values recorded for 5 months (August to December, 2010) in Etuno

River	Mar	Apr	May	Jun	Jul	Aug	Sept	Oct	Nov	Dec	Overall	Health status
Etuno	10	4	6	8	2	—	—	—	—	—	14	Moderate
Orlie	10	1	1	3	7	4	—	5	8	7	16	Moderate
Edion	3	4	6	7	6	5	6	6	3	6	12	Moderate
Omodo	1	7	8	3	8	5	7	5	4	11	11	Moderate

Table 5.
Monthly variation and overall pollution tolerance index (PTI) of the sampled rivers.

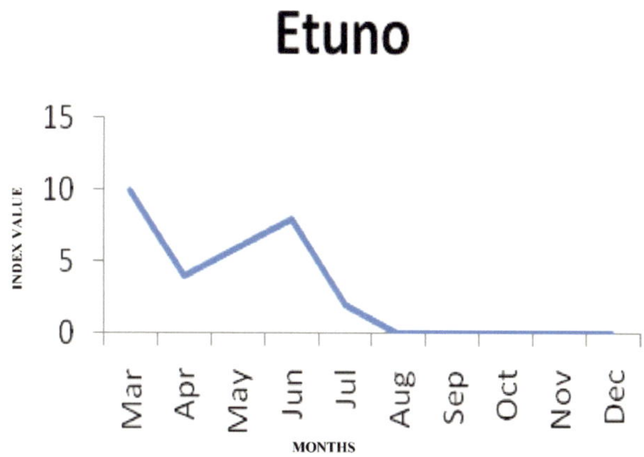

Figure 2.
The spatial and seasonal variation in pollution tolerance index (PTI) of the Etuno river.

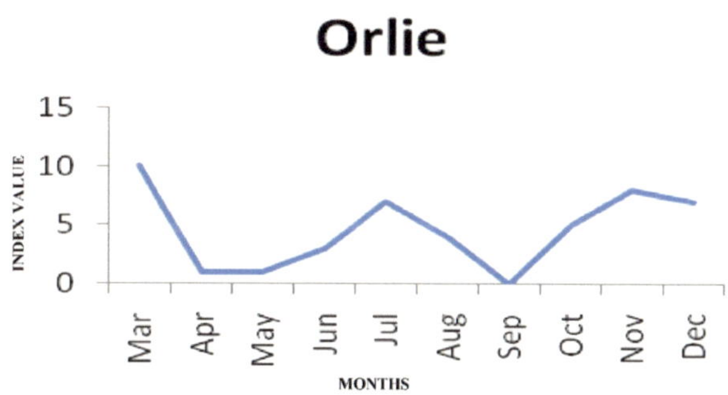

Figure 3.
The spatial and seasonal variation in pollution tolerance index (PTI) of the Orlie river.

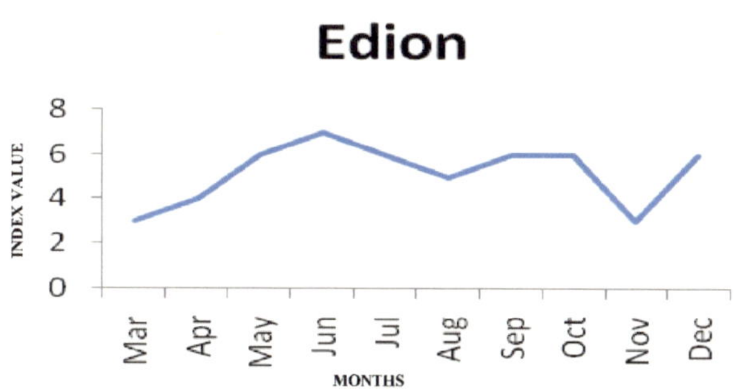

Figure 4.
The spatial and seasonal variation in pollution tolerance index (PTI) of the Edion river.

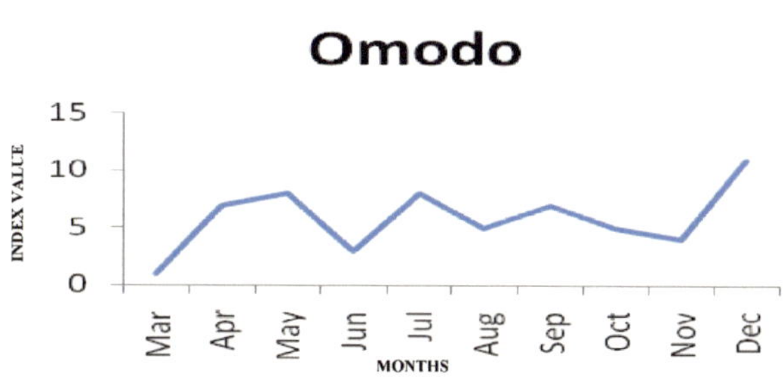

Figure 5.
The spatial and seasonal variation in pollution tolerance index (PTI) of the Omodo river.

River as the various PTI groups were completely absent, a situation that could not be explained. Though, surface run offs and inundation do greatly affect and destabilise the colonisation patterns of substrates by macrofauna, this was not the case as rainfall had diminished in those months. However, the water quality status were generally moderate, an indication that the water bodies had not been seriously compromised.

6. Conclusion

The presence of a species is more valuable than its absence in aquatic ecosystems. Therefore, factors limiting diversity of macrofauna should be discouraged around water bodies. It is important that the different sources of pollutants, ranging from industrial through municipal to domestic activities be controlled during siting in the Edo-North with particular reference to the small factories located on the banks of Etuno River as contaminants are eroded and transported downstream with ease.

Acknowledgements

Thanks to Mr. Festus Arijode for all the assistance he rendered during the field work. We are thankful to the then management of the National Centre for Energy and Environment (NCEE), Energy Commission of Nigeria, University of Benin, for funding the study from which this paper has been drafted.

Author details

Abdul-Rahman Dirisu* and John Ovie Olomukoro
Department of Animal and Environmental Biology, Faculty of Life Sciences, University of Benin, Benin City, Nigeria

*Address all correspondence to: dedonrahman10@yahoo.com

References

[1] Ekiye E, Lou Z. Water quality monitoring in Nigeria: Case study of Nigeria's industrial cities. Journal of American Science. 2010;**6**(4):22-28

[2] Dirisu AR, Olomukoro JO. Investigation of water quality of two rivers in Agbede wetlands in southern Nigeria. Global NEST Journal. 2015;**17**(3):451-462

[3] Kebede G, Mushi D, Linke RB, Dereje O, Lakew A, Hayes DS, et al. Macroinvertebrate indices versus microbial fecal pollution characteristics for water quality monitoring reveals contrasting results for an Ethiopian river. Ecological Indicators. 2019;**108**:105733. DOI: 10.1016/j.ecolind.2019.105733

[4] Adelegan JA. The History of Environmental Policy and Pollution of Water Sources in Nigeria (1960-2004): The Way Forward; 2004. Available from: http://www.nigerianlawguru.com/articles/environmental%20law/THE%20HISTORY%20OF%20ENVIRONMENTAL%20POLICY%20AND%20POLLUTION%20OF%20WATER%20SOURCES%20IN%20NIGERIA.pdf. [Accessed: 13 August 2020]

[5] Water Aid Nigeria. 2007. Available from: Wateraid.org/international/Nigeria

[6] Lear G, Dopheide A, Ancion P, Lewis GD. A comparison of bacterial, ciliate and macroinvertebrate indicators of stream ecological health. Aquatic Ecology. 2011;**45**(4):517-527. DOI: 10.1007/s10452-011-9372-x

[7] Odiete WO. Environmental Physiology of Animals and Pollution. Lagos, Nigeria: Diversified Resources; 1999. pp. 220-246

[8] Tanaka MO, de Souza ALT, Moschini LE, de Oliveira AK. Influence of watershed landuse and riparian characteristics on biological indicators of stream water quality in southeastern Brazil. Agriculture, Ecosystems and Environment. 2016;**216**:333-339. DOI: 10.1016/j.agee.2015.10.016

[9] Olomukoro JO, Egborge ABM. Hydrobological studies on Warri River, Nigeria: Part I. the composition, distribution and diversity of macrobenthic fauna. Biosciences. 2003;**15**(4):279-294

[10] Olomukoro JO, Dirisu AR. Biodiversity and Pollution Tolerance Index of selected Rivers in Derived Savannah Belt in Edo-North, Nigeria. In: Proceedings of the 4th Annual Biodiversity Conference of Nigerian Tropical Biology Association/Designing Research Capacity among African Environmental Scientist. Lagos: University of Lagos; 3rd-4th September, 2013. pp. 54-59

[11] Izaak Walton League of American. Creek Connections Aquatic Life Module: Aquatic Macroinvertebrate Sampling; 1994

[12] Hynes HBN. The ecology of stream insects. Annual Review of Entomology. 1970;**15**:25-42

[13] Paterson CG, Fernando CH. Benthic fauna colonization of a new reservoir with particular reference to the Chironomidae. Journal of the Fisheries Research Board of Canada. 1970;**27**:213-232

[14] Olomukoro JO. Macrobenthic fauna of Warri; River [Ph.D. thesis]. Nigeria: University of Benin; 1996. p. 244

[15] Umar DM, Harding JS, Winterbourn HM. Freshwater Invertebrates of the Mambilla Plateau, Nigeria. New Zealand: Canterbury Educational Printing Services, University of Canterbury; 2013. p. 86

[16] Hammer Ø, Harper DAT, Ryan PD. PAST: Paleontological statistics software package for education and data analysis. Palaeontologia Electronica. 2001;**4**(1):9. Available from: http://palaeo-electronica.org/2001_1/past/issue1_01.htm

[17] Klemm DJ, Philip AL, Florence F., Lazoreckak JM. Macroinvertebrate field and laboratory method for evaluating the biological integrity of surface waters. USEPA. EPA/600/4-90.030. XII; 1990. pp. 256

[18] Olomukro JO, Dirisu AR. Macroinvertebrate community and pollution tolerance index in Edion and Omodo rivers in derived Savannah Wetlands in Southern Nigeria. Jordan Journal of Biological Sciences. 2014;7:19-24

[19] Shimba MJ, Jonah FE. Macroinvertebrates as bioindicators of water quality in the Mkondoa River, Tanzania, in an agricultural area. African Journal of Aquatic Science. 2016;**41**:1-9. DOI: 10.2989/16085914.2016.1230536

Chapter 13

Freshwater Invertebrates of Southwestern South America: Diversity, Biogeography, and Threats

Claudio Valdovinos Zarges, Pablo Fierro and Viviana Olmos

Abstract

This chapter reviews the current state of knowledge of invertebrates of rivers, lakes, and wetlands in western South America, from southern Peru to the Strait of Magellan in southern Chile. A characterization of the diverse groups of insects, mollusk crustaceans, and other smaller groups is presented, and a biogeographic analysis of them is made with emphasis on their main forcing factors, ecology, and threats in the Anthropocene. This fauna presents Gondwanic characteristics, with clear North–South latitudinal patterns, covering from the Desert of Atacama in the North, one of the most arid deserts of the world, to the rainy and cold regions of the southern end of South America. The central zone of this territory includes one of the global biodiversity "hot spots," which currently presents serious threats associated with changes in habitat, introduction of invasive species, climate change, and overexploitation of aquatic resources.

Keywords: freshwater, invertebrates, South America, biodiversity, conservation

1. Introduction

Invertebrates represent the majority of the world's animal species, comprising a total of 32 phyla, of which 15 are present in freshwater [1]. These organisms all lack a vertebral column, are generally small in size, and present very diverse morphologies. Some have soft bodies such as worms and planarians, while others have hard bodies such as crustaceans, insects, and mollusks. Freshwater invertebrates offer the opportunity to contemplate the enormous diversity of forms and functions existing in the animal kingdom. It's precisely in this group of organisms where animal life is expressed without limits to particular forms or colors and is specialized to diverse forms of life. In the freshwater ecosystems of southwestern South America, there is a diverse fauna of invertebrates [2]. In this part of the world, around 1000 species of freshwater invertebrates are known. However, many scientists believe that the number of unknown species in this area could significantly increase that number. Knowledge of the diversity of these organisms is still fragmentary, despite the efforts of many researchers, especially over the last two centuries [3]. Some groups of insects, mollusks,

IntechOpen

and crustaceans are relatively well studied, but in most other groups, much remains to be done.

Knowledge of freshwater invertebrates in this area of South America has historically lagged far behind that available for vertebrates (e.g., fish; see [4–5]). This is explained by the fact that vertebrates are easier to study than invertebrates, as they are low in diversity, large in size, and easy to identify. In addition, there are taxonomic guides for most of them. In contrast, invertebrates tend to be very diverse and small in size, and for most of them, a stereomicroscope is required for correct identification. In addition to these disadvantages, there is a lack of identification guides for most taxonomic groups.

Invertebrates play a fundamental role in the function of inland aquatic ecosystems, as they allow the transfer of energy from producers (aquatic and terrestrial vegetation) to the upper trophic levels (fish and waterfowl) [1]. In this group of animals, there are herbivorous, omnivorous, carnivorous, and detrivorous species. These feed mainly on bacteria, fungi, microalgae, vascular plants, protozoa, invertebrates, and detritus. The latter may be of autochthonous origin (remains of dead aquatic organisms) or of allochthonous origin (from the terrestrial system, such as riparian tree leaves).

Depending on their way of life, two types of invertebrates can be identified in freshwater ecosystems: planktonic (those that live suspended in the body of water) and benthic (those that live associated with the substrates at the bottom). Zooplankton is composed mainly of Protozoa, Rotifera, Cladocera, and Copepoda, and sometimes we find other elements such as the crustaceans Ostracoda and Cnidaria [1]. Benthic invertebrates are generally more diverse than zooplankton and are composed of a large number of groups of Protozoa, Porifera, Cnidaria, Platyhelminthes, Nemertea, Aschelminthes, Annelida, Mollusca (Bivalvia and Gastropoda), Arthropoda (Chelicerata, Crustacea, and Insecta), Tardigrada, and Bryozoa, among other groups of free life.

This chapter presents a characterization of the diverse groups of insects, mollusks, crustaceans, and other smaller groups from Southwestern South America and a biogeographic analysis of them, with emphasis on their main forcing factors, ecology, and anthropogenic threats.

2. The hydrographic system

In the Southwest of South America (18–55°S), five freshwater ecoregions are recognized: Atacama, Altiplano, Mediterranean, Valdivian Lakes, and Patagonia, which are defined as large areas with homogeneous hydrological and climatic conditions [6–8]. The ecoregions of the Altiplano and Patagonia are shared with Bolivia and Argentina, while the other three ecoregions are only located in Chile. The ecoregions of Atacama (18–22°S) and Altiplano (18–23°S) are characterized by having an arid climate, being four subclimates: coastal desert climate at the coast; interior desert climate at the intermediate zone; the marginal desert climate, which is located between 2000 and 3500 masl; and the steppe climate, characterized by low temperatures and wide thermal amplitude between day and night, which is located at 3500 masl. In the Mediterranean ecoregion (23–33°S), there is a typical Mediterranean climate, with long dry periods of drought and very humid winters that are concentrated in a few months. The ecoregion of the Valdivian Lakes (35–39°S) is characterized by the presence of numerous lakes, mostly oligotrophic, with a typical rainy climate temperature. Finally, the Patagonian ecoregion (42–55°S) is characterized by a dry cold temperate climate, decreasing in temperature as the latitude increases.

3. Freshwater invertebrates

The freshwater invertebrates of the Southwest of South America have numerous singularities that highlight them compared to those existing in other regions of the world [1]. Among them are the following:

a. Very primitive fauna with ancestral Gondwanic-type relations in many of the taxonomic groups. As an example, the freshwater snails of the *Chilina* genus, which have presented their maximum evolutionary radiation in the Chilean territory and southern Argentina, correspond to one of the most primitive pulmonate gastropod groups known (Archaeopulmonata). They present evolutionary affinities with marine gastropods of the Cephalaspidea order (Opisthobranchia).

Additionally, many of the invertebrate groups have a typical Gondwanic geographic distribution. The fragmentation of the Gondwana supercontinent almost 100 million years ago (150–50 ma BP) caused geographic isolation of the ancestral biota. Each of the pieces of this giant mosaic continued to evolve in isolation but retained the signs of ancient connections. This is why, for many invertebrate groups, there is more affinity with the fauna of New Zealand than with that of the rest of South America (e.g., Brazil, northern Peru), a relationship that was already recognized by von Ihering in the late nineteenth century [9].

b. High diversity in a small geographical area and marked endemism: As in the case of terrestrial flora and terrestrial and freshwater vertebrates, in central-southern Chile there is a biodiversity "hot spot" (35° and 43° S), of freshwater invertebrates (these are territories that host a large number of endemic species and, at the same time, have been significantly impacted by human activities). This "hot spot" has been recognized as one of the 25 most important worldwide [10] and is clearly isolated from the rest of South America by a series of geographical barriers (e.g., arid diagonal, Andean mountain range, cold and dry southern zones). This biodiversity "hot spot" of freshwater invertebrates is part of the great Archiplata region [11, 12]. However, it could also be recognized as a sub-unit, called "Chilenia" [1], following the nomenclature used by some geologists to refer to a large part of this territory [13, 14].

Within this "hot spot," a study of the spatial pattern of species richness and indices of genetic and phylogenetic diversity of aeglids was performed [15]. Based on these indicators, they ordered the six hydrographic regions present along this territory according to their conservation priority. They concluded that the hydrographic region composed of the Tucapel (near Cañete), Imperial, and Toltén rivers is a priority for the conservation of aeglids, which can also be extended to many other invertebrate groups.

c. With cases of extremely small geographic ranges: There are examples of species with very small geographic ranges, for example, the "Desert Snail" (*Chilina angusta*), discovered by Rodulfo Amando Philippi on his exploratory trip to the Atacama Desert (1853 and 1854), which inhabits only the Aguada de Paposo. This is a spring with a surface of ca. 30 m^2, located in the coastal desert north of Taltal [1].

Another example is the "Cangrejo tigre" (*Aegla concepcionensis*) that lives in the small basin of the Andalién river [1]. This crab had been considered extinct up until less than a decade ago when it was rediscovered. In the case of aeglids, there are numerous other examples of very small geographic ranges where species are restricted to a limited portion within a given small basin [15–17]. An example of this is the case of the "Cangrejo Camaleón" (*Aegla hueicollensis*), which is found mainly in remote sectors of the Hueicolla and Pichihueicolla rivers, located in the Valdivian forest.

The best studied taxonomic groups in Southwestern South America are those of greatest relevance to the characterization of the structure and functioning of freshwater ecosystems, such as Insecta, Crustacea, Rotifera, Bivalvia, and Gastropoda [1]. Rotifera and Crustacea are well represented in lake zooplankton. Additionally, Crustacea, Insecta, Bivalvia, and Gastropoda constitute an important fraction of the zoobenthos, both lacustrine and fluvial (**Table 1**). Here the discussion will be limited to the large taxonomic groups present in this territory, emphasizing those most known or important in freshwater ecosystems. Most of the information has been obtained from literature regarding the Chilean territory [1–3, 15, 17–44] and southern Peru territory [3, 11, 45–48].

Phylum (subphylum) Class (order)	**Family**	**Genera (*and number of species*)**
Mollusca		
Bivalvia (Paleoheterodonta)	Hyriidae	*Diplodon 2*
	Sphaeriidae	*Pisidium 7, Sphaerium 2, Musculium 2*
Gastropoda (Mesogastropoda)	Cochliopidae	*Potamolithus 1, Helobia 21*
Gastropoda (Basommatophora)	Physidae	*Physa 4*
	Planorbidae	*Biomphalaria 7*
	Ancylidae	*Anisancylus 1, Uncancylus 3*
	Chilinidae	*Chilina 30*
	Lymnaeidae	*Lymnaea 5*
Arthropoda (Crustacea)		
Branchiopoda (Cladocera)	Bosminidae	*Bosmina 2, Eubosmina 1*
	Chydoridae	*Camptocercus 3, Alona 6, Leydigia 1, Alonella 2, Pleuroxus 5, Chydorus 4, Ephemeropus 1, Dunhevidia 1, Biapertura 2*
	Daphnidae	*Daphnia 7, Scapholeberis 2, Simocephalus 4, Ceriodaphnia 2*
	Macrothricidae	*Macrothrix 4, Echinisca 1, Cactus 1, Streblocerus 1*
	Moinidae	*Moina 1*
	Sididae	*Diaphanosoma 1, Latonopsis 1*
Copepoda (Calanoidea)	Boeckellidae	*Boeckella 17*
	Centropagidae	*Parabroteas 1*
	Diaptomidae	*Tumeodiaptomus 2*
Copepoda (Cyclopoidea)	Cyclopidae	*Acanthocyclops 3, Diacyclops 2, Metacyclops 1, Mesocyclops 2, Microcyclops 2, Tropocyclops 1, Eucyclops 7, Macrocyclops 1, Paracyclops 3*
Copepoda (Harpacticoidea)	Canthocamptidae	*Attheyella 33, Lofflerella 5, Antarctobius 8, Moraria 2*
	Harpacticidae	*Tigriopus 1*
Malacostraca (Decapoda)	Aeglidae	*Aegla 20*
	Palaemonidae	*Cryphiops 1*
	Parastacidae	*Parastacus 2, Samastacus 1, Virilastacus 1*
Malacostraca (Amphipoda)	Hyalellidae	*Hyalella 7*

Phylum (subphylum) Class (order)	Family	Genera (*and number of species*)
Malacostraca (Isopoda)	Janiridae	*Heterias 1*
Arthropoda (Insecta)		
Insecta (Ephemeroptera)	Ameletopsidae	*Chiloporter 2, Chaquihua 2*
	Baetidae	*Americabaetis 2, Andesiops 1, Callibaetis 3, Deceptiviosa 3, Camelobaetidius 1*
	Caenidae	*Caenis 3*
	Leptophlebiidae	*Archethraulodes 1, Atalophlebia 7, Atalophlebioides 1, Dactylophlebia 1, Demoulinellus 1, Gonserellus 1, Hapsiphlebia 1, Magallanella 1, Massartellopsis 1, Meridialaris 7, Nousia 6, Penaphlebia 5, Rhigotopus 1, Secochela 1, Thraulodes 1*
	Nesameletidae	*Metamonius 2*
	Oligoneuriidae	*Murphyella 1*
	Oniscigastridae	*Siphlonella 2*
Insecta (Plecoptera)	Austroperlidae	*Andesobius 1, Klapopteryx 2, Penturoperla 1*
	Diamphipnoidae	*Diamphipnoa 3, Diamphipnopsis 2*
	Eustheniidae	*Neuroperlopsis 1, Neuroperla 1*
	Gripopterygidae	*Andiperla 1, Andiperlodes 1, Antarctoperla 2, Araucanioperla 2, Aubertoperla 2, Ceratoperla 2, Chilenoperla 3, Claudioperla 1, Limnoperla 1, Megandiperla 1, Notoperla 2, Notoperlopsis 1, Pelurgoperla 1, Plegoperla 2, Potamoperla 1, Rhitroperla 2, Senzilloides 1, Teutoperla 3*
	Notonemouridae	*Austronemoura 9, Neofulla 3, Neonemoura 2, Udamocercia 3*
	Perlidae	*Inconeuria 1, Kempnyella 2, Nigroperla 1, Pictetoperla 2*
Insecta (Trichoptera)	Anomalopsychidae	*Anomalopsyche 1, Contulma 1*
	Calamoceratidae	*Phylloicus 1*
	Ecnomidae	*Austrotinodes 12, Chilocentropus 1*
	Glossosomatidae	*Mastigoptila 7, Scotiotrichia 1, Tolhuaca 1*
	Hydrobiosidae	*Amphichorema 3, Androchorema 1, Apatanodes 2, Australobiosis 2, Cailloma 3, Clavichorema 7, Heterochorema 1, Iguazu 1, Isochorema 2, Metachorema 2, Microchorema 4, Neoatopsyche 5, Neochorema 4, Neopsilochorema 1, Nolganema 1, Parachorema 1, Pomphochorema 1, Pseudoradema 1, Rheochorema 4, Stenochorema 1*
	Helicophidae	*Alloecentrelodes 2, Austrocentrus 3, Eosericostoma 2, Microthremma 7, Pseudosericostoma 1*
	Helicopsychidae	*Helicopsyche 2*
	Hydropsychidae	*Smicridea 15*
	Hydroptilidae	*Hydroptila 1, Oxyethira 4, Celaenotrichia 1, Neotrichia 1, Metrichia 5, Nothotrichia 2*
	Kokiriidae	*Pangullia 1*
	Philopotamidae	*Dolophilodes 20*
	Stenopsychidae	*Pseudostenopsichidae 3*
	Leptoceridae	*Hudsonema 1, Triplectides 3, Nectopsyche 2, Brachysetodes 10*
	Limnephilidae	*Austrocosmoecus 1, Metacosmoecus 1, Monocosmoecus 5, Platycosmoecus 1, Verger 19*
	Polycentropodidae	*Polycentropus 7*
	Phylorheithridae	*Mystacopsyche 2, Psylopsiche 3*
	Sericostomatidae	*Chiloecia 1, Myotrichia 1, Notidobiella 3, Parasericostoma 10*
	Tasimiidae	*Charadropsyche 1, Trichovespula 1*

Phylum (subphylum) Class (order)	Family	Genera (*and number of species*)
Insecta (Coleoptera)	Dytiscidae	*Rhantus 4, Lancetes 14, Leuronectes 2, Anisomeria 1, Megadytes 2, Laccophilus 2, Liodessus 4, Laccornellus 1, Platynectes 1, Desmopachria 1, Agabus 1*
	Elmidae	*Mycrocylloepus 1, Macrelmis 1, Austrolimnius 2, Austrelmis 8, Stenelmis 1, Neoelmis 1, Hydora 2*
	Gyrinidae	*Andogyrus 2, Gyrinus 2*
	Haliplidae	*Haliplus 3*
	Hydraenidae	*Ochtheosus 2, Gymnochthebius 7, Hydraenida 5*
	Hydrophilidae	*Andotypus 1, Dactylosternum 1, Cylorygmus 2, Stethoxus 2, Cercyon 2, Dibolocelus 2, Enochrus 5, Chaetarthria 1, Tropisternus 1, Hydrochus 1, Berosus 3, Hemiosus 2, Anticuar 1, Paracymus 3*
	Psephenidae	*Tychepsephus 1, Ectopria 1, Eubrianax 1*

Table 1.
Synoptic vision of the most well-known families and genera of freshwater macroinvertebrates of southwestern South America (based on [1–3, 11, 15, 17–48]).

3.1 Arthropods

The most frequent groups in freshwater ecosystems are crustaceans, insects, and chelicerates. Within crustaceans is a wide diversity of organisms, ranging from complex to very simple forms such as Copepoda, Branchiopoda, and Ostracoda. Copepoda is a very important component of lake zooplankton, while the others are benthic, and frequently associated with the bottom surface [1]. Amphipods are also common components in the benthos and in some isopod areas. Regarding higher crustaceans, there are three families in Chile: Palaemonidae, Parastacidae, and Aeglidae. The two last ones host very particular commensals, such as temnocephalous and histriobdelids (see below). Insects are notably more represented in freshwater environments than crustaceans. Thus, there are several orders whose larval or nymphal stages develop in water [1]: Ephemeroptera, Plecoptera, Trichoptera, and Odonata. Adults, on the other hand, live outside of water. Almost all the other orders of insects present families adapted for aquatic life, especially in the larval state. These are Diptera (Chironomidae, Culicidae, Tipulidae, Simuliidae, Athericidae, Blephariceridae), Coleoptera (Dytiscidae, Hydrophilidae, Psephenidae), and Hemiptera (Notonectidae, Belostomatidae). Within the aquatic insects, there are different forms of feeding. Some are microphages, equipped with sophisticated filtration systems (e.g., Diptera, Trichoptera), while others are efficient carnivores, located at the terminal levels of certain food chains (e.g., Megaloptera, Odonata). Some may even prey on freshwater vertebrates (e.g., Hemiptera Belostomatidae). The number of larvae or nymphs in freshwater under natural conditions is normally high, contributing significantly to the feeding of vertebrates and invertebrates.

Freshwater Chelicerata are not as diverse as the previous two groups. There are very few aquatic spiders, and only mites (Hydracarina) are a common component in this habitat. There are some lake ecosystems, such as the Quiñenco lagoon in the Biobío Region, whose use as a source of drinking water has been limited by the presence of high densities of Oribatida mites (*Scapheremaeus*, Galumnidae, Nothridae, Galumnoidea, Malaconothridae, Oribotuloidea) and Prostigmata (*Hygropatella*, Arrenouridae, Oxidae) [24].

Copepod, and cladoceran Crustaceans: As mentioned above, these are small fundamental organisms in lake zooplankton. The copepods are characterized by having their body divided into two regions, with the anterior region (cephalothorax) typically being elongated, with a nauplian eye and appendages. Three main groups are recognized: Calanoidea (mainly planktonic), Harpacticoidea (generally microbenthic littorals), and Cyclopoidea (littoral and only a few are typically limnetic). In contrast, cladocerans are typically planktonic organisms characterized by a thin bivalve shell (which does not cover the head) and with a reduced abdomen.

Knowledge of zooplankton in Chilean lake ecosystems has progressed significantly over recent decades [21, 42]. Southwestern South America is characterized by marked latitudinal and altitudinal gradients. In these gradients it is possible to find different types of lentic ecosystems, whose environmental diversity is clearly reflected by the composition of species of zooplanktonic crustaceans, and five zones can be recognized [42]:

a. Northern Chile corresponds to lakes and lagoons located in the Chilean-Peruvian Altiplano, where it is possible to register endemic species of the genus *Daphnia* and *Boeckella*, among others.

b. Central Chile brings together a series of aquatic bodies of low height and shallow depth. This area is characterized by the presence of *Diaptomus diabolicus* (synonym for *Tumeodiaptomus vivianae*). At this latitude there are also high mountain lakes of greater depth. These are characterized by the presence of species of the *Boeckella* genus, of which there are few records and taxonomic studies.

c. Central-South Chile includes the so-called Nahuelbutan Lakes, whose zooplanktonic fauna is just beginning to be studied.

d. Southern Chile and Chilean Patagonia include the lakes of the Magellanic region, which have a high diversity of species, especially those of the Torres del Paine area, characterized by its high endemism.

The Chilean freshwater zooplankton is composed mainly of 53 species of Cladocera and 73 species of Copepoda [42] (**Table 1**). Cladocera includes six families, of which Daphnidae and Chydoridae are the most diverse with 15 and 25 species, respectively. The Copepoda are made up of 20 species of Calanoidea, 22 of Cyclopoidea, and 49 of Harpacticoidea. The most diverse families are Cyclopidae and Canthocamptidae with 22 and 48 species, respectively. Within these taxonomic groups, the least studied are the Harpacticoidea and the Cladocera of coastal environments (e.g., Chidoridae), which require taxonomic revision.

Among the Calanoidea copepods, the genus *Boeckella* is commonly found throughout the Southern Hemisphere, in fresh and saline continental waters [42]. *Boeckella gracilipes* is one of the species with the widest geographic distribution in South America, with its presence being reported from Ecuador (Lake Mojanda) to Tierra del Fuego, although it presents morphologically differentiated populations, probably associated with temperature [42]. Among the copepods, Calanoidea also includes the endemic species of the extreme south of South America *Parabroteas sarsi*. This is a predatory copepod, which is widely distributed in the Chilean-Argentine Patagonia, which stands out for its large size, reaching up to 8 mm, a size that places it as the largest copepod in the world. Calanoid copepods in Chilean continental waters are characterized for being the main group in zooplanktonic assemblages, being represented by the genera *Boeckella, Parabroteas,* and *Tumeodiaptomus* [21]. The genus *Boeckella* is represented by three species of wide geographic distribution: *B. gracilipes,* which is found between 18 and 44°S; *B. poopoensis,* found mainly in saline lakes of northern Chile (14–27°S); *and B. michaelseni* which is found between 44 and 54°S. *Tumeodiaptomus,* represented *by T. diabolicus*, is distributed between 32 and 42°S. Finally, *Parabroteas*, with only one species, *P. sarsi*, is found in shallow lakes between 44 and 54°S. There are different species for the northern zone (*Boeckella occidentalis*, *B. gracilipes*, and *B. poopoensis*) central (*Tumeodiaptomus diabolicus*, *B. bergi*, *B. gracilipes*), and southern Chile (*T. diabolicus, B. michaelseni,* and *B. gracilipes*).

In regard to Cladocera, six species of *Daphnia* (*D. pulex, D. ambigua, D. obtusa, D. peruviana, D. commutate,* and *D. sarsi*) have been recorded, of which *D. obtusa* and *D. pulex* have a cosmopolitan distribution [42]. In Chile they are located from north to south and from coast to mountain range, representing excellent indicators of water quality and with great potential to be used in toxicity tests performed in the laboratory.

Malacostraca crustaceans: In the Chilean freshwater ecosystems, they are composed of the orders Decapoda, Isopoda, and Amphipoda. They are of particular relevance in freshwater ecosystems, because they are fundamental components in the diet of large fish and water birds and also because of the great commercial importance of the different species of the families Palaemonidae and Parastacidae. Another important aspect of this group is that it is the only one in which there is evidence of species extinction.

The knowledge of Malacostraca crustaceans in Chilean freshwater ecosystems has progressed remarkably in recent decades [16, 17]. The freshwater decapods are the largest group of Malacostraca, made up of shrimp of the families Palaemonidae (one species) and Parastacidae (four species; **Figure 1(26)**), plus Anomura crabs of the family Aeglidae (19 species and 2 subspecies; **Figure 1(27)**). Peracarids are represented by seven species of Amphipoda of the *Hyalella* genus, while only one species of Isopoda has been identified (*Heterias* (*Fritzianira*) *exul*). Thus, the limetic Malacostraca fauna in Chile is composed of 35 taxa, i.e., 33 species and 2 subspecies (**Table 1**) [17].

The geographical range occupied by the Malacostraca includes practically the entire continental territory in its latitudinal and altitudinal extension [17]. However, not all taxonomic groups are included in this range, nor is the distribution of their species continuous. On the contrary, most species have a more or less discontinuous distribution; on the one hand, they are related to the natural discontinuity of the hydrographic basins and, on the other, to the mosaic of habitats found in each basin.

The largest geographic amplitude is observed in the amphipods of the *Hyalella* genus, from Guallatire (Tarapacá) to Punta Arenas (Magallanes). In this latitudinal range, the seven *Hyalella* species are staggered in a North–South direction, with individual ranges that are very dissimilar in extension. The most widely distributed species is *H. costera*, which is recorded in sites as far away as Quebrada de Paposo (Antofagasta) and Isla Teja (Valdivia) [17]. A similar situation, of very remote limits marked by discontinuous populations, is that of *Cryphiops caementarius* ("Camarón de río del Norte"), registered between Arica and Valparaíso. Of the Parastacidae

DOI: http://dx.doi.org/10.5772/intechopen.89077

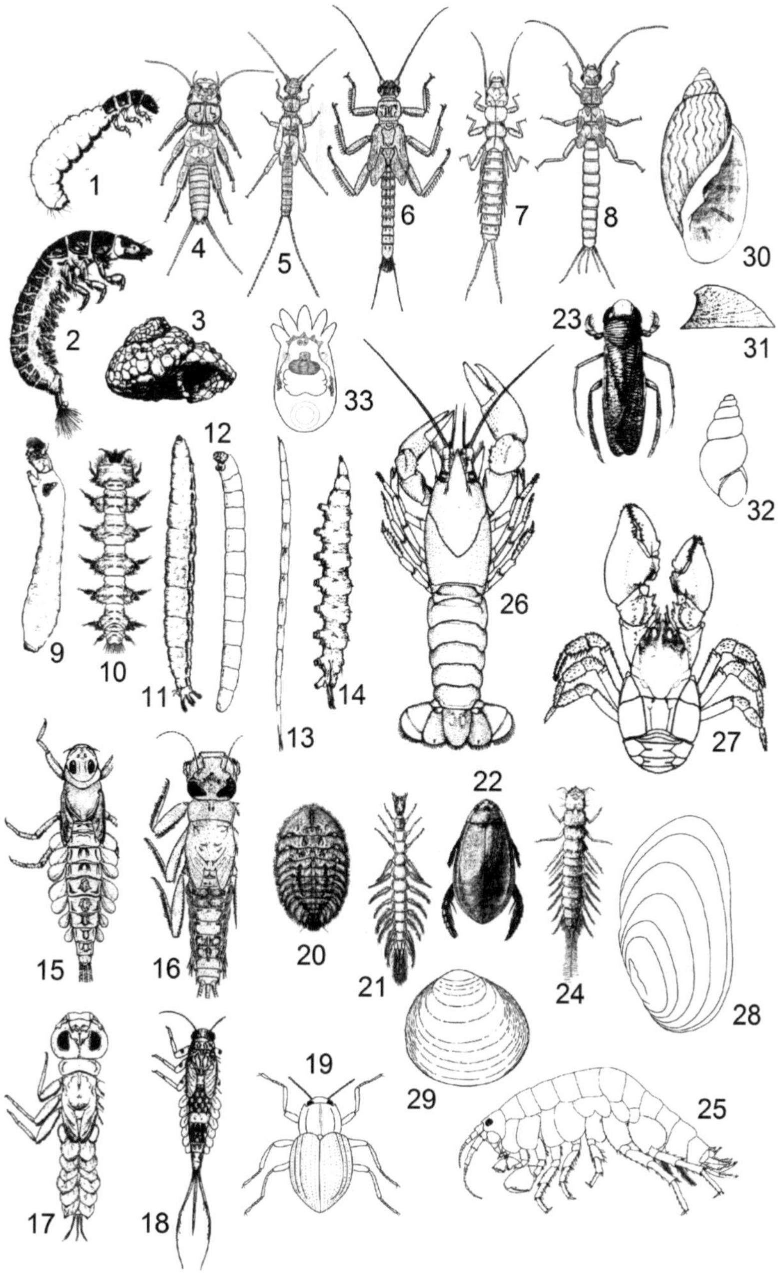

Figure 1.
Some families of macroinvertebrates found in rivers of the southwest of South America (based on [1–3, 16, 34]). 1–3 Trichoptera: (1) Hydroptilidae, (2) Hydropsychidae, (3) Helicopsychidae. 4–8 Plecoptera: (4) Perlidae, (5) Notonemouridae, (6) Gripopterygidae, (7) Eustheniidae, (8) Austroperlidae. 9–14 Diptera: (9) Simuliidae, (10) Blephariceridae, (11) Tipulidae, (12) Chironomidae, (13) Ceratopogonidae, (14) Athericidae. 15–18 Ephemeroptera: (15) Siphlonuridae, (16) Leptophlebiidae, (17) Ameletopsidae, (18) Baetidae. 19–22 Coleoptera: (19) Elmidae, (20) Psephenidae, (21) Gyrinidae, (22) Dytiscidae. 23 Hemiptera (Corixidae). 24 Megaloptera (Sialidae). 25–27 Crustacea: (25) Hyalellidae, (26) Parastacidae, (27) Aeglidae. 28–29 Bivalvia: (28) Hyriidae, (29) Sphaeriidae. 30–32 Gastropoda: (30) Chilinidae, (31) Ancylidae, (32) Cochliopidae. (33) Temnocephala.

shrimp species, the one with the widest range is *Samastacus spinifrons* ("Camarón de río del Sur"), distributed without interruption between Aconcagua and Chiloé. Its presence in islands south of Chiloé, up to Taitao, is not documented, but it can be presumed to be in at least the major islands of the archipelagos of Guaitecas and Chonos [17]. The remaining species of parastacids, of excavator habit, associated with wetlands of the Central Valley and the Coastal Mountain Range, have delimited and allopatric distributions, i.e., *Parastacus pugnax* to the north of the Toltén River and *Parastacus nicoleti* to the south of the same river. The known distribution of *Virilastacus araucanius* is clearly discontinuous, between Concepción and Hueyusca (near Osorno).

Among the Aeglidae species, *Aegla pewenchae* stands out as having the broadest latitudinal range, followed by *Aegla papudo.* The remainder of the species have narrow latitudinal ranges, such as *Aegla concepcionensis*, which is restricted to a single hydrographic basin [17]. Another example is the distribution of *Aegla neuquensis neuquensis,* whose known distribution in Chile is restricted to one sector of the Aysén River basin [38]. Recently, the distribution of freshwater decapods in rivers and lakes in Patagonia was mapped, identifying Duque de York Island (50°75′S) as the southernmost habitat of these organisms in the world (*Aegla alacalufi*) [27].

According to the available antecedents, the highest density of Malacostraca crustacean taxa occurs between 32° and 43°S [17]. All species of parastacids, 16 of the 20 *Aegla* species and 3 of the 7 *Hyalella* species, are found exclusively in Chilean territory [17]. In Chile, *Aegla affinis* is found extralimitally. It is found in Laguna del Maule, introduced by Argentine sport fishermen from the Rio Grande basin (in the south of the Mendoza Province). On the other hand, Chilean species are generally endemic to restricted sectors of the national territory. Apart from the narrow distributional range of *Aegla concepcionensis* and *Aegla hueicollensis*, which qualifies them as extremely endemic species, there are other cases in which a species is known only from one basin or from some adjacent basins. Such is the case of *Aegla spectabilis* in the Chol Chol River basin and of *Aegla bahamondei* and *Aegla occidentalis* in the adjacent basins of the Tucapel-Paicaví and Lleu Lleu rivers, in the coastal strip of the Biobío Region [17]. On the western slope of the Cordillera de la Costa, south of Corral and up to the mouth of the river Bueno, there is *Aegla hueicollensis* distributed in a series of small individual basins isolated from each other. *Parastacus pugnax* and *Parastacus nicoleti* species, digging species associated with the coastal wetlands and the intermediate depression to the north of Temuco, are endemic in their respective dispersion areas, separated by the Toltén River basin.

The conservation status of the freshwater invertebrate species in this territory was initially established for shrimps and anomurans, based on a classification using IUCN criteria and expert opinion [49]. Subsequently, the Chilean *Aegla* species were reclassified, relying on phylogenetic and genetic diversity arguments combined with the criteria proposed by IUCN [15]. The initial proposals determined that three of the four species of parastacids (*P. nicoleti*, *P. pugnax*, and *S. spinifrons*) are "vulnerable" in much or all of their geographic range [49]. The status of *Cryphiops caementarius* was recognized as "endangered" in the Valparaiso and Metropolitan regions and "vulnerable" in the rest of its geographic range. The status of the *Aegla* species was less compromised, although *Aegla laevis laevis* and *Aegla papudo* were "endangered" in the Valparaiso and Metropolitan regions. In addition, *Aegla laevis talcahuano* was described as "vulnerable" throughout its range. The remaining *Aegla* species were classified as "insufficiently known" or "less concern" [49]. Subsequent classifications subscribed only partially to the above qualifications, establishing that *Aegla concepcionensis* and *Aegla expansa* were "extinct in the wild" and that *Aegla papudo, Aegla spectabilis*, and *Aegla laevis laevis* qualified as "critically endangered" [15]. More recently, in 2014 the Chilean state approved and made the classification of 24 species of freshwater decapod crustaceans official.

There are two species classified as "critically endangered," corresponding to *Aegla affinis* and *Aegla denticulata lacustris*. In addition, there are eight "endangered" and five "vulnerable." The remaining nine qualify in the category of "less concern."

Of the Chilean groundwater crustaceans, relatively little is known. A synthesis has been made for the South American continent, indicating that groups such as Amphipoda (e.g., *Ingolfiella*, *Bogidiella*) and several Isopoda (e.g., *Microcerberus*), are preferably found in the Chilean-Argentine area (previously described) [11]. There are also groups of Syncarida, such as *Bathynella*, *Leptobathynella*, *Parabathynella*, *Chilibathynella*, and *Stygocaria*, registered along the territory [11].

Insects: The larval stages of Ephemeroptera, Plecoptera, and Trichoptera, as primary consumers, are a relevant component of the benthic freshwater fauna, both in abundance and in biomass. They process a significant amount of periphytic microalgae and organic matter (autochthonous and allochthonous), either by triturating large particles or filtering small ones. On the other hand, during the adult stage, in some cases they return a significant amount of energy to the terrestrial environment. Many terrestrial predators, such as spiders, insects, insectivorous birds, and bats, consume a large number of adults during the periods of emergency, nuptial flight, and oviposition. These aquatic insects are one of the most important in river trophic nets, since spawning, larvae, and adults are a fundamental part of the diet of fish and amphibians or participate in some of the stages that end in them. Due to their abundance and ubiquity, as well as the differential tolerance of different species to different degrees of pollution or environmental impact, they have been used for some time as biological indicators of water quality. In particular, Plecoptera, preferably living in fast, turbulent, cold, and highly oxygenated waters, are considered excellent indicators of water quality.

The knowledge of insects in freshwater ecosystems has progressed significantly in recent decades [1]. It is relatively easy to identify the families and genera of these invertebrates due to the existence of identification guides, for example, the book *Macroinvertebrados bentónicos sudamericanos: Sistemática y biología* [3]. However, identification at the species level of many genera is difficult and in some cases still impossible.

Ephemeroptera: These are elongated body organisms whose adult states have reticulated nervation wings. The first pair of wings is larger than the second, and when at rest, the wings are left in an upright position. Both adult aerial and aquatic nymphal states are recognized by the presence of three (or two) filiform caudal appendages (**Figure 1(15–18)**). Globally, this is a rather small group in terms of number of species. However, they are conspicuous components of freshwater benthos in their immature stages.

In Chile, a total of 57 species belonging to 7 families have been described (**Table 1**) [18]. The most diverse family is Leptophlebiidae, with 36 species belonging to 15 genera. Of the 25 genera existing in Chile, the genera *Nousia*, *Meridialaris*, and *Penaphlebia* are the most diverse, with 7 and 6 species, respectively. One of the problems with the identification of the aquatic states of Ephemeroptera is that most of the descriptions are based on the diagnostic characteristics of adults [18]. Only 40% of the species have been described at both the adult and nymph stage, 12% only in nymphs, and 47% only in adult males and/or females [18]. From the point of view of their endemicity in Chilean territory, 56% of species would be exclusive to Chilean territory, 33% shared with Argentina, and 11% of them a wider distribution [18]. Below are general comments for each family based on [18]:

a. *Baetidae*: only three of its four genera and four of the nine species can be identified by their nymphs (*Andesiops peruvianus*, *Andesiops torrens*, *Andesiops ardua*, and *Americabaetis alphus*).

b. *Oniscigastridae*: of the two species of this group, only *Siphlonella ventilans* is known to be a nymph. This species is known as the male imago and subimago but not the female.

c. *Nesameletidae*: only the nymph of *Metamonius anceps* has been identified.

d. *Ameletopsidae*: only one of the two genera of this family is known to have nymphs—*Chiloporter penai* and *Chiloporter eatoni*. *Chaquihua*, the other genus of this interesting Gondwanic family, characterized by having carnivorous nymphs, is practically known only by its female imagos.

e. *Oligoneuriidae*: the *Murphyella needhami* nymph is known, which is very curious due to the absence of abdominal gills.

f. *Caenidae*: this family has unstudied species at the nymph level and is even unidentifiable at the imago level.

g. *Leptophlebiidae*: of the 36 species described in Chile, 14 could be recognized by their adult stage and 17 by adults and/or nymphs, and only 5 are known only in their nymph stage (*Hapsiphlebia anastomosis*, *Nousia delicata*, *Massartellopsis irarrazavali*, *Meridialaris laminata*, *Nousia maculata*, *Nousia grandis*, *Nousia minor*, *Meridialaris diguillina*, *Meridialaris biobionica*, *Penaphlebia vinosa*, *Penaphlebia chilensis*, and *Penaphlebia fulvipes*).

Plecoptera: The adult stages of these organisms have long antennas and two pairs of generally well-developed membranous wings. Both aerial and aquatic nymphal states are recognized by the presence of two caudal appendages, which are multisegmented and variable in length (**Figure 1(4–8)**). In this territory, there are Plecoptera that can live in extreme temperature environments. For example, it has been observed that the "Dragon of Patagonia" (*Andiperla willinki*) may inhabit glacial areas, where temperatures may reach freezing point.

A total of 63 species belonging to 6 families have been identified in the territory. Gripopterygidae corresponds to the most diverse family, with 28 species included in 18 genera [39]. Like most freshwater invertebrates, the greatest diversity of the Plecoptera species is located in central-southern Chile.

In general, most plecopteran genera include no more than two species, with the exception of *Chilenoperla*, *Diamphipnoa*, and *Teutoperla,* which have three species. From a point of view of its endemicity in the Chilean territory, it covers 57% of species. The following are specific comments for each family based on [18]:

a. Eustheniidae: has predatory nymphs present only in Oceania and Chile, and it is represented in Chile by two monospecific genera exclusive to our country (*Neuroperla schedingi* and *Neuroperlopsis patris*).

b. Diamphipnoidae: this family of detritivorous nymphs lives exclusively in South America. Of the five Chilean species, *Diamphipnoa helgae* and *Diamphipnopsis samali* are also found in Argentina.

c. Austroperlidae: this family has detritivore nymphs, whose representatives are found in South America and Australia. Of the four Chilean species, only *Klapopteryx armillata* also inhabits Argentina.

d. Gripopterygidae: this is the most diversified Plecoptera family in Chile. It has mostly detritivorous nymphs, distributed in South America, Australia, and New Zealand. Despite being a well-studied family, 55% of species with unknown or well-described nymphs persist but have only been assigned to a genera [39]. The case of *Araucanioperla* is notable, of which two species are known based on their imagos, and three different nymphs have not been assigned to any of them. In this family only the *Claudioperla tigrina* species has been identified and is shared with Peru and Bolivia.

e. Notonemouridae: This family is distributed in South Africa, Madagascar, Australia, New Zealand, and South America, and in Chile it is the second most important with 16 species (unknown diet). According to the previously cited authors, in general immature states only allow recognition of the genera.

f. Perlidae has predatory nymphs represented in Chile by five species.

In regard to the conservation status of aquatic insect species, the only taxa officially classified in Chile correspond to *Andiperla willinki* "Dragón de la Patagonia," which has been classified in the "endangered" category. This species is highly unique worldwide. It is a species native to Chile, found in the Aysén and Magallanes Regions and in Argentina (only one record) [40]. They are extremophilic and psychrophilic insects that live exclusively on ice in the northern, southern, and Darwin mountain ranges. Their habitat is not shared with other aquatic insects or vertebrates [31, 32]. Adults and nymphs may feed on other psychrophilic species, such as microalgae and Collembola, which also support the food web of glacial communities [31, 32]. In this way, *A. willinki* would be the top predator in these systems. This species is a potential resource for obtaining biotechnological products, especially those associated with enzymatic processes carried out under freezing conditions.

Trichoptera: correspond to insects of soft body, whose adult aerial phases have two pairs of undeveloped hairy membranous wings. The larvae are aquatic (**Figure 1(1–3)**) and build "little houses" of diverse organic materials (e.g., fragments of leaves and logs) and inorganic materials (e.g., grains of sand; **Figure 1(3)**), depending on the taxonomic group involved.

There are nearly 200 species described in Chile, with the Hydrobiosidae family being the most diversified, with 47 species belonging to 20 genera [29]. However, it is the Limnephilidae family, due to its size and abundance, which is almost emblematic in Chile, particularly distributed in the watercourses of Patagonian forests. Of the total species described, mostly based on adult states, aquatic immature states are known for only 45 of them. In other words, so far it is possible to identify only 21% of the aquatic larvae of Trichoptera up to the species level [29]. For each family, a summary is given below of the number of species per genus for which the aquatic immature states are known in relation to the total of species (immature aquatic states/known species) based on [29]:

a. Hydrobiosidae: *Cailloma* 3/3, *Neopsilochorem*a 1/1, *Apatanodes* 1/2, *Neoatopsyche* 5/5, *Iguazu* 1/1, *Rheochorema* 4/4, and *Stenochorem*a 1/1.

b. Glossosomatidae: *Mastigoptila* 1/7.

c. Hydroptilidae (**Figure 1(1)**): *Neotrichia* 1/1, *Celaenotrichia* 1/1, and *Metrichia* 1/5.

d. Ecnomidae: *Austrotinodes* 1/12.

e. Limnephilidae: *Austrocosmoecus* 1/1, *Monocosmoecus* 3/5, *Platycosmoecus* 1/1, *Metacosmoecus* 1/1, and *Verger* 6/19.

f. Leptoceridae: *Triplectides* 1/3 and *Hudsonema* 1/1.

g. Anomalopsychidae: *Contulma* 1/1 and *Anomalopsyche* 1/1.

h. Helicophidae: *Eosericostoma* 1/2 and *Austrocentrus* 1/3.

i. Sericostomatidae: *Parasericostoma* 2/10.

j. Tasimiidae: *Trichovespula* 1/1 and *Charadropsyche* 1/1.

In Chile, the known geographic distribution of the Trichoptera ranges from the Coquimbo Region to the Magallanes Region (30–55°S), bounded to the north by aridity, although intrusions have also been observed in the Brazilian subregion (e.g., species of the family Hydroptilidae in the Loa River) [29]. The greatest diversity of species is located in the regions of Biobío and Los Lagos (37–42°S), dominating in the Biobío Region, with a percentage that exceeds half of all species recorded in Chile [1]. On the other hand, the fundamentally endemic condition of the species, added to the fact that a group of genera has been included in exclusive families of the Australo-zelandic area (Philorheithridae, Helicophidae, Kokiriridae, Tasimiidae) and Australasian area (Stenopsychidae), has provided the foundation to distinguish and characterize the Chilean-Patagonian subregion within the Neotropical Area. Because such a distinction goes beyond the political boundaries of the territory, such endemism does not prevent species from being shared with Argentina. Only Helicophidae has presented an important diversification of genera (=5) and species (=15), among the families that have been assigned a Gondwanic origin.

Coleoptera: The adults of aquatic or aerial life are characterized by the presence of two pairs of wings, of which the previous pair has been modified as solid protective covers (elytra), the posterior pair being membranous. These insects, and mainly the larval stage, are part of the benthic macroinvertebrate fauna, participating in multiple food webs where they act as predators, detritivores, or herbivores (**Figure 1(19–22)**). From a taxonomic point of view, aquatic Coleoptera constitute a heterogeneous group that includes taxa belonging to different lineages of the Adephaga and Polyphaga sub-orders. In Chile, almost a hundred species belonging to seven families are known [23]. Three families of Hidradephaga are present, of which Dytiscidae presents the greatest richness at a generic and specific level with 11 genera and 34 species. Gyrinidae is represented by two genera and four species and Haliplidae with one genus and three species. **Table 1** summarizes the taxonomy of the group, indicating the number of known species based on [23]. Among the Dytiscidae genera, *Lancetes* is the most diversified with a total of 14 species, followed by *Rhantus* and *Liodessus* with 4 species each. The remaining genera are monospecific, except *Megadytes*, *Laccophilus*, and *Leuronectes*, each with two species. The Polyphaga are represented by four families, of which Hydrophilidae is the most diversified with 13 genera and 26 species. The Elmidae family presents seven genera, highlighting *Austrelmis* with eight species and *Austrolimnius* with two species. Hydraenidae presents only three genera, of which *Gymnochyhebius* is the most diversified with seven species. Finally, the Psephenidae family is the least diversified with three monospecific genera.

Chile does not have endemic families of aquatic Coleoptera, unlike other Mediterranean regions [23]. However, this fauna shows South American elements of tropical and Australian origin. This is the case of the *Tropisternus* genus, which is widely distributed in the Neotropical and *Lancetes* regions with links to Australia, New Zealand, and Tasmania, as well as the genus *Tychepsephus* (Psephenidae) and *Austrolimnius*, a taxa described as the dominant genus of freshwater elmids, which is also found in Central and South America. Most of the genera are not very diversified, and many of them are monotypic, a situation mainly related to the isolation of the territory since the Tertiary Period.

There are some taxa that have restricted geographic distributions, a situation that would give rise to a certain degree of endemism [23]: This is the case of the following species [1]:

a. Hydrophilidae: *Enochrus concepcionensis* (Concepción and Biobío and Puyehue National Park)

b. Dytiscidae: Platynectes magellanicus (Magallanes), Rhantus obscuricollis (Aysén in Chile and Neuquén in Argentina), Rhantus antarcticus (provinces of Concepción and Cautín), Lancetes towianicus (Magallanes), Lancetes flavipes (Magallanes), and *Lancetes kuscheli* (Province of Antofagasta)

c. Elmidae: *Microcylloepus chilensis* (Quebrada de Camarones en Tarapacá); *Austrelmis chilensis, A. scissicollis*, *A. trivialis*, and *A. nyctelioides* (all three in the province of Quillota); and *Austrelmis elegans* and *A. costulata* (Tumbre, Province of Antofagasta)

In the Juan Fernández Archipelago, the fauna of Dytiscidae would be composed of only three species, all of the Colymbetinae subfamily. Of these, *Anisomeria bistriata* and *Lancetes bäckstromi* are endemic to Masafuera Island and Masatierra, respectively. For Easter Island, *Bidessus skottsbergi* has been described as endemic, and *Rhantus signatus* was described on Mocha Island.

The following are general comments for each family based on [23]:

a. Gyrinidae: Larvae and adults are of aquatic habits and predators; they frequent bodies of water of scarce current.

b. Haliplidae: Their habitat consists of lentic water bodies with abundant filamentous algae and underwater plants and detritus-rich bottoms. The larvae are not swimmers and are shredders of the phytophagous regime. Adults move through water by alternating movements of mesothoracic legs.

c. Dytiscidae: They are present in all types of freshwater ecosystems. Larvae and adults have aquatic and carnivorous habits (adults also have a great flight capacity). They are distributed throughout Chile, and the central zone reaches up to 2300 masl (they are also found in salt flats in northern Chile).

d. Hydrophilidae: detritivorous adults (dead animals and decaying plants) and predatory larvae, of aquatic or semiaquatic habits. *Tropisternus setiger* has been reported as a predator of mosquitoes (Culicidae). The adults of the subfamily Sphaeridiinae are the only ones that are present in terrestrial habitats.

e. Hydraenidae: They are small insects, not swimmers, and move by walking on rocks and algae in the banks of bodies of water or in the margins of clear water currents and sandy bottom. Some species are associated with mosses and most are found under boulders.

f. Elmidae: They are aquatic in both adult and larval stages and are generally found in sandy, gravelly, or submerged bottoms among vegetation. They feed on algae and detritus and can also feed on microorganisms and small aquatic invertebrates.

g. Psephenidae: Aquatic larvae that live attached to boulders in corrugated sectors, presenting the appearance of crustaceans. They are typical inhabitants of rithron areas that correspond to sectors of high slope, with high speeds of currents, low and stable temperatures, and high concentrations of oxygen (e.g., *Tychepsephus felix*).

Diptera: Correspond to insects whose adult aerial phases have two pairs of simple wings (they also have two pairs of strongly modified vestigial wings). Many of the Diptera groups have aquatic immature phases, with an enormous variety of shapes and adaptations to the aquatic environment (**Figure 1(9–14)**). Among the most common families in Chilean rivers and lakes, the following stand out: Chironomidae, Athericidae, Simuliidae, Ceratopogonidae, Blephariceridae, Psychodidae, Empididae, Culicidae Tabanidae, Ephydridae, and Tipulidae [1]. Within the Diptera, the Chironomidae family deserves special attention because it is one of the most important groups of insects in aquatic ecosystems (**Figure 1(12)**), due to its abundance, species richness, and wide ecological spectrum, being found in a range of natural conditions greater than any other group of insects. The bibliographic information available suggests that this group has a great taxonomic diversity in the Southwest of South America. However, the dispersion of the literature, the scarcity of specialists, and the lack of reference collections do not allow us to provide a list of the genera and number of species present. As an example of its great diversity, in a stretch of only 5 km along the middle course of the Biobío River, 18 morphospecies have been observed, belonging to the subfamilies Orthocladiinae, Podonominae, Chironominae, and Tanyponidae [1]. Of these, the Orthocladinae and Chironominae families were the most represented in terms of specific wealth, with eight and seven taxa, respectively. The most common genera in this area are *Eukiefferiella*, *Thienemanniella*, *Demycryptochironomus*, *Cricotopus*, *Oliveridia*, *Pentaneura*, *Nanocladius*, *Parasmittia*, *Dicrotendipes*, *Cryptochironomus*, *Stictochironomus*, *Chironomus*, and *Parakiefferiella*.

Despite their great diversity, enormous abundance, and great ecological relevance, these small aquatic larvae are one of the least studied groups in the territory [1]. The larvae of this group of insects play an environmentally important role, as they are sensitive bioindicators of aquatic ecosystem conditions, such as temperature, pH, dissolved oxygen, and nutrients, in addition to a wide range of toxic substances. It is important to note that the effects of pollution on chironomid communities have been widely reported in the literature, showing differential taxa tolerances to specific pollution sources. As a consequence of the different tolerances to eutrophication and organic enrichment of sediments, chironomids are used to study the trophy level of aquatic ecosystems. Furthermore, the potential use of chironomids as indicators of heavy metal contamination has been reported, based on the study of deformations of their buccal structures or due to the increase in dominance of groups capable of inhabiting environments with high concentrations of metals. Additionally, the habitat preference of the different groups of chironomids can provide information on the particular eco-hydraulic characteristics of an aquatic ecosystem [1].

Chironomids have acquired particular relevance over the last decades from an ecotoxicological point of view due to malformations produced in the structures of the cephalic capsule of larvae (antennas and buccal parts and others). These morphological alterations have been observed in the genera *Chironomus*,

Procladius, and *Cryptochironomus*. Moreover, this group of insects is of great importance in paleolimnological studies, since the cephalic chitinous capsules of the larvae are preserved in lacustrine sediments, allowing them to be used in paleoenvironmental reconstitutions. Additionally, this group has been greatly relevant in biogeography studies, being used as reference elements in intercontinental faunal relations.

Other orders of common insects: There are three other orders of insects that are frequently found in freshwater ecosystems, although with less abundance and diversity than those described above. These are [1]: (a) Odonata, represented mainly by the families Lestidae, Gomphidae, Coenagrionidae, Cordulidae, Calopterygidae, Aeshnidae, Libellulidae, and Petaluridae; (b) Megaloptera, represented by the predatory families Sialidae (*Sialis*; **Figure 1(24)**), and Corydalidae (*Protochauliode*); (c) Hemiptera, with the families Gerridae, Corixidae (**Figure 1(23)**), Notonectidae, and Belostomatidae; and (d) Lepidoptera with the family Pyralidae.

3.2 Molluscs

This group is represented by several species of the Bivalvia and Gastropoda classes, which are very common in different types of freshwater habitats.

a. *Bivalvia*: These organisms are characterized by the soft parts of their bodies (e.g., visceral mass and foot), enclosed by two calcareous shells connected dorsally by a flexible ligament. In Chile, 13 species belonging to 2 families and 4 genera have been described [28]. Numerous studies have demonstrated the important role they play in the ecosystems they integrate. For example, the large bivalves of the Hyriidae family, due to their feeding by suspension and because they are long-lived organisms, can significantly influence the abundance of phytoplankton communities, water quality, and nutrient recycling [28]. In addition, they are an important component of energy flow and nutrient cycling, as they constitute a significant portion of freshwater macrobenthic biomass. These organisms have been used as sentinel organisms and have potentially been considered as biomonitors of the health of freshwater ecosystems. The clams of the Sphaeriidae family have been less studied because of their small size, hidden way of life, and difficulty in being identified. However, since they can inhabit environments where no other bivalve can live, they can serve as biomonitors of the environmental conditions of their habitat.

The species described to date for Chile belonging to the Hyriidae family are represented only by the *Diplodon* genus, with two subgenera [28], *Diplodon* and *Australis*, each with its respective species *D. (D.) chilensis* and *D. (A.) solidulus*, and the family Sphaeriidae represented by three genera: (a) *Pisidium*, with the species *P. lebruni*, *P. observationis*, *P. chilense*, *P. meierbrooki*, *P. magellanicum*, *P. huillichum*, and *P. llanquihuense*; (b) *Sphaerium*, with species *S. forbesi* and *S. lauricochae*; and (c) *Musculium*, with the species *M. patagonicum* and *M. argentinum*.

An analysis of the geographic distribution of the species reported for Chile allows us to propose the existence of three zoogeographic areas and the postulation of four species of Sphaeriidae and one of Hyriidae as endemic species of Chile [28]: (a) High Andean region, *Sphaerium lauricochae*, *Sphaerium forbesi*, and *Pisidium meierbrooki* are species specific to this region, sharing geographical areas with Peru and Bolivia; (b) Central-southern region of Chile, characterized by *Pisidium llanquihuensis*, *Pisidium huillichum*, *Pisidium chilense*, and *Musculium argentinum* (the three species of *Pisidium* have only been recorded in Chile, but not *M. lauricochae* and *Sphaerium forbesi*); and (c) Patagonian region, the species *P. magellanicum*,

P. observationis, and *Musculium patagonicum* are shared with Argentina. *Pisidium lebruni* is also a Patagonian species but is currently registered only in Chile. The authors mentioned above indicated that this biogeographic proposal is preliminary and that future studies are required for its validation. In regard to Hyriidae, endemism is at the subfamily Hyriinae level, which is endemic to South America. The species *Diplodon chilensis* is widely distributed in Argentina, but *D. solidulus* is not, a fact that allows it to be considered endemic, along with the subgenus *Australis*. However, like *Pisidium lebruni*, its presence outside Chile must be corroborated [28].

So far there are no reports of the introduction of exotic bivalve species, as is happening in Argentina with *Corbicula* and *Limnoperna* or with *Dreissena polymorpha* in the northern hemisphere [1]. The great competitive capacity of these exotic species is causing the decline of native populations, especially Hyriidae, since these are used as substrates to settle, with consequent death by asphyxia.

b. *Gastropoda*: In the freshwater ecosystems of the extreme south of South America, there are gastropod species with a high degree of endemism, which present archaic zoogeographical relations of the Gondwanic type and constitute functionally relevant elements in the benthic communities of such ecosystems [33, 36] (e.g., Chilinidae (**Figure 1(30)**), Cochliopidae (**Figure 1(32)**), Planorbidae, Lymnaeidae, Physidae, and Ancylidae (**Figure 1(31)**). Chile has described 73 species belonging to six families and eight genera. However, many of these groups must be revised [36].

Although the inventory of Chilean freshwater gastropods began in the early eighteenth century and continued into the nineteenth century, it was not until the beginning and middle of the twentieth century that the number of new species described began to stabilize [36]. Subsequently, from the last half of the twentieth century until now, the number of new species described has been remarkably low. Since the compilation of freshwater gastropod mollusks from Chilean territory, the taxonomy of a few families has progressed considerably (e.g., Ancylidae, Planorbidae), while others (e.g., Chilinidae, Cochliopidae) still require the attention of researchers [36]. Cochliopidae species have been mostly included in the genus *Heleobia* or *Littoridina* [19]. However, it has recently become evident from penile morphology that many of the species traditionally assigned to *Littoridina* in Chile actually belong to *Heleobia* [19, 20]. According to a review of the existing literature on the group, inland water gastropod mollusks in Chile involve representatives of the subclasses Prosobranchia (one family) and Pulmonata (five families) (**Table 1**). The Pulmonata constitute the largest group, integrated by "snails" of the families Chilinidae (30 species of the genus *Chilina*), Lymnaeidae (five species of the genus *Lymnaea*), Physidae (four species of the genus *Physa*), and Planorbidae (seven species of the genus *Biomphalaria*), plus "limpets" of the Ancylidae family (four species of the *Anisancylus* and *Uncancylus* genera). Additionally, the prosobranchs are represented only by the family Hydrobiidae (one species of the *Potamolithus* genus and 21 species of the *Littoridina* genus (*Helobia*)).

The taxonomy of the group has been quite abandoned, so information on the species remains incomplete [36]. In this regard, a critical taxonomic review of the species of the six families described for Chile is urgent. Most of them have been described on the basis of shell characters, which, due to their strong intra- and interpopulation variability, must be validated with taxonomically more conservative characters, referring to the protoconch, radula, and the anatomy of the soft parts, i.e., penile complex and lung. Additionally, it is evident that the application of molecular taxonomy techniques is a requirement for the definitive validation of species. Preliminary studies carried out by the first author, of the penean complex of Chilean species, suggest that several of them would be synonymous and others not yet described. In this regard, it is expected that the number of valid species described for Chile could be reduced between 10 and 20%.

The geographic range occupied by freshwater gastropods covers the entire territory, in its latitudinal extension, and from the coastal boundary (and in many cases estuarine) to the Andean highlands in the north or to the Andes Cordillera in the rest of the area. However, not all groups are included in this global range, nor is the distribution of their species continuous. On the contrary, most species have a more or less discontinuous distribution, associated on the one hand with the location of the hydrographic basins and on the other with the mosaic of habitats found within each of the basins.

One of the problems facing the analysis of the geographic distribution patterns of freshwater gastropod species is the scarcity of sampling and lack of data specific to the collection site. In fact, most of the published records correspond to the "type location" [36]. Another obvious problem is that several species have imprecise localities. At the supra-specific level, the Cochliopidae, Physidae, and Lymnaeidae families are the most widely distributed in Chile, from the basins of the extreme north of the Atacama Desert to the Magellanic Region. On the contrary, Chilinidae species are mainly distributed between Valparaíso and Tierra del Fuego. An exception within this Family is the *Chilina angusta,* which has a different distribution compared to the rest of the family, living in springs of the Quebrada de Paposo on the coast of the Atacama Desert (25°S). Most species of the Planorbidae family are restricted to northern and central Chile (e.g., *Biomphalaria atacamensis*, *B. termala, B. montana, B. Schmiererianus*, *B. costata*, and *B. aymara* [34]). Only one species of Planorbidae extends in central and southern Chile to the Puelo River (*B. chilensis*). The Ancylidae family is the one with the most restricted geographic distribution in Chile, covering from Valparaíso to Chiloé, being a very abundant group in the stony coastal rivers of the VIII Region. According to the available antecedents, the highest density of freshwater gastropod taxa is located between regions VII and X, the latter being the one that concentrates the greatest diversity of species. To the south of Chiloé and to the north of the Choapa River, the number of species clearly tends to decrease. Of the total of 72 species described for Chile, 91.7% are endemic to the country [36]. In this regard, all the species of the families Cochliopidae, Chilinidae, Physidae, and Planorbidae are endemic. Of the Lymnaeidae family, only *Lymnaea lebruni* is endemic, having only been cited for Punta Arenas. In the case of Ancylidae, only *Uncancylus foncki* is endemic, having been cited for the Maullín River and Llanquihue Lake.

As for most Chilean freshwater invertebrates, the proposal of conservation categories for gastropod species is a difficult task given the lack of information [36]. On the other hand, no specific criteria and parameters have yet been developed for the classification of freshwater mollusks in the different conservation categories proposed by the World Conservation Union (IUCN) and established in Article 37 of Law No. 19,300 on the General Bases of the Environment (Chile). As a result, there is no general picture of the conservation status of Chilean species. Valdovinos et al. [37] tentatively proposed a classification of freshwater mollusks of the Chilean Coastal Range, mainly between 36° 50′S and 39° 26′S (following the IUCN "B criterion" (1994), which classifies a species as threatened when its geographical distribution is very restricted, and there are other factors that allow us to suspect that it is endangered. According to these authors, their proposal should be considered with caution as it is based on fragmentary information and general observations made by the author over the last 20 years. They considered all the representatives of the Chilinidae family within the "vulnerable" category, since, although their species still have relatively high occupation areas, it is evident that there is a continuous decline in the availability of their habitats. In contrast, given their large areas of occupation, along with their high dispersal and colonization capacity, and their abundance in different habitat types, the following species were considered to be at a "lower risk": *Physa chilensis*, *Lymnaea viator*, *Littoridina cumingi*, and *Biomphalaria chilensis.* Due to the scarcity of information, species such as *Littoridina pachispira* were classified within the category of "insufficient data." Following the same criteria, apart from these species from southern Chile, the *Chilina angusta* species from the Quebrada de Paposo (Taltal) should be considered "critically endangered." Regarding the conservation status of mollusk species, the only taxa officially classified are *Biomphalaria costata* and *Heleobia atacamensis*, both classified as "critically endangered," and *Heleobia chimbaensis*, considered "vulnerable."

The exotic freshwater mollusks present in Chile have been analyzed on the basis of specimens collected in wetlands and in commercial aquariums or intercepted in customs barriers, as well as bibliographical references. A total of seven species belonging to six genera were recorded, i.e., *Melanoides maculata*, *Thiara (Melanoides) tuberculata*, *Helobia* sp., *Pomacea bridgesii*, *Physella venustula*, *Physa* sp., and *Biomphalaria* sp. (*M. maculata* was collected in the Lluta River and classified as a cryptogenic species).

3.3 Other invertebrate groups

Protozoa (several phyla): Within the protozoa of free life, only the Sarcomastigophora and Ciliophora are normally found in freshwater. The first group is mainly represented by heliozoa, amoebas, and various flagellates and the second, by planktonic ciliates such as *Ophrydium* and *Stentor* and benthic ciliates such as suctors and hypotrichs. A review has been made of the ciliates present in lake ecosystems of central and southern Chile [43]. The ecological importance of the species of the orders Prostomatida (*Urotricha* spp. and *Balanion planctonicum*) and Haptorida (*Lacrymaria* sp., *Askenasia* spp.), Peritrichida (*Ophrydium naumanni*, *Vorticella* sp., *Vaginicola* sp.), Heterotrichidae (*Stentor araucanus*, *Stentor amethystinus*), Oligotrichida (*Strobilidium viride*, *Strombidium spp.*, *Halteria* spp., *Strobilidium* spp.), and Scuticociliatida (*Uronema* spp., *Cyclidium* spp.). The taxonomic knowledge of this group of typed amoebas has progressed substantially in recent decades, with a large number of species having been described, especially in freshwater ecosystems in central and southern Chile [44]. In addition, a complete taxonomic listing of Chilean tecamebas (sensu lato) was recently published [22]. The authors point out that the known diversity includes 416 taxa (64 genera and 352 taxa), 24 of which were reported for the first time.

Porifera: The members of the Spongillidae family are the only freshwater organisms. Species diversity is proportionally low relative to other groups of invertebrates. However, they are usually very abundant in the benthos of some lakes (e.g., Lake Lleu Lleu). Some species live in shady habitats, but those of the *Spongilla* genus, due to their association with zooxanthellae, live in illuminated areas. As an adaptation to the highly variable conditions of abiotic factors in freshwater, these sponges lack larvae and produce resistance structures.

Cnidaria: Like porifers, Cnidaria are poorly represented in Chile's freshwater ecosystems. In some lacustrine systems of central Chile, which show clear eutrophication symptoms (e.g., Laguna Grande de San Pedro and Lago Lanalhue, Biobío Region), the presence of the invasive jellyfish *Craspedacusta sowerbii* from Asia has been recorded [1]. This organism is an active predator of zooplankton, which can alter the planktonic communities of the ecosystems it invades. This small jellyfish is shaped like a bell, has between 50 and 500 tentacles, and does not usually exceed 25 millimeters in length. Some of its tentacles are long and allow it to maintain its position in the water while favoring movement, while the others are short and have a food function. It is in the latter where the nematocysts are housed, which include small harpoon-shaped cells (cnidocytes) that shoot when they come into contact with prey. As far as coloring is concerned, the hydromedusa is translucent, although with certain whitish or greenish tones. In addition to these jellyfish, the polyps also include *Hydra*, which are organisms that can become very abundant in palustrine systems.

Platyhelminthes: Of the six classes of this group, only Turbellaria and Temnocephala are found in freshwater ecosystems, as free living organisms or commensals (other groups are parasites; see below) [1]. The first group is composed of the genera *Dugesia* (*D. anceps*, *D. rincona*, *D. chilla*, *D. titicacana*, *D. sanchezi*, *and D. dimprpha*) and *Curtisia* (*C. michaelseni*), benthic organisms still imperfectly known in the area. The Temnocephala are a very interesting group that live as ectocommensals on the Parastacidae and Aeglidae crustaceans. Temnocephala were originally discovered in Chile and have a clear Gondwana-type distribution, restricted to the southern continents. The most common species of this group found in central-southern Chile are *Temnocephala chilensis* and *T. tumbesiana*.

Nemertea: Only a few nemertines have been found in freshwater worldwide. Most South American freshwater nemertes belong to the genus hoplonemertes, *Prosoma* (e.g., Venezuela, Brazil, and Argentina) [1]. The other type of South American nemerte is the heteronemerte, *Siolineus turbidus*, found in Brazil. In Chile, bdelonemertean *Malacobdella auriculae* has been described as an ectoparasite of a pulmonary gastropod [30]. These authors described a new genus and species for Chile, *Koinoporus mapochi*, which has been recorded in central Chile (Melipilla, Talagante, San Javier, Angostura de Paine, Pelarco, and Concepción). This species lives in low velocity waters, in both lotic and lentic ecosystems, associated with areas containing abundant aquatic macrophytes such as *Hydrocotyle ranunculoides*.

Aschelminthes: The Rotifera, Nematomorpha, Nematoda, and Gastrotricha are very common groups in freshwater environments of Southeastern South America [1]. Rotifers are an important part of lake zooplankton, although many are benthic. Nematodes are a very diverse group, but little is known about their taxonomy, despite being very common in practically all types of freshwater environments. Gastrotriches are a group of organisms that in freshwater ecosystems are apparently not very diverse and are frequently associated with muddy bottoms and the roots of aquatic plants (e.g., *Ichthydium*, *Polymerurus*, *Lepidodermella, Chaetonotus*, *Heterolepidoderma*, and Aspidiophorus). Nematomorpha are quite common in seasonal pools, which are represented by *Gordius paranensis*, *Gordius chilensis*, and *Beatogordius latastei*.

Annelida: Of this group of segmented worms, oligochaetes are undoubtedly the most common in freshwater ecosystems, especially in the low-moving water environments of lakes and rivers [1]. In benthic environments with a high organic load, Oligochaeta Tubificidae (*Tubifex*, *Limnodrilus*, *Potamothrix*, *Bothrioneurum*, *Epirodrilus*, *Isochaetides*) are usually abundant, while in less extreme environments, Naididae (*Nais*, *Chaetogaster*, *Schmardaella*, *Paranais*, *Pristinilla*, *Dero*, *Pristina)* and Lumbriculidae (*Lumbriculus*) are frequent [1]. Not much is known about this group in the area, and its scarce knowledge has been derived fundamentally from studies carried out by Argentine researchers. Other common annelids, especially in marshy environments, are the Hirudinea or "leeches" of which *Mesobdella gemmata* is perhaps one of the most common in central-southern Chile.

Another quite frequent group, especially in the soft bottoms of the estuary of some rivers, is the Archiannelida (e.g., Biobío River). Within this group, those belonging to the Histriobdellidae family are of great evolutionary and biogeographic importance, as they are commensals of decapod crustaceans (they live in their gill chambers) and because of their typical Gondwanic distribution (e.g., Madagascar, Tasmania, New Zealand, and the southern tip of South America). This is a group, represented in Chile by two species of the *Stratiodrilus* genus, which have highly specialized hosts [41].

For example, *Stratiodrilus aeglaphilus* inhabits the "crab" *Aegla laevis* (e.g., Río Maipo, in central Chile), and *S. pugnaxi* inhabits the "Camarón de Vega" *Parastacus pugnax* (e.g., Reumén, southern Chile).

Tardigrada: These small invertebrates are present in almost all types of freshwater environments, forming part of the benthos located on submerged plants and also in humid areas outside the water, for example, among mosses [1].

Bryozoa: The six species of Phylactolaemata bryozoans present in Chilean freshwater ecosystems belong to the cosmopolitan genera, *Fredericella* (*F. sultana*) and *Plumatella* (*P. repens*, *P. mukaii*, *P. patagonica*, *P. casmiana*). Although many authors acknowledge that freshwater bryozoans are common and abundant organisms in all freshwater bodies around the world (e.g., lakes, rivers, temporary pools), the development of knowledge regarding these organisms in Latin America is scarce and far from being a line of research that has persisted over time [26].

Parasitic invertebrate metazoa: Until now mainly free life forms have been studied, but there are many other species of parasitic invertebrates associated with freshwater organisms, at least for some part of their life cycle. In the case of metazoa parasites of aquatic and semiaquatic organisms, there are approximately 60 taxa described in Chile [25]. 47% of them have been identified at the species level and 53% as a genus or family. These parasites are composed of five phyla (Arthropoda, Nematoda, Acanthocephala, Platyhelminthes, and Myxozoa), between 1 and 3 classes per phylum, with a total of 8 classes, 19 orders, and 31 families. Phylum Platyhelminthes is the most diverse and is composed of 3 classes, 11 orders, and 19 families. Within this group, Digenea has the highest number of species. Like most of the Chilean freshwater invertebrate groups, the greatest diversity is found in central-southern Chile. The study of the ecological aspects of the digenees present in freshwater organisms of our country is of great importance, because many of them negatively affect man (e.g., *Fasciola hepatica*, which affects livestock, and Furcocercaria, which produce swimmer itch) [1]. Swimmer itch is caused by exposure to cercarias that have birds or mammals as definite hosts. In the summer of 2004, public alarm was generated when this parasitic disease was registered for the first time in central-southern Chile. It was found that the culprit was *Trichobilharzia* sp., whose intermediate host was the freshwater snail *Chilina dombeyana* [35]. This is clearly an emerging phenomenon, which is expected to intensify with increased eutrophication of ecosystems and climate change.

4. Threats to the biodiversity

As mentioned above, in central-southern Chile there is a biodiversity "hot spot" of global interest, which includes benthic freshwater invertebrates. Given the climatic, geographic, and hydrological conditions of this territory, located approximately between 35° and 43° S, this zone also corresponds to a "hot spot" of economic activities which place a lot of pressure on aquatic biodiversity [1]. Paradoxically, the Biobío Region, which corresponds to the heart of this territory for its valuable heritage of freshwater fauna (fish and invertebrates), is one of the most intervened and unprotected in the country. For example, in the region's

36,929.3 km^2 of territory, only 843.6 km^2 correspond to "protected wild areas" (2.28%), which are outside areas of interest for the conservation of freshwater biota [1]. The most intense impact within this zone is concentrated in the middle and lower parts of watersheds, particularly in areas formerly occupied by coastal forests, whose importance has already been mentioned [50]. At present, these forests have consisted in a high proportion of pine and eucalyptus plantations, which support cellulose production at a global scale.

There are numerous factors that put freshwater invertebrate communities at risk (as well as fish), many of which work together to enhance the effects on biota [17]. In this regard, the following factors are of particular importance:

a. Loss and degradation of freshwater habitats: Important "pharaonic works" carried out in northern and central Chile, such as the construction of hydroelectric power plants (reservoir or run-of-river), irrigation works (ponds and canals), and mining operations, are currently endangering the survival of many freshwater invertebrate species, especially those with benthic habits. For example, there has been a notable loss of diversity and local extinctions (e.g., *Chilina dombeyana*) in lakes regulated by hydroelectric generation activities (e.g., Laja Lake). Massive local extinctions of *Aegla pewenchae* have also been observed, associated to the strong hourly changes in the level of the Biobío River, product of the activity of the plants of the Biobío [1]. Similarly, the "minimum ecological flows" considered in many hydroelectric and irrigation projects are insufficient for the conservation of potentially threatened macroinvertebrate species. For example, the Maipo River basin is fragmented by 10 hydroelectric plants (Queltehues, Alfalfal, Carena, Maitenes, Puntilla, El Volcán, Los Bajos-Caemsa, Los Morros, La Florida, Planchada-La Ermita), the Maule River by eight plants (Loma Alta, Cipreses, Curillinque, Pehuenche, Isla, Colbún, Machicura, San Ignacio), and the Biobío River by nine plants (El Toro, Abanico, Mampil, Pangue, Antuco, Peuchén, Ralco, Rucúe, and Angostura).

As mentioned above, not only the "pharaonic works" are a serious threat to the conservation of Chile's freshwater invertebrates. Deforestation of native forest basins and its transformation into areas dedicated to forestry, agriculture, and urbanization are generating alterations of great magnitude to freshwater ecosystems of central-southern Chile. Although many aspects of the patterns and processes that occur in riparian environments have been studied in recent years, little is known about the effect of vegetation type on Chilean fluvial communities [1]. However, replacement of native forest with introduced species and deforestation are common practices in many regions. Currently, in forested areas of central Chile, deciduous native components are being replaced on a large scale by pine and eucalyptus, suggesting that this process has an important impact on freshwater invertebrates and hence on the energetic characteristics of river communities. This situation reaches its greatest impact in the Chilean regions Maule, BioBío, and Araucanía, extending progressively towards the south. Within these territories, the areas of the coastal mountain range have been almost completely destroyed and are now occupied by forest plantations of exotic species.

There are numerous other factors that cause the degradation of aquatic habitats; among them it is necessary to highlight the erosion of basins, which produces high sedimentations in lakes and rivers [1]. In this regard, it is estimated that 70% of rivers in central Chile are affected by this process (e.g., rivers Aconcagua, Cachapoal, Maipo, and many others). The effect of large sediment loads on rivers is multiple. However, its effect on photosynthesis and the development of microalgae limits the penetration of light and the filling of microhabitats that are occupied by benthic invertebrates. Another direct effect on aquatic habitats is the extraction of aggregates directly in river beds.

b. Invasive species: The introduction of exotic fish species produces significant negative effects on Chilean freshwater invertebrate populations, the magnitude of which is just beginning to be discovered. Among the most common species are "rainbow trout" (*Oncorhynchus mykiss*) and "brown trout" (*Salmo trutta*), which actively prey on benthic macroinvertebrates in the upper and middle part of rivers, juvenile states especially on Chironomidae diptera; intermediate states on Ephemeroptera, Plecoptera, and Trichoptera; and adult stages on Aeglidae crustaceans. It is very probable that one of the main causes of the extinction of local Aeglidae populations is associated with predation by these two trout species. In the lower part of rivers and in many mesotrophic and eutrophic lakes, the "carp" (*Cyprinus carpio*) and the "mosquito fish" (*Gambusia holbrooki*) have relevant effects on invertebrate communities, either by resuspending sediment produced by the former (which affects suspended species such as the mussel *Diplodon chilensis*) or by predation of planktonic and benthic organisms, which produces the latter. Another species, the "chanchito" (*Cichlasoma facetum*), which is very common in lakes of central Chile (e.g., in the area of Concepción), is extremely voracious and generates large modifications in invertebrate populations. As previously mentioned, in the zooplankton of lentic ecosystems located in central Chile, with clear symptoms of eutrophication, it is possible to find the hydromedusa *Craspedacusta sowerbi*, originating from Asia. This organism is an active predator of zooplankton, which can alter the planktonic communities of the ecosystems it invades.

There are also other invasive benthic invertebrates, such as the mollusks mentioned above. Exotic freshwater mollusks are reported in local wetlands, particularly *Melanoides maculata* and *Physella venustula* [1]. Species of the genera *Helobia* and *Physa* have also been recorded. However, they have been classified as cryptogenic, species not defined as native or exotic, since *Physa* is a genus widely distributed in the Pacific basin of North and Central America and its current taxonomic knowledge in South America is limited. The main pathways of voluntary and involuntary introduction of exotic freshwater molluscs are trade. Given the increase in interregional trade, appropriate ecological and taxonomic data must be collected to evaluate their eventual establishment in local ecosystems. In addition, the advance of the invasive diatom *Didymosphenia geminata* known as "didymo" has recently been recorded. This is a highly invasive algae with a high capacity to affect aquatic ecosystems into which it is introduced. This microalgae is present from the Biobío River basin to the Patagonian ecosystems, where it is generating important changes to fluvial benthic macroinvertebrates that are only now beginning to be studied.

c. Aquatic pollution: Pollution of rivers and lakes is one of the most visible threats affecting the survival of the area's freshwater invertebrates. The nature of the compounds that affect aquatic biota varies from one basin to another, depending on the productive activities developed there and the presence of human settlements. For example, in sectors with a high population density, the main factors are associated with nutrients and organic matter (e.g., the Aconcagua and Maipo River basins, among others); in areas with strong mining activity, the main factors are metals and pH (e.g., acid drains in northern and central Chile); in agricultural areas, fertilizers and pesticides (e.g., central-southern Chile); and in areas with cellulose and paper industries, a large diversity of persistent organic compounds (e.g., Biobío River basin [51]). A relevant aspect in Chile regarding this disturbance factor is the absence of a policy that regulates the quality of surface waters of rivers and lakes. It is evident that the almost total absence of this regulation has a negative impact on aquatic biota, as it does not set limits regarding the concentration and pollutant loads contributed to these systems, either from specific or diffuse sources.

d. Overexploitation: With the exception of "Camarón de río del norte" (*Cryphiops caementarius*), present in the rivers of northern Chile, and the species "Camarón de vega" (*Parastacus pugnax*) and "Camarón de río del Sur" (*Samastacus spinifrons*), present in central-southern Chile, there are no other species that are strongly affected by overexploitation [1].

e. Global climate change: The effect of global climate change is the least predictable of the five factors considered. This lack of prediction is associated with the uncertainty of future climate scenarios and the difficulty of anticipating their ecological consequences [52]. However, it is predicted that changes in water availability and thermal regime will be more serious at medium and high latitudes, where important changes in the latitudinal and altitudinal distribution of species will occur. The biota of these areas is particularly vulnerable due to their thermal regime (many species associated with low water temperatures, e.g., Plecoptera) and the lack of adequate escape routes to more suitable habitats. It is evident that the most vulnerable species to extinction within freshwater invertebrates are those of large size (>10 mm), without the capacity to fly, stenohalines, and with narrow geographical distribution ranges, such as many crustaceans (e.g., Aeglidae) and mollusks (e.g., Chilinidae).

5. Conclusions

The invertebrates of rivers, lakes, and wetlands of the Southwest of South America are one of the oldest testimonies of the great climatic changes suffered by the landscape, especially during the Tertiary and Quaternary periods. In particular, the freshwater ecosystems of the mountainous territories and plains, ranging from the Maule to the Aysén Regions of Chile, which were once covered by forests in virtually all their extension. Today, many invertebrate species in the northern and central parts of this territory are threatened, particularly those with small populations and low dispersal capacity, such as mollusks and decapod crustaceans. In this territory of high diversity and endemism, the rivers are severely fragmented as a result of the presence of hydroelectric plants and irrigation infrastructures. In addition, water quality shows a progressive deterioration, associated with strong industrial and urban growth. Also, deforestation and substitution of native forest for pine and eucalyptus plantations, especially in the central valley and the coastal mountain range, are generating profound changes in this group of invertebrates. However, in these zones it is still possible to find small remnants of rivers associated with native vegetation, although markedly isolated from each other. These remnants of great past climatic and geological changes are today highly threatened.

Acknowledgements

We are grateful to Dr. Oscar Parra and Amber Philp for the revision of the manuscript. This work was supported by DAND Codelco-Andina.

Author details

Claudio Valdovinos Zarges[1*], Pablo Fierro[2] and Viviana Olmos[1]

1 Departamento de Sistemas Acuáticos, Facultad de Ciencias Ambientales y Centro EULA, Universidad de Concepción, Chile

2 Instituto de Ciencias Marinas y Limnológicas UACh, Universidad Austral de Chile, Chile

*Address all correspondence to: cvaldovi@udec.cl

References

[1] Valdovinos C. Invertebrados dulceacuícolas. In: Rovira J, Ugalde J, Stutzin M, editors. Biodiversidad de Chile Patrimonio y Desafíos. 3rd ed. Santiago de Chile: Ocho Libros Eds; 2008. pp. 202-223

[2] Dos Santos D, Milineri C, Nieto C, Zúñiga M, Emmerich D, Fierro P, et al. Cold/Warm stenothermic freshwater macroinvertebrates along altitudinal and latitudinal gradients in Western South America: A modern approach to an old hypothesis with updated data. Journal of Biogeography. 2018;**45**: 1571-1581. DOI: 10.1111/jbi.13234

[3] Domínguez E, Fernández H, editors. Macroinvertebrados bentónicos sudamericanos. Sistemática y biología. 1st ed. Tucumán, Argentina: Fundación Miguel Lillo; 2009. 654 p

[4] Habit E, Dyer B, Vila I. Estado de conocimiento de los peces dulceacuícolas de Chile. Gayana. 2006;**70**:100-113. DOI: 10.4067/S0717-65382006000100016

[5] Vila I, Pardo R, Dyer B, Habit E. Peces límnicos: diversidad, origen y estado de conservación. In: Vila I, Veloso A, Schlatter R, Ramírez C, editors. Macrófitas y vertebrados de los sistemas límnicos de Chile. Santiago, Chile: Editorial Universitaria; 2006. pp. 73-102

[6] Abell R, Thieme M, Revenga C, Bryer M, Kottelat M, Bogutskaya N, et al. Freshwater ecoregions of the world: A new map of biogeographic units for freshwater biodiversity conservation. Bioscience. 2008;**58**: 403-414. DOI: 10.1641/B580507

[7] Fuster R, Escobar C, Lillo G, de la Fuente A. Construction of a typology system for rivers in Chile based on the European Water Framework Directive (WFD). Environmental Earth Sciences. 2015;**73**:5255-5268. DOI: 10.1007/s12665-014-3772-x

[8] CEPAL. Latin America and the Caribbean: Inventory of water resources and their use. South America. United Nations Economic Commission for Latin America and the Caribbean – ECLAC; 1991. 333 p

[9] von Ihering H. On the ancient relations between New Zealand and South America. Transactions and Proceedings of the New Zealand Institute. 1891;**24**:431-445

[10] MyersN,MittermeierR,MittermeierC, da Fonseca G, Kent J. Biodiversity hotspots for conservation priorities. Nature. 2000;**403**:853-858

[11] Noodt W. Die grundwasserfauna Südamerikas. In: Fittkau EJ, Illies J, Klinge H, Schwabe GH, Sioli H, editors. Biogeography and Ecology in South America. The Hague: Dr. W. Junk N.V. Publishers; 1969. pp. 659-684

[12] Illies J. Biogeography and ecology of Neotropical freshwater insects, especielly those from running waters. In: Fittkau EJ, Illies J, Klinge H, Schwabe GH, Sioli H, editors. Biogeography and Ecology in South America. The Hague: Dr. W. Junk N.V. Publishers; 1969. pp. 685-707

[13] Chernicoff C, Zappettini E. Delimitación de los terrenos tectonoestratigráficos de la región centro-austral Argentina: evidencias aeromagnéticas. Revista Geologica de Chile. 2003;**30**:299-316. DOI: 10.4067/S0716-02082003000200009

[14] Rapalini AE. The accretionary history of southern South America from the latest Proterozoic to the late Palaeozoic: Some palaeomagnetic

constraints. In: Vaughan ARM, Leat PT, Pankhurst RJ, editors. Terrane Processes at the Margins of Gondwana. London: Geological Society, Special Publications; 2005. pp. 305-328. DOI: 10.1144/GSL.SP.2005.246.01.12

[15] Pérez-Losada M, Jara C, Bond-Buckup G, Crandall K. Conservation phylogenetics of Chilean freshwater crabs *Aegla* (Anomura: Aeglidae): Assigning priorities for aquatic habitat protection. Biological Conservation. 2002;**105**:345-353. DOI: 10.1016/S0006-3207(01)00218-X

[16] Jara C. Crustáceos del género *Aegla* (Decapoda: Anomura) en la Cordillera de la Costa: su importancia para la conservación de la biodiversidad de aguas continentales en Chile. In: Smith C, Armesto J, Valdovinos C, editors. Historia, biodiversidad y ecología de los bosques costeros de Chile. Santiago, Chile: Editorial Universitaria; 2005. 708 p

[17] Jara C, Rudolph E, González E. Estado de conocimiento de los crustáceos malacostracos dulceacuícolas de Chile. Gayana. 2006;**70**:40-49. DOI: 10.4067/S0717-65382006000100008

[18] Camousseight A. Estado de conocimiento de los efemerópteros de Chile. Gayana. 2006;**70**:50-56. DOI: /10.4067/S0717-65382006000100009

[19] Collado G. Nuevo registro de distribución geográfica y antecedentes de historia natural de *Heleobia chimbaensis* (Biese, 1944) (Caenogastropoda: Cochliopidae) en la costa del desierto de Atacama: Implicancias para su conservación. Amici Molluscarum (Chile). 2012;**20**:13-18

[20] Collado G, Méndez M. Estrategias reproductivas y tipos de desarrollo en especies endémicas del género *Heleobia* Stimpson, 1865 (Caenogastropoda: Cochliopidae) de Chile. Amici Molluscarum (Chile). 2011;**19**:67-71

[21] De los Ríos P, Rivera R, Morrone J. Calanoids (Crustacea: Copepoda) reported for Chilean inland waters. Boletín de Biodiversidad de Chile. 2010;**3**:9-23

[22] Fernández L, Lara E, Mitchell E. Checklist, diversity and distribution of testate amoebae in Chile. European Journal of Protistology. 2015;**51**:409-424. DOI: 10.1016/j.ejop.2015.07.001

[23] Jerez V, Moroni E. Estado de conocimiento de los coleópteros dulceacuícolas de Chile. Gayana. 2006;**70**:72-81. DOI: 10.4067/S0717-65382006000100012

[24] Muñoz S, Mendoza G, Valdovinos C. Evaluación rápida de la biodiversidad de cinco lagos costeros de la VIII Región de Chile. Gayana. 2001;**65**:256-262. DOI: 10.4067/S0717-65382001000200009

[25] Olmos V, Muñoz G. Estado de conocimiento de los parásitos eumetazoos de organismos dulceacuícolas de Chile. Gayana. 2006;**70**:112-139. DOI: 10.4067/S0717-65382006000100018

[26] Orellana M. Estado de conocimiento de los briozoos dulceacuícolas de Chile. Gayana. 2006;**70**:96-99. DOI: 10.4067/S0717-65382006000100018

[27] Oyanedel A, Valdovinos C, Sandoval N, Moya C, Kiessling G, Salvo J, et al. The southernmost freshwater anomurans of the world: Geographic distribution and new records of Patagonian aeglids (Decapoda: Aeglidae). Journal of Crustacean Biology. 2011;**31**:396-400. DOI: 10.2307/41315729

[28] Parada E, Peredo S. Estado de conocimiento de los bivalvos dulceacuícolas de Chile. Gayana.

2006;**70**:82-87. DOI: 10.4067/S0717-65382006000100013

[29] Rojas F. Estado de conocimiento de los tricópteros de Chile. Gayana. 2006;**70**:65-71. DOI: 10.4067/S0717-65382006000100011

[30] Sánchez M, Moretto H. A new genus of freshwater hoplonemertean from Chile. Zoological Journal of the Linnean Society. 1988;**92**:193-207. DOI: 10.1111/j.1096-3642.1988.tb00102.x

[31] Takeuchi N, Kohshima S, Shiraiwa T, Kubota K. Characteristics of cryoconite (surface dust on glaciers) and surface albedo of a Patagonian glacier, Tyndall glacier. Southern Patagonia Icefield. Bulletin of Glaciological Research. 2001;**18**:65-69

[32] Takeuchi N, Koshima S. A snow algal community on a Patagonian glacier, Tyndall glacier in the southern Patagonia Icefield. Arctic, Antarctic, and Alpine Research. 2004;**36**:91-98. DOI: 10.1657/1523-0430(2004)036[0092:ASACOT]2.0.CO;2

[33] Valdovinos C. Biodiversidad de moluscos chilenos: Base de datos taxonómica y distribucional. Gayana. 1999;**63**:111-164

[34] Valdovinos C, Stuardo J. Planórbidos Altoandino del Norte de Chile y *Biomphalaria aymara spec. nov.* (Mollusca: Basommatophora). Studies on Neotropical Fauna and Environment. 1991;**26**:213 224. DOI: 10.1080/01650529109360854

[35] Valdovinos C, Balboa C. Cercarial dermatitis and lake eutrophication in south-Central Chile. Epidemiology and Infecctions. 2007;**136**:391-394. DOI: 10.1017/S0950268807008734

[36] Valdovinos C. Estado de conocimiento de los gastrópodos dulceacuícolas de Chile. Gayana. 2006;**70**:88-95. DOI: 10.4067/S0717-65382006000100014

[37] Valdovinos C, Olmos V, Moya C. Moluscos terrestres y dulceacuícolas de la Cordillera de la Costa Chilena. In: Smith C, Armesto J, Valdovinos C, editors. Historia, biodiversidad y ecología de los bosques costeros de Chile. Santiago, Chile: Editorial Universitaria; 2005. pp. 292-306

[38] Valdovinos C, Kiessling A, Mardones M, Moya C, Oyanedel A, Salvo J, et al. Distribución de macroinvertebrados (Plecoptera y Aeglidae) en ecosistemas fluviales de la Patagonia chilena: Muestran señales biológicas de la evolución geomorfológica postglacial? Revista Chilena de Historia Natural. 2010;**83**:267-287. DOI: 10.4067/S0716-078X2010000200008

[39] Vera A, Camousseight A. Estado de conocimiento de los plecópteros de Chile. Gayana. 2006;**70**:57-64. DOI: 10.4067/S0717-65382006000100010

[40] Vera A, Zuñiga-Reinoso A, Muñoz-Escobar C. Perspectiva histórica sobre la distribución de *Andiperla willinki* "Dragón de la Patagonia" (Plecoptera: Gripopterygidae). Revista Chilena de Entomología. 2012;**37**:87-93

[41] Vila I, Bahamonde N. Two new species of *Stratiodrilus, S. aeglaphilus* and *S. pugnaxi* (Annelida: Histriobdellidae) from Chile. Proceedings of Biological Society of Washington. 1985;**98**(2):347-350

[42] Villalobos L. Estado de conocimiento de los crustáceos zooplanctónicos dulceacuícolas de Chile. Gayana. 2006;**70**:31-39. DOI: 10.4067/S0717-65382006000100007

[43] Woelfl S. Notas sobre protozoos ciliados dulceacuícolas de Chile. Gayana. 2006;**70**:24-26. DOI: 10.4067/S0717-65382006000100005

[44] Zapata J. Estado de conocimiento de los tecamebianos dulceacuícolas de Chile. Gayana. 2006;**70**:27-30. DOI: 10.4067/S0717-65382006000100006

[45] Villanueva M, Chamamé F. Análisis de la biodiversidad de macroinvertebrados bentónicos del río Cunas mediante indicadores ambientales, Junín-Perú. Scientia Agropecuaria. 2016;**7**:33-44. DOI: 10.17268/sci.agropecu.2016.01.04

[46] Trama F, Mejía J. Biodiversidad de macroinvertebrados bentónicos en el sistema de cultivo de arroz en el sector muñuela margen derecho en Piura, Perú. Ecología Aplicada. 2013;**12**:147-162

[47] Huamantico A, Ortiz W. Clave de géneros de larvas de Trichoptera (Insecta) de la Vertiente Occidental de los Andes, Lima, Perú. Revista Peruana de Biología. 2010;**17**:75-80

[48] Iannacone J, Mansilla J, Ventura K. Macroinvertebrados en las lagunas de puerto viejo, Lima – Perú. Ecología Aplicada. 2003;**2**:116-124

[49] Bahamonde N,Carvacho A, Jara C, López M, Ponce F, Retamal M, et al. Categorías de conservación de decápodos nativos de aguas continentales de Chile. Boletín del Museo Nacional de Historia Natural (Chile). 1998;**47**:91-100

[50] Parra O, Valdovinos C, Urrutia R, Cisternas M, Habit E, Mardones M. Caracterización y tendencias tróficas de cinco lagos costeros de Chile Central. Limnética. 2003;**22**:51-83

[51] Habit E, Belk M, Tuckfield R, Parra O. Response of the fish community to human-induced changes in the Biobío River in Chile. Freshwater Biology. 2006;**51**:1-11. DOI: 10.1111/j.1365-2427.2005.01461.x

[52] Allan JD, Flecker AS. Biodiversity conservation in running waters. Bioscience. 1993;**43**:32-43